Application of Plasticity and Generalized Stress-Strain in Geotechnical Engineering

Proceedings of the Symposium on
Limit Equilibrium, Plasticity and Generalized
Stress Strain Applications in Geotechnical Engineering
held in conjunction with the 1980 ASCE Annual Convention
and Exposition
Hollywood, Florida
October 27-31, 1980

Sponsored by the Geotechnical Engineering Division
of the
American Society of Civil Engineers

R.N. Yong and E.T. Selig, Editors

Published by the
American Society of Civil Engineers
345 East 47th Street
New York, New York 10017

TA
710
.A1
S87
1980

The Society is not responsible for any statements made or opinions expressed in its publications.

Copyright © 1982 by the American Society of Civil Engineers,
All Rights Reserved.
Library of Congress Catalog Card No. 81-71796
ISBN 0-87262-294-0
Manufactured in the United States of America.

CONTENTS

INTRODUCTION *Soil Constitutive Model Assessment*
by R. N. Yong and E T. Selig 1

SESSION 1 *Limit Equilibrium Plasticity Applications in Geotechnical Engineering*

STATE-of-the-ART REPORT
by G. G. Meyerhof 7

A Plasticity Model for the Load Unload Behavior of Sand,
by R. Baker, S. Frydman and J. Galil 25

A Reassessment of Limit Equilibrium Concepts in Geotechnique,
by R. N. Chowdhury 53

A Generalized Bounding Surface Constitutive Model for Clays,
by Y. F. Dafalias and L. R. Herrmann 78

Soil as an Anisotropic Kinematic Hardening Solid,
by W. D. L. Finn and G. R. Martin 96

Analysis of Soil Structures with Different Plasticity Models,
by E. Mizuno and W. F. Chen 115

Shallow Penetration of Marine Foundations.
by J. D. Murff and T. W. Miller 139

Expansion of Spherical Cavity in Sand,
by H. B. Poorooshasb and B. Lievre 153

Plastic-Limit Equilibrium States in Soil Media
by J. H. Prevost and B. E. Hjorth 166

SESSION 2 *Generalized Sress-Strain Applications in Geotechnical Engineering*

STATE-of-the-ART REPORT
by J. T. Christian 182

Deformation Analysis for Braced Excavation in Clay,
by C. S. Chang and M. H. B. Abbas 205

Re-Evaluation of Work Hardening Model,
by E. Evgin and Z. Eisenstein 226

Verification of Non-Linear Effective Stress Model in Simple Shear,
by W. D. L. Finn and S. K. Bhatia 240

Characterization of the Undrained Anisotropy of Clays,
by L. A. Hansen and G. W. Clough 253

Solution to a Certain Class of Problems in Soil Mechanics,
by H. B. Poorooshasb 277

Yielding Load of Anchor in Sand,
by M. C. Wang and A. H. Wu 291

Stress-Strain Influence on Soil Compactability by Rollers,
by R. N. Yong and E. A. Fattah 308

Panel Members' Discussions

J. M. Duncan 333

Zdenek Eisenstein 336

George Y. Baladi 341

Subject Index 353

Author Index 355

SOIL CONSTITUTIVE MODEL ASSESSMENT

by

R.N. Yong,[1] M.ASCE and E.T. Selig,[2] F.ASCE

The need to identify stress-strain and strength characteristics of soil has always existed in geotechnical engineering practice. Traditionally, the problems involving stability and especially deformation of soil have been handled with simple models or with empirical methods based on field experience. However, the need to solve increasingly sophisticated geotechnical engineering problems has placed demands on engineering practitioners which require better analytical tools. The availability of large, high-speed computers has made possible the application of analytical modelling to a degree that was not previously feasible. Satisfactory use of these new tools requires appropriate soil constitutive models, which in general must be more elaborate than the simple models used in the past.

A great deal of research has been carried out in recent years with the objectives of developing analytical models for representing the stress-strain behaviour of soils and establishing the techniques for using these models in computer methods for solving geotechnical engineering problems. Various kinds of yield and failure criteria have also been proposed for soil. Although the incentive for such research has often resulted from the identification of problems by practitioners, the suitability of the constitutive models and computer solutions for practical problems has not been fully documented or established. In addition, the requirements for physical tests needed to produce the material parameters for the constitutive models have not been clearly established, nor has it been determined whether standard tests can be used routinely for quantifying the soil parameters for a particular application.

Vastly improved capabilities for prediction of soil deformation and stability have been achieved through this research. However, opinions differ widely within the profession on such basic questions as which models are best, what the capabilities of the models are, and what the expected reliability of these models is in design applications.

Soil is known to be a nonlinear, inelastic, anisotropic, and nonhomogeneous material with stress-dependent and time-dependent properties. Yielding and plastic deformation take place at stress states that vary with stress history and stress path. The controlling stress states are three-dimensional, although traditionally soil is usually analyzed by two-dimensional models. Recognition of these realities leads to pertinent questions such

---

[1] William Scott Professor of Civil Engineering and Applied Mechanics, McGill University, Montreal, Quebec, Canada. H3A 2K6.

[2] Professor of Civil Engineering, University of Massachusetts, Amherst, Mass. 01003. U.S.A.

as the following that need to be answered:

1. To what degree can real soil behaviour, as identified above, be represented by a constitutive model and yet require only soil parameters which can be readily obtained?

2. How simple can a constitutive model be and yet satisfy the essential requirements of the problem?

3. How well can any of these models be applied in practice and what is the expected accuracy?

4. How detailed or sophisticated should the models be for application to practical problems?

To obtain answers to such questions and to fully identify how all the models presently available can be used and under what conditions they are suitable, three companion programs were established. These were:

1. A three-day workshop in Montreal at McGill University in May 1980, which concentrated on the understanding and evaluation of the soil constitutive models.

2. An ASTM one-day symposium in Chicago in July 1980, dealing with test methods for obtaining the stress-strain and strength parameters of soil.

3. An ASCE one-day symposium in Hollywood, Florida in October 1980, to assess applications of soil constitutive models in geotechnical engineering practice.

The scope of these programs was specifically limited to "static" problems, that is, problems involving a single load cycle which might include both loading and unloading stress paths.

The Montreal workshop was designed to bring together the developers of the constitutive models, as well as researchers and practitioners in geotechnical engineering. The sponsors were the U.S. National Science Foundation and the Natural Sciences and Engineering Research Council of Canada. The co-organizers of the workshop were Professor H.Y. Ko at the University of Colorado and Professor R.N. Yong at McGill University.

As a focus for the workshop, a representative group of international experts, identified as predictors, were invited to participate. Each predictor was first asked to explain the features of his model. The predictors had been supplied with soil information and test data in advance of the workshop to which they were to fit their soil models. This exercise provided a common basis for showing how the model parameters are obtained and demonstrated how the models could be used. The predictors were then asked to calculate the expected soil behaviour for stress paths other than those from which they established their parameters. For the first time at the workshop, comparisons were made between the predictions and the measured results for these stress paths in an attempt to provide an unbiased assessment of the suitability of the models.

The following persons participated as predictors:

T. Adachi, Kyoto University
G.Y. Baladi, Waterways Experiment Station, and I. Sandler, Weidlinger Associates
Z.P. Bazant, Northwestern University, and A. Ansal, Istanbul Technical University
W.F. Chen, Purdue University
J.M. Duncan, University of California at Berkeley
Y.F. Dafalias, and L.R. Herrmann, University of California at Davis
G. Gudehus, Karlsruhe University
P.V. Lade, University of California at Los Angeles
J.K. Mitchell, University of California at Berkeley, and E. Kavazanjian, Stanford University
J.H. Prevost, Princeton University
C.P. Wroth, Oxford University

The results are presented in the workshop proceedings (Ref. 1).

The ASTM symposium following the workshop was intended to highlight the advances made in soil testing techniques and methods for reduction of the test data. The program sessions and participants were:

Session I -- Strength Testing Methods and Requirements

Chairman	E.T. Selig, University of Massachusetts at Amherst
State-of-the-Art Speaker	A.S. Saada, Case Western Reserve University, Cleveland
Panel Moderator	F.C. Townsend, University of Florida at Gainesville
Panelists	C.C. Ladd, Massachusetts Institute of Technology, Cambridge
	P. LaRochelle, Laval University, Quebec City
	S. Wright, Texas University at Austin
	S. Poulos, Geotechnical Engineers, Inc., Winchester, Mass.

Session II -- Data Reduction and Application of Measurements for Analytical Modelling

Chairman	E.T. Selig, University of Massachusetts at Amherst
State-of-the-Art Speaker	H.Y. Ko, University of Colorado at Boulder
Panel Moderator	R.N. Yong, McGill University, Montreal
Panelists	F. Tavenas, Quebec City
	J.H.A. Crooks, Golder Associates, Mississauga, Ontario
	G.Y. Baladi, Waterways Experiment Station, Corps of Engineers, Dept. of the Army, Vicksburg, Miss.
	R.J. Krizek, Northwestern University, Evanston, Ill.

The proceedings (Ref. 2) include the two state-of-the-art reports, 30 technical papers, discussions and panel presentations.

The symposium on applications of soil stress-strain and strength models held at the ASCE conference in Hollywood, Florida was divided into two sessions, one for each of the following two broad categories:

1. Stability - problems involving limit equilibrium of soil which are not usually concerned with deformation prior to failure.

2. Deformation - problems involving stress-strain behaviour with stress conditions that do not involve failure states.

However, it was recognized that many of the stress-strain or constitutive models have failure as a limiting condition or state. Thus the historical separation of stability and deformation problems may no longer be necessary if the new constitutive models are used.

The organization of the ASCE symposium sessions consisted of a state-of-the-art speaker on each of the two topics, followed by panel presentations by experienced professionals and finally, panel discussion incorporating questions and comments from the audience. The participants of the two sessions are as follows:

SESSION 1
Limit Equilibrium Plasticity Applications in Geotechnical Engineering

    Presiding: Ernest T. Selig, University of Massachusetts at Amherst
    Session Chairman: Harold W. Olsen, U.S. Geological Survey, Denver
    State-of-the-Art Presentation: George G. Meyerhof, Technical University of Nova Scotia, Halifax
    Panel Discussion Moderator: Raymond N. Yong, McGill University
    Panelists: W.F. Chen, Purdue University
              W.D.L. Finn, University of British Columbia
              H.B. Poorooshasb, Concordia University
              C.P. Wroth, Oxford University

SESSION 2
Generalized Stress-Strain Applications in Geotechnical Engineering

    Presiding: Woodland G. Shockley, U.S. Army Corps of Engineers
    Session Chairman: Harold W. Olsen, U.S. Geological Survey, Denver
    State-of-the-Art Presentation: John T. Christian, Stone and Webster Engineering Corp., Boston
    Panel Discussion Moderator: Ernest T. Selig, University of Massachusetts at Amherst
    Panelists: Ralph E. Brown, Law Engineering Testing Company, Atlanta
              J. Michael Duncan, University of California at Berkeley
              Z. Eisenstain, University of Alberta
              George Y. Baladi, Waterways Experiment Station

These proceedings provide results of this ASCE symposium.

The following proceedings of this symposium contain the state-of-the-art papers and contributions from the panel, as well as technical papers submitted by the profession to provide examples of applications of stress-strain and strength relationships to geotechnical engineering problems.

A number of basic questions were addressed by the symposium in assessing the applicability of soil constitutive models to geotechnical engineering practice. The first was - Has a particular model been used in practice and, if so, what has been the experience, and to what extent is the model documented? The identified purposes for using the models, other than for research, were categorized as: 1) design prior to construction, 2) guide to instrumentation, 3) assessment of a problem that had developed in construction to seek a remedy, and 4) obtain a better understanding of a geotechnical engineering problem in order to achieve improved design methods. Most of the applications related to the fourth of these categories, and a few represented the first.

The second question was - What tests are required to determine the model parameters? The categories considered were: 1) empirical, estimated from past experience, 2) field tests, such as standard penetration test and cone penetrometer, 3) conventional triaxial test, and 4) special tests like cubical triaxial test. A distinction was made between the minimum essential tests and the desired tests. The parameters for most models could be derived from conventional triaxial tests. Although empirical or field test methods were generally insufficient, in some cases approximate values of the parameters could be estimated from prior triaxial testing using soil classification and physical state information.

A third question was - What is the cost of performing analyses with the models? The basic cost categories concerned: 1) computer time, 2) consulting time from model developer, and 3) engineering time to obtain the soil parameters, set up and debug the finite element computer program and, finally, checking and interpreting the results. By far the largest cost factor was the third of these.

The fourth question was - How accurate can the computer solutions be which incorporate the soil models? The large variability of natural soil deposits and the uncertainty of field conditions severely limit the accuracy with which the soil properties can be represented. Thus it is likely that precision, or the ability to examine effects of parameters, rather than accuracy, is a more promising expectation in most applications.

Geotechnical problems can be approached on several levels of sophistication. These progress from using empirical methods to using the most elaborate constitutive models. The fifth question then, is - What is the basis for selection of the level, and when is it necessary to go beyond the standard methods? There is no simple answer to this question. The proper level depends on the importance of the problem, as well as the availability of funds, experienced personnel and needed soil information.

Based on the results of the workshop and ASTM and ASCE symposia, an assessment may be made of the status and application of soil constitutive models. First, none of the models can completely represent the complex behaviour of soil. Second, deficiencies exist in all of the commonly available tests for measuring the soil properties, and often the desired data are not even available. Third, some constitutive models are too complex or too difficult to use in solving geotechnical problems.

Nevertheless, the use of constitutive models in the solution of geotechnical engineering problems through finite element methods has advanced beyond the research stage and has many practical advantages. As long as the practitioner does not permit the computer to substitute for engineering judgement, the use of soil constitutive models can be a valuable aid in design and analysis. With the accumulation of application experience, continued advances in the future can be anticipated.

References

1. Yong, R.N. and Ko, H.-Y., Editors, *LIMIT EQUILIBRIUM, PLASTICITY AND GENERALIZED STRESS-STRAIN IN GEOTECHNICAL ENGINEERING*. Proceedings, NSF/NSERC North American Workshop, American Society of Civil Engineers, New York, November 1981. 880 p.

2. Yong, R.N. and Townsend, F.C., Editors, *LABORATORY SHEAR STRENGTH OF SOIL*, ASTM Special Technical Publication 740, American Society for Testing and Materials, Philadelphia, Penn., September 1981. 717 p.

State-of-the-Art Report

LIMIT EQUILIBRIUM PLASTICITY IN SOIL MECHANICS

by G.G. Meyerhof, F.ASCE[1]

Introduction

Limit plasticity applications in geotechnical engineering include both the ultimate limit state at collapse and the serviceability limit state under working conditions of earth structures. Recently two symposia on the topic have been held (Cambridge, 1971 and 1973), which included measurements of soil strength parameters, methods of solution of boundary value problems and applications of plasticity in soil mechanics.

The present report will only consider the following main areas of the ultimate static soil behaviour at collapse: - failure conditions, plasticity theory, and its application to problems of the stability of slopes, active and passive earth pressures and the ultimate bearing capacity of foundations. Deformation behaviour of soils and earth structures under working conditions will not be included since it is part of another report (12).

Failure Conditions

Soil masses in the ultimate limit state usually fail in shear, althougth tension failure may occur in some cases. It is generally recognized that the strength properties of a soil mass should be determined by in-situ methods of testing or laboratory tests on representative undisturbed soil samples, which simulate field conditions of the particular stability problem considered. Thus, plane-strain or triaxial tests may be required on soils under compression or extension with appropriate drainage conditions, stress system and history, loading path and rate of loading. To facilitate stability analyses the complex real strength and plastic deformation behaviour of soils must be idealized. Thus, soil having both peak and residual strength may be represented roughly as an ideal simple elastic-perfectly plastic or rigid-perfectly plastic material with a constant peak, residual or average strength failure surface, while a better model would be an ideal strain-softening material (Fig. 1 a). Similarly, the plastic strain increments can be characterized by a dilatancy angle ψ (28, 65), which for a drained soil at the peak strength is considerably smaller than the corresponding friction angle $\emptyset$. Moreover, the dilatancy angle decreases during strain-softening and becomes $\psi = 0$ at the residual strength and critical void ratio with constant volume behaviour, while for soil under high confining pressure the value of ψ may be negative at failure (Fig. 1 b).

Failure lines for peak and residual strengths of anisotropically-consolidated soil in compression and extension tests provide the corresponding drained shear strength parameters c and $\emptyset$ in Coulomb's equation using effective stresses. For an effective stress path the yield surface is roughly elliptical in shape, and strain-hardening, strain-softening and time-dependent behaviour of soils by consolidation or creep can be included by a corresponding expansion or contraction of the yield surface (76). Typical yield and failure lines for sand under triaxial compression are shown in Fig. 2a together with plastic potential curves, which define by the gradient the corresponding strain increment directions (60).

---

[1] Professor, Department of Civil Engineering, Technical University of Nova Scotia, Halifax, Nova Scotia, Canada.

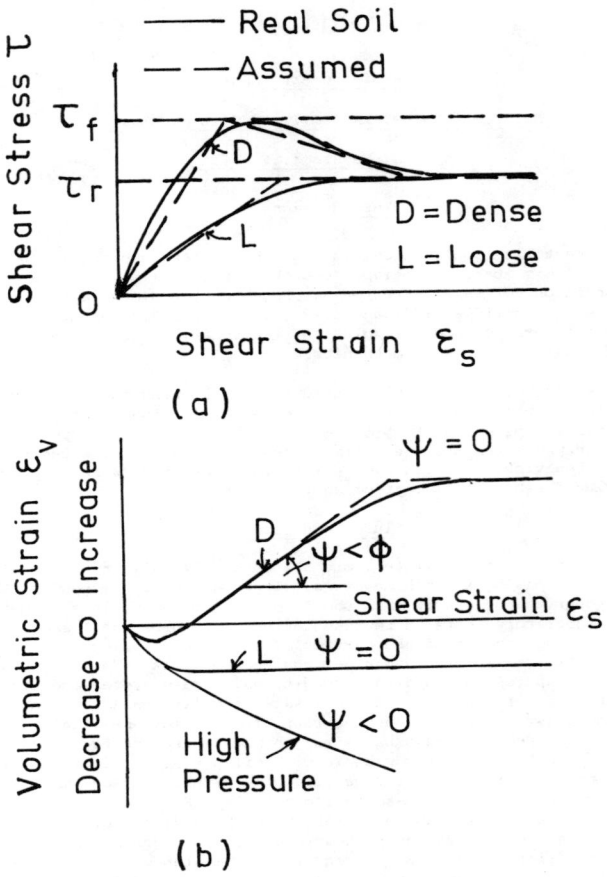

1. Typical Relationships between (a) Shear Stress and Shear Strain and (b) Volumetric Strain and Shear Strain.

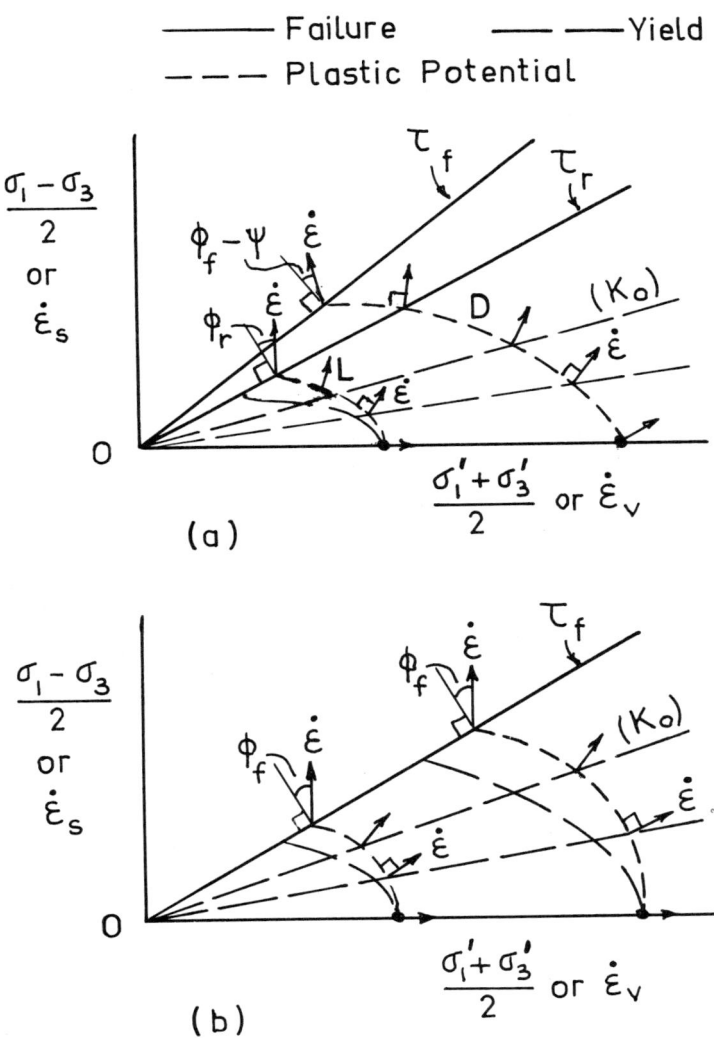

2. Typical Yield and Failure Lines and Plastic Potential Curves for (a) Sand and (b) Normally-Consolidated Clay.

The plastic potential and yield surface of sand do not coincide, and the direction of the plastic strain increments deviates by an angle of $\emptyset_f - \psi$ from the normal to the peak failure surface and the deviation increases to $\emptyset_r$ at residual failure, so that a non-associated flow rule would be required in plasticity theory. Normality to the yield surface would lead to an associated flow rule for an ideal material with $\psi = \emptyset$, which in practice applies only to saturated clay failing under undrained conditions with $\emptyset = 0$.

For cohesive-frictional soils a general plastic potential including anisotropy is difficult to establish, and the direction of the plastic strain increments may not be unique and depend on many different factors. However, for normally-consolidated and lightly over-consolidated clays of low sensitivity the plastic potential and yield curves are similar, and typical curves for saturated normally-consolidated clay under triaxial compression are shown in Fig. 2 b together with corresponding strain increment directions (41). The direction of the plastic strain increments deviates by an angle of $\emptyset_f$ from the normal to the failure surface, thus requiring a non-associated flow rule.

Plasticity Theory

Classical plasticity methods of analysis (29, 31, 61, 70) are based on associated flow rule materials with $\psi = \emptyset$ and can provide a unique collapse load for plane-strain problems of an ideal simple plastic material independent of the loading path. Some analyses are also available for axial-symmetric problems on the assumption that the intermediate principal stress is equal to the major or minor principal stress (4, 13, 14). A theoretically exact solution requires compliance with equilibrium, compatibility and boundary conditions and it must be both statically and kinematically admissible. In the absence of a rigorous analysis a correct solution can be bracketed by a limit analysis using a statically admissible solution for a lower bound and a kinematically admissible solution for an upper bound. Lower bound solutions can readily be obtained by combining regions of constant stress separated by allowable discontinuities in the stress field (11). Upper bound solutions can be improved by variational calculus methods based on the classical Euler equations to determine the critical kinematic composite rupture surface for given boundary conditions (37).

By the use of an incremental or numerical approach, classical plasticity theory can be applied to variable shear strength and dilatancy in stress and strain increment fields, strain-hardening and strain-softening plastic materials, including progressive failure of a soil mass by reduction of the shear strength from peak to residual value in portions of the mass. In these cases the theoretical unique collapse load depends on the loading path, stress history and other factors, and a solution can readily be obtained by a finite element method of analysis (76). This method provides information about the complete load-deformation behaviour of the soil mass to collapse and the corresponding increase of plastic zones, in which a zero or positive rate of plastic work must be ensured for a correct solution under the assumed input conditions (17).

For an ideal plastic material with a non-associated flow rule ($\psi < \emptyset$) classical plasticity analyses have to be modified (15). The theoretical collapse load may not be unique and depend on the loading path, and it can be determined by numerical and finite element methods of analysis. The corresponding stress and strain increment fields at the collapse load are usually interrelated by assuming that the principal axes of stress and of strain increment coincide. This assumption has been confirmed experimentally for monotonic loading to the peak collapse load, but not during a subsequent unloading to the residual strength nor for significant changes of the overall geometry (63, 64) when the principal directions diverge. Under the condition of coincidence of the principal axes Mohr's circles of stress and of strain increment can readily be used to determine the stresses and strain increments at any point A in the soil mass with given values of c, ∅ and ψ (Fig. 3 a and b). Furthermore, these circles show that the stress characteristics make angles of 45° - ∅/2 to the direction of the major principal stress σ_1, while the strain increment characteristics or zero-extension lines make angles of 45° - ψ/2 to the direction of the major principal strain increment $\dot{\varepsilon}_1$ and hence of σ_1 (Fig. 3 c). Moreover, the angles between the stress and strain increment characteristics are (∅ - ψ)/2, and on the strain increment characteristics the mobilized shear strength parameters are c_o and $∅_o$ (Fig. 3a).

Recent mathematical analysis of plane-strain problems of ideal elastic-plastic materials with a non-associated flow rule shows that simultaneous solution of the complete system of stress and strain increment equations for a given loading path leads to a unique solution for the collapse load (2). This analysis also predicts a corresponding unique rupture surface in the soil mass, which may coincide with either a stress or a strain increment characteristic, whichever is the innermost limiting line encountered during the mathematical solution process of the particular problem. This location of the rupture surface would also be expected from considerations of the minimum total plastic work of the particular system considered.

Limit theorems do not apply to plastic materials with a non-associated flow rule. However, an upper bound for such a material is given by the unique solution for an associated flow rule material having the same shear strength parameters c and ∅ (15). Further, a lower bound solution for a material with a non-associated flow rule can be obtained from a statically admissible stress field, which must be based on the given loading path and not produce plastic regions in the soil mass. Moreover, a safe lower bound solution would be obtained from a statically admissible solution with the shear strength parameters c_o and $∅_o$, mentioned previously. Thus, for a constant volume material with ψ = 0, the corresponding shear strength parameters would be c_o = c.cos∅ and $∅_o$ = $\tan^{-1}$ (sin ∅), which compares with the statically admissible solution using $∅_o$ = $\sin^{-1}$ (sin∅.cos∅) (21) for a relative rotation of the principal axes of stress and strain increment.

Stability of Slopes

The stability of slopes of homogeneous ideal rigid-plastic soil is commonly analysed by assuming plane or cylindrical failure surfaces (71, 74), while composite failure surfaces with interslice forces are used for non-homogeneous soil (36, 58), as shown in Fig. 4. These kinematic analyses of associated flow rule materials provide upper bound solutions, which have recently been enhanced by application of the calculus of variations (10, 26). For simple two-dimensional slopes of homogeneous soil the results of these analyses can be expressed in terms of a stability factor N_s = γH/c, where γ = unit weight of soil and H = height of slope with slope angle β (Fig. 5). The stability factors N_s shown as given by the variational method have been corrected for the influence of the interslice forces, which increase the original factors by up to about 10%. It is of interest to note that the corresponding values of N_s agree well with those deduced from statically admissible stress fields, which form a lower bound solution (23, 70). These practically unique slope stability factors for an associated flow rule material are found to be smaller than those based on the conventional circular arc method except for a vertical bank (β = 90°). In the latter case 3.4 < N_s < 3.8 for purely cohesive soil with ∅ = 0 (22).

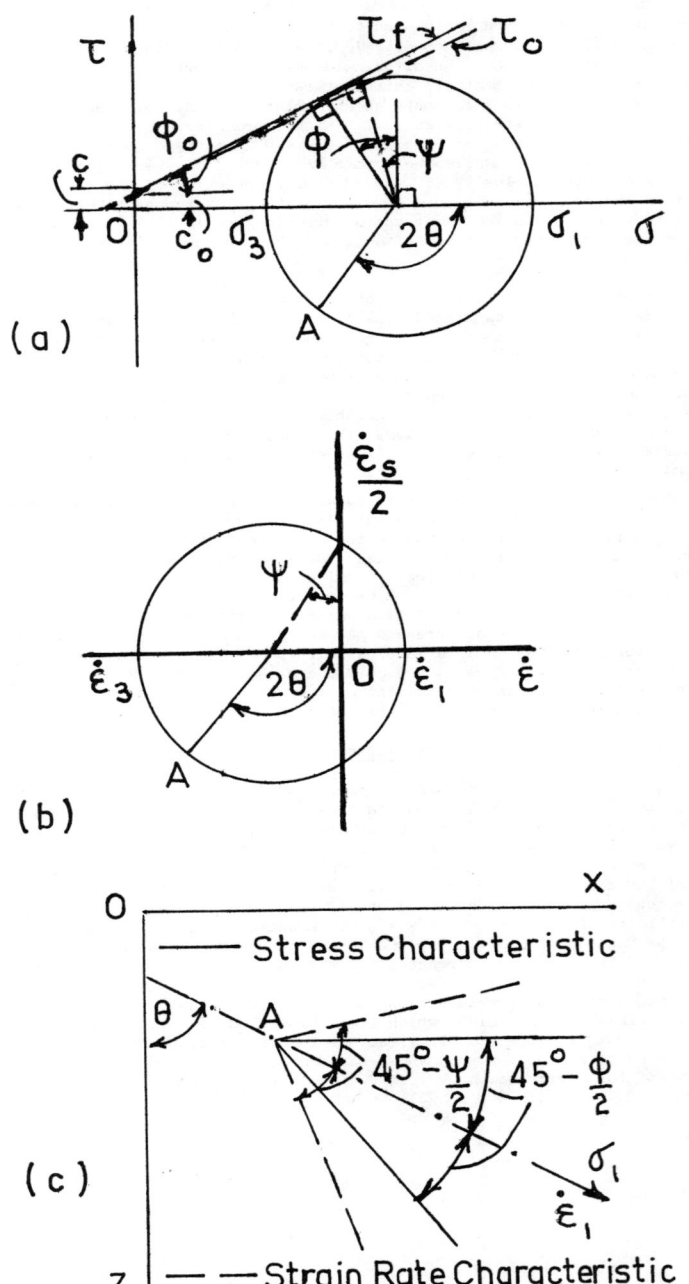

3. Mohr's Circles of (a) Stress and (b) Strain Increment and (c) Directions of Stress and Strain Increment Characteristics.

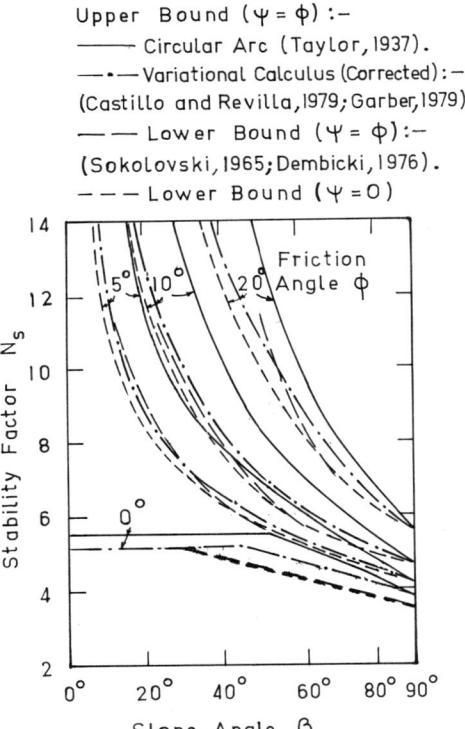

5. Stability Factors for Simple Slopes.

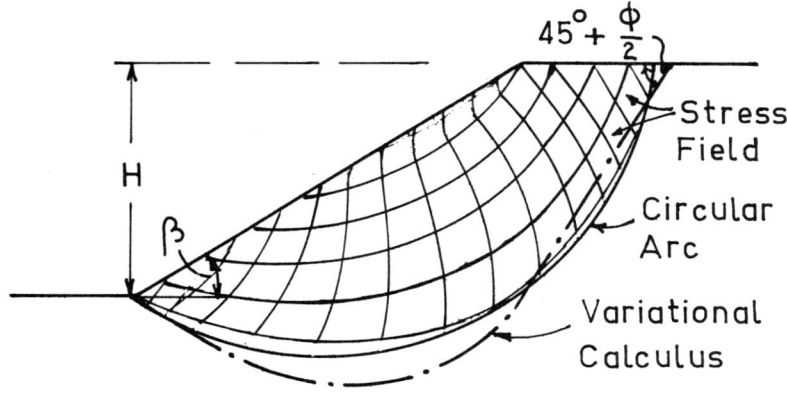

4. Comparison between Assumed Theoretical Failure Mechanisms of Simple Slope.

Unique stability factors for slopes of soil with a non-associated flow rule material ($\psi < \emptyset$) are not yet available. However, as mentioned before, the stability factors for an associated flow rule material are upper bounds, and the use of a friction angle $\emptyset_o$ would provide a lower bound analysis for flat slopes with $\beta = \emptyset_o$ while for a vertical bank the friction angle $\emptyset$ applies. Assuming a linear variation of the effective friction angle between these limits for intermediate slope angles, the lower bound values of N_s have been evaluated by the writer for simple slopes of a constant volume material with $\psi = 0$ (Fig. 5). For friction angles of $\emptyset < 20°$ these lower bound factors differ by no more than about 5% from the corresponding upper bound values so that the range of non-uniqueness has no engineering importance for cohesive soils, which has been confirmed by finite element analysis (76). However, for soils with $\emptyset > 30°$ the difference becomes more significant, as indicated by finite element analysis for a particular slope geometry (17). In accordance with recent mathematical analysis (2) the theoretical location of the rupture surface would be expected to change from the stress characteristics in the active state near the top of the slope to the strain increment characteristics in the passive state near the bottom of the slope.

The effects of surcharge on slopes, anisotropy and non-homogeneity of associated flow rule materials on the stability of slopes have been analysed by limit methods for lower bounds (70) and upper bounds (6, 42). These analyses indicate that for flat slopes with $\beta < 45°$ anisotropy of purely cohesive soil ($\emptyset = 0$) can, approximately, be taken into account by using an average cohesion $c = (c_1 + c_2)/2$ with the stability factor N_s for isotropic soil, where c_1 and c_2 = cohesion for major principal stress during shear in vertical and horizontal directions, respectively, relative to direction of soil deposition. Similarly, for strain-softening of such soil an average cohesion of peak and residual strengths may be used for rough preliminary estimates of short-term slope stability. Finite element solutions have been obtained for progressive failure of slopes by strain-softening (38, 43), which is particularly important for slopes of heavily overconsolidated clay when gradual softening of the soil and line rupture along shear bands may occur (57).

Active and Passive Earth Pressures

The lateral earth pressures on rigid walls at soil failure are usually determined by limit analyses for homogeneous ideal rigid-plastic materials with an associated flow rule using lower bound stress fields (9, 70) or using upper bound kinematic analyses with plane or curved failure surfaces (74). Upper bound solutions based on composite soil failure surfaces have also been obtained for the collapse load of walls under various types of movement and rotation (7, 29). Recently unique solutions of stress and strain increment fields for active and passive earth pressures have been derived from rigorous plasticity methods of analysis for walls which are translated or rotated about the base (39, 40), as shown in Figs. 6a and b.

The corresponding coefficients of active and passive earth pressures of cohesionless (K_a and K_p) and cohesive (K_{ac} and K_{pc}) associated flow rule materials with a horizontal surface for vertical two-dimensional walls are shown in Fig. 7, together with the lower bound values based on statically admissible stress fields for a constant volume material with $\psi = 0$. It is found that the effect of dilatancy on the earth pressure coefficients for rough walls increases rapidly with the friction angle $\emptyset$. However, in practice the value of ψ of soil generally increases with $\emptyset$ only to a maximum of about $\psi = \emptyset/2$ for very dense soil,

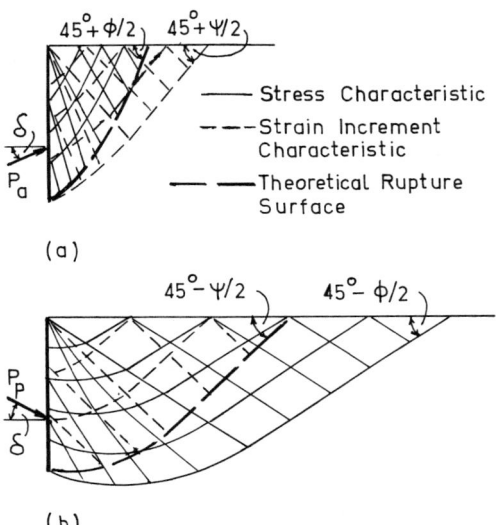

6. Fields of Stress and Strain Increments for Rough Vertical Wall Under (a) Active Earth Pressure and (b) Passive Earth Pressure.

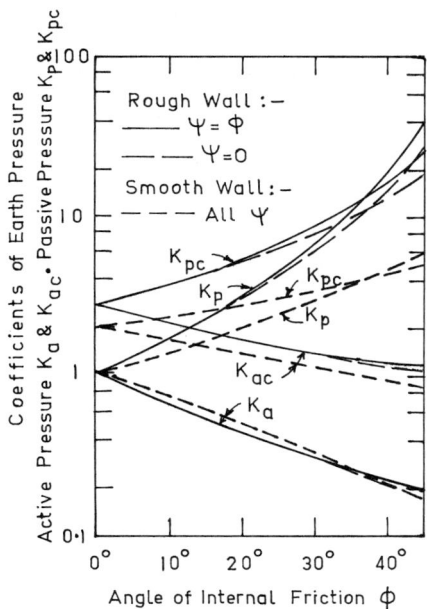

7. Coefficients of Active and Passive Earth Pressures on Vertical Wall.

when the earth pressure coefficients are roughly midway between the lower limit for $\psi = 0$ and the upper limit for $\psi = \emptyset$. It is also of interest to note that the observed rupture surface in sand for the active case of a wall translating or rotating near the base agrees well with stress characteristics corresponding to the peak friction angle (56, 66, 69). Further, the corresponding observed passive rupture surface is with some exceptions (32, 67) generally close to the strain increment characteristics (3, 63, 72). The difference of the rupture surface locations in these passive earth pressure tests may be explained by the effects of progressive failure and overall change of geometry. Moreover, the different characteristic behaviour of active and passive rupture surfaces would be expected theoretically (2).

Plasticity analyses have also been developed for walls rotating about the top for both active and passive earth pressures of cohesionless soil with a non-associated flow rule ($\psi < \emptyset$), and they are supported by the results of some model tests on sand (34, 35, 63). The axial symmetric problem of lateral earth pressures on cylindrical walls has been solved by lower bound stress fields for associated flow rule materials, which show that the active earth pressure is decreased and the passive earth pressure is considerably increased by the induced circumferential stresses (4, 24).

Ultimate Bearing Capacity of Foundations

The ultimate bearing capacity of shallow strip foundations on homogeneous ideal rigid-plastic soil is conveniently expressed by bearing capacity factors N_c, N_q and N_γ (73), which have been evaluated by plasticity theory for associated flow rule materials by unique solutions for N_c and N_q and by closely bracketed upper and lower bound limit solutions for N_γ (8, 47). Lower bound values of the bearing capacity factors for a non-associated flow rule material ($\psi < \emptyset$) can be determined from the vertical component of the corresponding passive earth pressure on the sides of a rigid soil wedge below the footing (Fig. 8). The sides of this wedge, which are inclined at an angle θ to the horizontal, have shear strength parameters c_o and $\emptyset_o$. The corresponding values of the bearing capacity factors have been calculated by the writer for a rough strip foundation on a constant volume material with $\psi = 0$ by setting $\theta = 45°$ for N_c and N_q, and finding the critical θ for the minimum N_γ (47). These lower bound values are shown in Fig. 9 together with the above-mentioned factors for $\psi = \emptyset$. For two particular cases the present bearing capacity factors for $\psi = 0$ compare well with those obtained by a rigorous solution based on stress and strain increment fields (16, 17, 76). The effect of dilatancy on the bearing capacity factors increases rapidly with $\emptyset$, but for common values of ψ of soils in practice the difference does not exceed about 10%, as found above for the passive earth pressure coefficients on rough walls.

The rupture surface for a foundation on an ideal plastic material is theoretically expected to coincide with strain increment characteristics, as found for model strip footing tests on a bed of rigid rollers (1, 5). However, load tests on large footings on dense sand (59) as well as numerous model footing tests on sand (1, 30, 44, 45, 75) have shown that the observed rupture surface is generally close to the stress characteristics corresponding to an angle $\emptyset$ between the peak and residual values. This difference between the theoretical and observed location of the rupture surface in sand can be explained largely by the effects of change of overall geometry and the strain-softening of the failed soil. Thus, the rupture surface generally develops fully only after the residual bearing capacity has been reached at a footing settlement which is several times greater

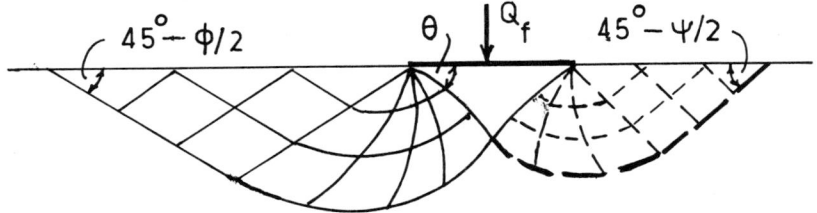

— Stress Characteristic
— — — Strain Increment Characteristic
— — Theoretical Rupture Surface

8. Fields of Stress and Strain Increments for Rough Strip Footing.

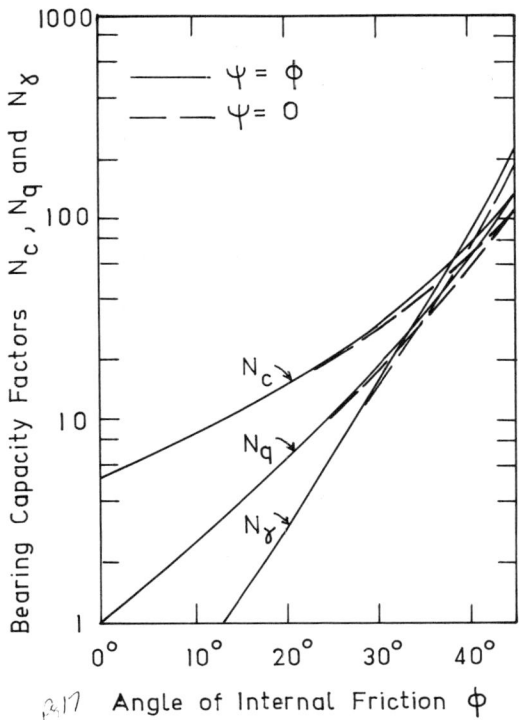

9. Bearing Capacity Factors for Shallow Rough Strip Foundation.

than that of the peak bearing capacity. The corresponding significant change of the initial geometry is accompanied by substantial progressive post-peak soil failure during which the planes of principal strain increment near the potential rupture surface rotate relative to the principal stress planes. It should be noted that for a corresponding rotation by an angle of $(\phi - \psi)/2$ the stress and strain increment characteristics would overlap along the rupture surface, which has also been observed in pure shear tests on dense sand at the residual strength (64).

The ultimate bearing capacity of strip footings on anisotropic flow rule materials has been analysed for purely cohesive soil (19) and for cohesionless soil (52). These analyses show that anisotropy of purely cohesive soil can roughly be taken into account by using an equivalent cohesion $c' = (c_1 + 2c_2)/3$ with the bearing capacity factor N_c for isotropic soil, while an equivalent friction angle $\phi' = (2\phi_1 + \phi_2)/3$ may be used in estimating the factors N_γ and N_q of cohesionless soil, where ϕ_1 and ϕ_2 = friction angles for major principal stress during shear in vertical and horizontal directions, respectively, relative to direction of soil deposition.

Unique solutions of the ultimate bearing capacity of clay have been obtained for the effects of non-homogeneity (18) and of strain-softening of the soil using finite element methods (33). Lower bound (70) and upper bound (54) methods have been used to obtain solutions for the ultimate bearing capacity of foundations on layered associated flow rule materials, including punching failure when a relatively thin strong soil layer tests on a weak stratum. Similar theoretical results are available for eccentric and inclined loading on horizontal and inclined foundations, foundations on slopes, uplift of foundations (53), and other special cases (8, 27, 46, 48, 49, 50).

The bearing capacity factors for rough circular footings on homogeneous associated flow rule materials have been obtained by lower bound solutions based on statically admissible stress fields (4, 68). The results can be expressed in terms of shape factors s_c, s_q and s_γ, which are the ratios of the corresponding bearing capacity factors of circular to strip foundations (Fig. 10). It is found that these shape factors are, except for the $\phi = 0$ case, considerably greater than those deduced from the results of load tests on circular and strip footings on sand (20, 45, 47). This difference is due to the fact that the ultimate bearing capacity of circular footings is largely governed by friction angles of triaxial compression tests, while the greater plane-strain angles govern the ultimate bearing capacity of strip footings. Corresponding bearing capacity factors for circular footings on non-associated flow rule materials ($\psi < \phi$) are not yet available, but would lie between the upper bound values given above and the lower bound values obtained by using the shear strength parameters c_o and ϕ_o, mentioned previously. These approximate shape factors for a constant volume material with $\psi = 0$ are shown in Fig. 10.

Bearing capacity factors for deep foundations in homogeneous soil have been obtained by extending the methods used for shallow foundations (45). However, due to the important effects of the method of installation of deep foundations, soil compressibility and arching and other factors a semi-empirical approach is necessary at present to determine the base resistance bearing capacity factors and the skin friction factors for driven and bored piles (51, 55) and for pier foundations (62).

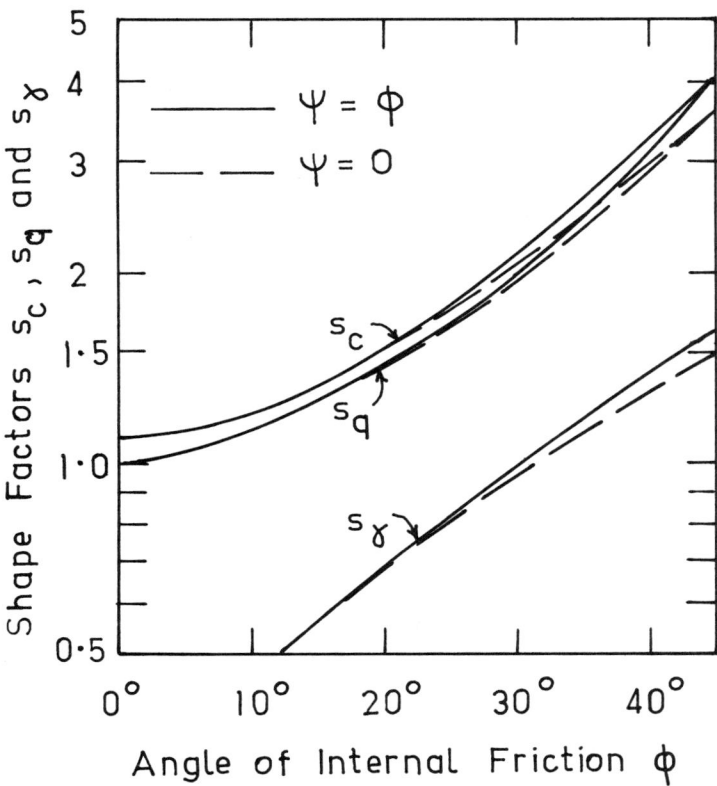

10. Shape Factors for Shallow Rough Circular Foundation.

Conclusions

Using idealized failure conditions classical plasticity theory can provide a unique collapse load for plane-strain and axial symmetric problems of associate flow rule materials. Alternatively, the solution can be bracketed by limit analyses of statically admissible lower bounds and of kinematically admissible upper bounds which may be improved by variational calculus methods. For non-associated flow rule materials classical plasticity analyses have to be modified, and safe upper and lower bound solutions can be obtained in many cases within reasonable limits. Path dependent unique collapse loads can readily be determined by numerical and finite element methods of analysis.

The stability of slopes based on recent limit methods of analysis is generally found to be smaller than that using the conventional circular arc method for associated flow rule materials. Safe lower bound values of slope stability of non-associated flow rule soil with constant volume behaviour have been calculated. Recently unique solutions have been derived for active and passive earth pressures of associated flow rule materials on rough walls which are translated or rotated about the base, and general plasticity analyses have been developed for non-associated flow rule materials. The ultimate bearing capacity of shallow strip foundations on associated flow rule materials has been evaluated by plasticity theory, and lower bound values have been determined for non-associated flow rule soil with constant volume behaviour. Lower bound solutions have also been obtained for circular footings on associated flow rule materials, and the corresponding problem on non-associated flow rule materials has been analysed to derive approximate shape factors.

It is found that the effect of soil dilatancy on the stability of slopes, active and passive earth pressures and the ultimate bearing capacity of foundations increases rapidly with the friction angle of the soil. However, for values of the angle of dilatancy in practice this effect is relatively small, as shown by the corresponding moderate range of non-uniqueness of the limit methods of analysis of non-associated flow rule materials.

References

1. Arens, E. "Ebene Grundbruchversuche mit Lotrecht und Schräg Belasteten Streifengründungen," Soil Mechanics Research Report No. 3, Tech. Univ., Aachen, 1975.

2. Baker, R., and Frydman, S. "On the Solution of Plane Strain Problems for Non-Associated Soils," Civil Engg. Faculty Publ. No. 257, Technion, Haifa, Israel, 1980.

3. Bassett, R.H., "Stability of Rigid Structures," Discussion, Proc. 5th Europ. Conf. Soil Mech. and Foundt. Engg., Madrid, 1972, Vol.2, pp. 152-155.

4. Berezantsev, V.G., "An Axially Symmetric Problem of the Theory of Limiting Equilibrium of a Granular Medium." Gostekhizdat, Moscow, 1952.

5. Biarez, J., "General Theories of Earth Pressure," Discussion, Proc. 5th Europ. Conf. Soil Mech. and Foundt. Engg., Madrid, 1972, Vol. 2, pp. 53-58.

6. Booker, J.R., and Davis, E.H., "A Note on Plasticity Solutions to the Stability of Slopes in Inhomogeneous Clays," Geotechnique, 1972, Vol. 22, pp. 509-513.

7. Brinch-Hansen, J., "Earth Pressure Calculation." Inst. Danish Civil Engrs., Copenhagen, 1953.

8. Brinch-Hansen, J., "A Revised and Extended Formula for Bearing Capacity," Danish Geotech. Inst., Bull. No. 28, Copenhagen, 1970.
9. Caquot, A., and Kerisel, J., "Traité de Méchanique des Sols." Gauthier-Villars, Paris, 1949.
10. Castillo, E., and Revilla, A., "The Calculus of Variations and the Stability of Slopes," Proc. 9th Int. Conf. Soil Mech., Tokyo, 1977, Vol.2,pp. 25-30.
11. Chen, W.F., "Limit Analysis and Soil Plasticity." Elsevier, New York, 1975.
12. Christian, J.T., "Generalized Stress-Strain Applications in Geotechnical Engineering," Annual Meeting, ASCE, Geotech. Engg. Div., 1980, Hollywood, Florida.
13. Cox, A.D., "Axially Symmetric Plastic Deformation in Soils," Int. Jl. Mech. Science, 1962, Vol. 4, pp. 371-380.
14. Cox, A.D., Eason, G., and Hopkins, H.G., "Axially Symmetric Plastic Deformations in Soils," Phil. Trans. Royal Soc., Series A, London, 1961, Vol.254, pp. 1-45.
15. Davis, E.H., "Theories of Plasticity and the Failure of Soil Masses," Soil Mech. Selected Topics, Butterworths, London, 1968, pp. 341-380.
16. Davis, E.H., and Booker, J.R., "The Bearing Capacity of Strip Footings from the Standpoint of Plasticity Theory," Proc. 1st Aust. N.Z. Conf. on Geomechanics, Melbourne, 1971, pp. 276-282.
17. Davis, E.H., and Booker, J.R., "Some Applications of Classical Plasticity Theory for Soil Stability Problems," Symp. on Plasticity and Soil Mechanics, Cambridge, 1973, pp. 24-41.
18. Davis, E.H., and Booker, J.R., "The Effects of Increasing Strength with Depth on the Bearing Capacity of Clays," Geotechnique, 1973, Vol. 23, pp. 551-564.
19. Davis, E.H., and Christian, J.T., "Bearing Capacity of Anisotropic Cohesive Soil," Jl. Soil Mech. and Foundt. Div., ASCE, Vol. 97, No. SM5, 1971, pp.753-769.
20. De Beer, E.E., "Experimental Determination of the Shape Factors and Bearing Capacity Factors of Sand," Geotechnique, 1970, Vol. 20, pp. 387-411.
21. De Jong, G.J., "Lower Bound Collapse Theorem and Lack of Normality of Strain Rate to Yield Surface for Soils," Proc. Symp. Rheology and Soil Mechanics, UTAM, Grenoble, 1964, pp. 69-75.
22. De Jong, G.J., "Improvement of Lower Bound Solution for the Vertical Cutoff in a Cohesive Frictional Soil," Geotechnique, 1978, Vol. 28, pp. 197-201.
23. Dembicki, E., "Stabilité d'un Talus à l'Etat Plastique," Tech. Univ. Turin, 1976.
24. Dembicki, E., "Active, Passive Pressure and Bearing Capacity of Soil," Arkady, Warsaw, 1979.
25. Dubrova, G.A., "Interaction of Soil and Structures," Izd. Rechnoy Transport, Moscow, 1963.
26. Garber, E., Discussion of "The Calculus of Variations Applied to Stability of Slopes," Geotechnique, 1978, Vol. 28, pp. 223-227.
27. Giroud, J.P., Tran-Vo-Nhiem, and Obin, J.P., "Tables pour le Calcul des Foundations," Dunod, Paris, 1973.
28. Hansen, B., "Line Rupture Regarded as Narrow Rupture Zones," Proc. Conf. Earth Pressure Problems, Brussels, 1958, pp. 39-48.
29. Hansen, B., "A Theory of Plasticity for Ideal Frictionless Materials," Teknisk Forlag, Copenhagen, 1965.

30. Heinz, W.F., "Zur Tragfähigkeit von Flachgründungen auf Rolligem Boden," Bull. No. D83, Tech. Univ., Berlin, 1970.

31. Hill, R., "The Mathematical Theory of Plasticity," Clarendon Press, Oxford, 1950.

32. Hettiaratchi, D.R.P., and Reece, A.R., "Boundary Wedges in Two-Dimensional Passive Soil Failure," Geotechnique, 1975, Vol. 25, pp. 197-220.

33. Hoeg, K., "Finite Element Analyses of Strain-Softening Clay," Jl. Soil Mech. and Foundt. Div., ASCE, Vol. 98, No. SM1, 1972, pp. 43-58.

34. James, R.G., and Bransby, P.L., "Experimental and Theoretical Investigations of a Passive Earth Pressure Problem," Geotechnique, 1970, Vol. 20, pp.17-37.

35. James, R.G., and Lord, J.A., "An Experimental and Theoretical Study of an Active Earth Pressure Problem Relevant to Braced Cuts in Sand," Proc. 5th Europ. Conf. Soil Mech. and Foundt. Engg., Madrid, 1972, Vol. 1, pp. 29-38.

36. Janbu, N., "Slope Stability Computations," Embankment-Dam Engineering, Casagrande Volume, J. Wiley, N.Y., 1973, pp. 47-86.

37. Kopaczy, J., "Uber die Bruchflächen und Bruchspannungen in den Erdbauten," Jaky Memorial Volume, Akad. Kiado, Budapest, 1955, pp. 81-99.

38. Law, K.T., and Lumb, P., "A Limit Equilibrium Analysis of Progressive Failure in the Stability of Slopes," Can. Geotech. Jl., 1978, Vol. 15, pp. 113-122.

39. Lee, I.K., and Herington, J.R. "Effect of Wall Movement on Active and Passive Pressures," Jl. Soil Mech. and Foundt. Div., ASCE, Vol. 98, No. SM6, 1972, pp. 625-640.

40. Lee, I.K., and Herington, J.R., "A Theoretical Study of the Pressures Acting on a Rigid Wall by a Sloping Earth or Rock Fill," Geotechnique, 1972, Vol. 22, pp. 1-26.

41. Lewin, P.A., "The Influence of Stress History on the Plastic Potential," Symp. on Plasticity and Soil Mechanics, Cambridge, 1973, pp. 96-106.

42. Lo, K.Y., "Stability of Slopes in Anisotropic Soils," Jl. Soil Mech. Foundt. Div., ASCE, Vol. 91, No. SM4, 1965, pp. 85-106.

43. Lo, K.Y., and Lee, C.F., "Stress Analysis and Slope Stability in Strain-Softening Material," Geotechnique, 1973, Vol. 23, pp. 1-11.

44. Meyerhof, G.G., "An Investigation of the Bearing Capacity of Shallow Footings on Dry Sand," Proc. 2nd Int. Conf. Soil Mech., Rotterdam, 1948, Vol. 1, pp.237-243.

45. Meyerhof, G.G., "The Ultimate Bearing Capacity of Foundations, Geotechnique, 1951, Vol. 2, pp. 301-332.

46. Meyerhof, G.G., "The Bearing Capacity of Foundations Under Eccentric and Inclined Loads," Proc. 3rd Int. Conf. Soil Mech., Zurich, 1953, Vol. 1, pp. 440-445.

47. Meyerhof, G.G., "Influence of Roughness of Base and Ground-Water Conditions on the Ultimate Bearing Capacity of Foundations," Geotechnique, 1955, Vol. 5, pp. 227-242.

48. Meyerhof, G.G., "The Ultimate Bearing Capacity of Foundations on Slopes," Proc. 4th Int. Conf. Soil Mech., London, 1957, Vol. 1, pp. 384-386.

49. Meyerhof, G.G., "The Ultimate Bearing Capacity of Wedge-Shaped Foundations," Proc. 5th Int. Conf. Soil Mech., Paris, 1961, Vol.2, pp. 105-109.

50. Meyerhof, G.G., "Uplift Capacity of Foundations Under Oblique Loads," Can. Geotech. Jl., 1973, Vol. 10, pp. 64-70.

51. Meyerhof, G.G., "Bearing Capacity and Settlement of Pile Foundations," Jl. Geotech. Engg. Div., ASCE, Vol. 102, No. GT3, 1976, pp. 195-228.

52. Meyerhof, G.G., "Bearing Capacity of Anisotropic Cohesionless Soils," Can. Geotech. Jl., 1978, Vol. 16, pp. 592-595.

53. Meyerhof, G.G. and Adams, J.I., "The Uplift Capacity of Foundations," Can. Geotech. Jl., 1968, Vol. 5, pp. 225-244.

54. Meyerhof, G.G. and Hanna, A., "Ultimate Bearing Capacity of Foundations on Layered Soils under Inclined Load," Can. Geotech. Jl., 1978, Vol.15, pp. 565-572.

55. Meyerhof, G.G. and Valsangkar, A.J., "Bearing Capacity of Piles in Layered Soils," Proc. 9th Int. Conf. Soil Mech., Tokyo, 1977, Vol. 2, pp. 645-650.

56. Milligan, G.W.E., and Bransby, P.L., "Combined Active and Passive Rotational Failure of a Retaining Wall," Geotechnique, 1976, Vol. 26, pp. 473-494.

57. Morgenstern, N., Blight, G.E., Janbu, N., and Resendiz, D., "Slopes and Excavations," Proc. 9th Int. Conf. Soil Mech., Tokyo, 1977, Vol. 2, pp. 605-650.

58. Morgenstern, N., and Price, V.E., "The Analysis of the Stability of General Slip Surfaces," Geotechnique, 1965, Vol. 15, pp. 79-93.

59. Muhs, H., and Weiss, K., "Die Grenztragfähigkeit von Flach Gegründeten Streifenfundamenten unter Geneigter Belastung nach Theorie und Versuch," DEGEBO, Bull. No. 31, Tech. Univ., Berlin, 1975.

60. Poorooshasb, H.B., Holubec, I., and Sherbourne, A.N., "Yielding and Flow of Sand in Triaxial Compression," Can. Geotech. Jl., 1966, Vol. 3, pp. 179-190, and 1967, Vol. 4, pp. 376-397.

61. Prager, W., and Drucker, D.C., "Soil Mechanics and Plasticity Analysis or Limit Design," Quart. Appl. Math, 1952, Vol. 10, pp. 157-165.

62. Reese, L.C., "Design and Construction of Drilled Shafts," Jl. Geotech. Engg., Div., ASCE, Vol. 104, No. GT1, 1978, pp. 91-116.

63. Roscoe, K.H., "The Influence of Strains in Soil Mechanics", Geotechnique, 1970, Vol. 20, pp. 129-170.

64. Roscoe, K.H., and Bassett, R.H., and Cole, E.R.L., "Principal Axes Observed During Simple Shear of a Sand," Proc. Europ. Geotech. Conf., Oslo, 1967, Vol. 1, pp. 231-237.

65. Rowe, P.W., "The Stress-Dilatancy Relation for Static Equilibrium of an Assembly of Particles in Contact," Proc. Royal Soc., Series A, London, 1962, Vol. 269, pp. 500-527.

66. Rowe, P.W.,"Progressive Failure and Strength of a Sand Mass," Proc. 7th Int. Conf. Soil Mech., Mexico, 1969, Vol. 1, pp. 341-349.

67. Rowe, P.W., and Peaker, K., "Passive Earth Pressure Measurements," Geotechnique, 1965, Vol. 15, pp. 57-79.

68. Salencon, J., Croc, M., Michel, G., and Pecker, A., "Plasticité," Compt. Rend. Acad. Science, Series A, Paris, 1973, Vol. 276, pp. 1569-1572.

69. Smith, I.A.A., "Stress and Strain in a Sand Mass Adjacent to a Model Wall", Ph.D. Thesis, Cambridge Univ., Cambridge, England, 1972.

70. Sokolovski, V.V.,"Statics of Granular Media," Pergamon Press, London, 1965.

71. Taylor, D.W.,"Stability of Earth Slopes," Jl. Boston Soc. Civil Engrs.,1937, Vol. 24, pp. 197-246.

72. Tcheng, Y., "Stability of Rigid Structures," Discussion, Proc. 5th Europ. Conf. Soil Mech. and Foundt. Engg., Madrid, 1972, Vol. 2, pp. 131-134.

73. Terzaghi, K., "Theoretical Soil Mechanics, J. Wiley, N.Y., 1943.

74. Terzaghi, K., and Peck, R.B., "Soil Mechanics in Engineering Practice," J. Wiley, N.Y., 1967.

75. Vesic, A.S., "Analysis of Ultimate Loads of Shallow Foundations," Jl. Soil Mech. and Foundt. Div., ASCE, Vol. 99, No. SMI, 1973, pp. 47-73.

76. Zienkiewicz, O., Humpheson, C., and Lewis, R.W., "Associated and Non-Associated Visco-Plasticity and Plasticity in Soil Mechanics," Geotechnique, 1975, Vol. 25, pp. 671-689.

A PLASTICITY MODEL FOR THE LOAD UNLOAD BEHAVIOUR OF SAND

By

Rafael Baker[1], Sam Frydman[1], A.M. ASCE and J. Galil[2]

ABSTRACT

The development of an analytical model for the load-unload behaviour of sand is important, particularly as a means for providing an understanding of the liquefaction phenomenon. This paper describes the first stage of an attempt to develop such a model within the framework of plasticity theory.

The model developed consists of a non-linear elastic part together with a plastic part made up of two independently hardening yield functions (spherical and deviatoric). An associative flow rule was used for the spherical part and a non associative rule for the deviatoric part. It was found that the model reasonably predicted the triaxial behaviour of dense sand during one undrained compression-extension cycle, using parameters derived from drained triaxial tests. No agreement could be obtained, however, in further cycles. This is seen as being due to the difficulty in estimating the effect of stress reversal and cyclic loading on the parameters of the model. Further investigation of the form of the hardening characteristics of the model is required in order to extend its use into a multi cycle regime.

KEYWORDS: Cyclic loads; Liquefaction; Plasticity; Pore water pressure; Sand; Triaxial Tests.

---
1. *Senior Lecturers, Technion - Israel Institute of Technology*, Haifa, ISRAEL
2. *Formerly Research Student, Technion - Israel Institute of Technology*, Haifa, ISRAEL

INTRODUCTION

The need to design major structures on sites which may be subjected to earthquake effects requires an understanding of the response of soils to repeated loading. It has been observed that such loading applied to saturated, cohesionless soils leads to progressive increase in both pore water pressure and strain, possibly resulting in failure. Failure under these conditions has loosely been given the name "liquefaction". In spite of the immense practical importance of this phenomenon, no satisfactory basic model has yet been developed to explain and predict the build up of pore pressure in a cohesioless soil during undrained, cyclic loading.

This paper presents the first stage of an attempt to describe the undrained, cyclic behaviour of sand in terms of an elastic-plastic model.

BASIC CONSIDERATIONS

Continium mechanics provides a wide range of constitutive models which may be used to approximately describe the behaviour of real materials. The choice of a particular model depends both on the material being considered, and the phenomenon being described. In seeking a model to describe the behaviour of granular soils under undrained cyclic loading, a number of observations, and their implications with regards the choice of a suitable constitutive model, will be considered.

(a) It has been commonly observed that other than for inertia effects, the rate of loading has insignificant effect on the behaviour of granular soils in undrained conditions. This is manifested, for example, by the insensitivity of sand behaviour to the frequency of loading (eq. Wong (15)). An immediate consequence of this observation is that viscous models are unsuitable describers of sand behaviour, and commonly used procedures for liquefaction analysis, such as the SHAKE program (Schnabel (13)), can be at best considered empirically useful tools. It follows that a physically appropriate model for the description of the above behaviour must contain elastic and/or plastic components. Such a model conventionally admits three types of loading - first loading, unloading and reloading. During the first type both reversible (elastic) and irreversible (plastic) strains are generated, while in the latter two, only elastic strains develop.

(b) Finn and Coworkers (3,11), made the important observation that "In general we have observed in laboratory cyclic simple shear tests that most of the volume changes in dry sands, and the increases in pore-water pressure in undrained saturated sands occur during the unloading portions of the load cycle".

A study of test results published by other investigators shows that a similar phenomenon occurs under different test conditions, such as conventional triaxial liquefaction testing conditions in which the radial stress remains constant, and has been found by the authors to occur in undrained cyclic triaxial tests in which the mean total principal stress was held constant (this test type is referred to hereafter as an undrained pure deviatoric test). Results of one such test on a dense sand specimen are shown in Figure 1.

Fig. 2 shows results of a drained pure deviatoric test on a saturated dense sand specimen, showing the decrease in volume which was observed during unloading. Since behaviour during unloading is assumed to be governed by the elastic portion of the constitutive model, as suggested previously, it follows that the pore pressure build up during liquefaction is essentially a manifestation of the elastic behaviour of the soil. The simplest elastic model, which is specified in terms of two constants (eq. shear and bulk modulii), shows no cross effect - i.e. a change in deviatoric stress only will result in no volume change; this is true even if the two constants are strain dependent. A convenient way of introducing the cross effects is to formulate the elastic model in terms of the elastic strain energy function (eq. Desai (1)) as is done below. This cross effect between deviatoric and spherical regimes is generally a second order effect which can be reasonably ignored.

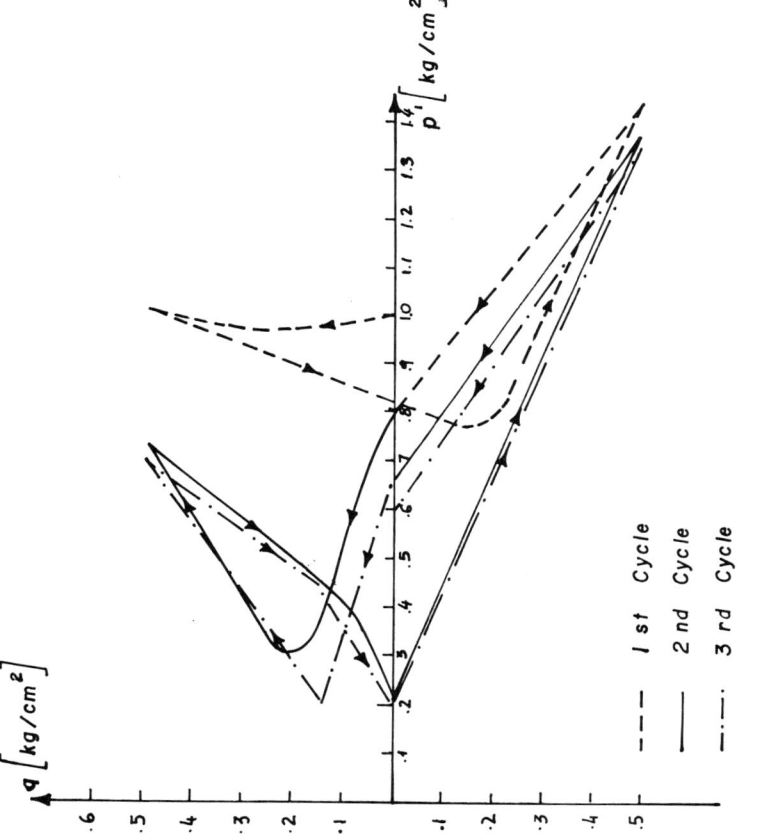

Fig l. Effective Stress Path

Figure 2, for example shows that the volumetric strain which occurred during pure deviatoric unloading was insignificant compared to that which occurred during loading. Only during undrained unloading, when plastic strains are absent and the incompressibility of water magnifies the effect does it became of major significance.

(c) Plastic models are specified in terms of three functions - a yield function, a plastic potential function and a hardening function. In addition to these three functions, a full definition of the plastic model requires a description of the nature of hardening, whether isotropic, kinematic or a combination thereof. As long as only monotonic loading conditions are being considered, the nature of hardening is irrelivent, and this has been the case in most published work on soil plasticity. On the other hand, the nature of hardening must be specified if the cyclic behaviour of soil is being investigated. Useful information on this point may be obtained from the observations of Ishihara and coworkers (7,8,9,14) from cyclic triaxial tests on sand specimens that as long as loading in any direction (compression or extension) is kept below a certain value, subsequent loading in the other direction shows almost no memory of this previous loading history, and the specimen behaves during the subsequent loading in a manner similar to that under virgin loading. While this observation is not sufficient to enable a formulation of the nature of hardening, it is in clear contradiction to isotropic hardening, and implies that when the soil is loaded into extension during the first load cycle, having undergone previous load-unload in compression, plastic strains will be initiated exactly as if no previous loading had occurred. This observation is therefore sufficient to specify the behaviour of the soil during one complete compression-extension loading cycle, without a full description of the hardening law. Study of the behaviour of sand under multiple cycles, however, must await the development of a more complete hardening law.

In the following, the elastic and plastic components of a constitutive model are developed in detail.

THE ELASTIC MODEL

Consider an ideal elastic material whose constitutive equation is defined in terms of a strain energy function, G. It is assumed that G depends on the first two strain invariants, taken as ε (volumetric strain, positive in compression) and γ (octahedral shear strain).

It is possible to write :

$$p' = \frac{\partial G}{\partial \varepsilon^e} \quad (1a)$$

and

$$q = \frac{\partial G}{\partial \gamma^e} \quad (1b)$$

where, for triaxial conditions p' is the effective spherical stress, $\dfrac{\sigma'_a + 2\sigma'_r}{3}$ and q is the octahedral shear stress, $\sqrt{\dfrac{2}{9}(\sigma_a - \sigma_r)^2}$, σ'_a = axial effective stress and σ'_r = radial effective stress.

From eqns. (1a) and (1b):

$$dp' = \frac{\partial^2 G}{(\partial \varepsilon^e)} d\varepsilon^e + \frac{\partial^2 G}{\partial \gamma^e \partial \varepsilon^e} d\gamma^e = G_{\varepsilon\varepsilon} d\varepsilon^e + G_{\gamma\varepsilon} d\gamma^e \quad (2a)$$

and

$$dq = \frac{\partial^2 G}{\partial \varepsilon^e \partial \gamma^e} d\varepsilon^e + \frac{\partial^2 G}{(\partial \gamma^e)^2} d\gamma^e = G_{\varepsilon\gamma} d\varepsilon^e + G_{\gamma\gamma} d\gamma^e \quad (2b)$$

In the case of linear elastic material, there is no crosseffect between the deviatoric and spherical stress and strain components, and it is possible to write:

$$p' = \frac{\partial G}{\partial \varepsilon^e} = K\varepsilon^e \quad (3a)$$

and

$$q = \frac{\partial G}{\partial \gamma^e} = G\gamma^e \quad (3b)$$

Consequenctly, for linear elastic material:

$$dp' = Kd\varepsilon^e \quad (4a)$$

and

$$dq = Gd\gamma^e \quad (4b)$$

If such a material is tested under pure deviatoric loading, without drainage, then $dp = d\varepsilon^e = 0$ and $dp' = dp - du$, so that $du = 0$.

PLASTICITY MODEL FOR SAND

Consequently, it is seen that for a linear elastic material no pore pressure change of volume change can occur under pure deviatoric loading conditions. However the pore pressure increases during unloading observed by Finn et al. and seen in Figure 1, represent an elastic unloading effect (since plastic strains are irreversible). Therefore it is clear that a linear elastic model is inconsistent with observed behaviour of sand. A cross effect is evident and the simple linear equations (3) and (4) must be replaced by equations (2), in which the function $G_{\gamma\epsilon} = G_{\epsilon\gamma}$ describes the cross effect, while functions $G_{\gamma\epsilon}$ and $G_{\epsilon\gamma}$ describe the direct efects between volumetric and deviatoric regimes.

Applying equations (2) to a drained, pure deviatoric test, in which $dp' = 0.$:

$$0 = G_{\epsilon\epsilon} d\epsilon^e + G_{\gamma\epsilon} d\gamma^e \quad (5a)$$

$$dq = G_{\epsilon\gamma} d\epsilon^e + G_{\gamma\gamma} d\gamma^e \quad (5b)$$

Solving for $d\gamma^e$ in (5a) and substituting in (5b) :

$$dq = G_{\epsilon\gamma} d\epsilon^e - G_{\gamma\gamma} \cdot \frac{G_{\epsilon\epsilon} d\epsilon^e}{G_{\gamma\epsilon}} = \frac{G_{\gamma\epsilon} - G_{\gamma\gamma} G_{\epsilon\epsilon}}{G_{\gamma\epsilon}} \cdot d\epsilon^e$$

or $\quad \left(\dfrac{d\epsilon^e}{dq}\right)_{p'} = \dfrac{G_{\gamma\epsilon}}{G_{\gamma\epsilon}^2 - G_{\gamma\gamma} G_{\epsilon\epsilon}} \quad (6a)$

where the subscript p' on the left hand side signifies constant p'. Equation (6a) describes the change of volumetric strain as a function of deviatoric stress. Obviously, if the material is linear and $G_{\gamma\epsilon} = 0$, then $d\epsilon^e/dq = 0$.

At the present stage, the rate of change of ε^e with q will be assumed linear, and equation (6a) will be assumed to be of the form:

$$\left(\frac{d\varepsilon^e}{dq}\right)_{p'} = \frac{G_{\gamma\varepsilon}}{G_{\gamma\varepsilon}^2 - G_{\gamma\gamma}G_{\varepsilon\varepsilon}} = a \tag{6b}$$

where a may be dependent on plastic strain level, but is constant with respect to elastic strain.

Now applying equations (2) to an undrained pure deviatoric test, in which $dp = d\varepsilon^e = 0$,

$$dp' = dp - du = -du = G_{\gamma\varepsilon} d\gamma^e \tag{7a}$$

and

$$dq = G_{\gamma\gamma} d\gamma \tag{7b}$$

Therefore

$$\left(\frac{du}{dq}\right)_{p_\varepsilon} = -\frac{G_{\gamma\varepsilon}}{G_{\gamma\gamma}} \tag{8}$$

and $G_{\gamma\varepsilon} = -G_{\gamma\gamma} \cdot \left(\frac{du}{dq}\right)_{p_\varepsilon}$ \hfill (9)

Substituting eqn. (9) into eqn. (6):

$$\left(\frac{d\varepsilon^e}{dq}\right)_{p'} = \frac{\left(\frac{du}{dq}\right)_{p_\varepsilon}}{G_{\varepsilon\varepsilon} - G_{\gamma\gamma}\left(\frac{du}{dq}\right)_{p_\varepsilon}^2} \tag{10}$$

Equation (10) defines the relationship between volumetric strain in a drained deviatoric test and pore pressure development in an undrained deviatoric test.

Consideration will now be given to experimentally observed behaviour of granular soil. El Sohby (2) and others found that the elastic, volumetric strain which results from spherical loading may be expressed by an equation of the form:

$$\varepsilon^e = c\left(\frac{p'}{p_o}\right)^m \tag{11}$$

where c, m are factors depending on the properties of the sand particles, and porosity of the mass, and p_o is a reference spherical stress such as atmospheric pressure.

Differentiating equation (11) :

$$d\varepsilon^e = \left(\frac{cm}{p_o^m}\right)(p')^{m-1} \cdot dp' \tag{12}$$

Although equation (11) was suggested on the basis of results of tests performed with deviatoric stress q=0, the authors have carried out a series of tests with q=const and p' increasing (Fig.3) which suggest that equation (12) may also be applicable to this condition.

In the case of drained, pure deviatoric tests, it is common to relate the shear stress increment to the elastic shear strain increment through a shear modulus G, such that :

$$dq = G \, d\gamma^e \tag{13}$$

For granular material, G is found to be dependent on effective spherical stress, through an expression of the form :

$$G = G_o\left(\frac{p'}{p_o'}\right)^{1/2} \tag{14}$$

where G_o is a dimensionless stiffness parameter, dependent on the shear strain level. Equation (13) can then be rewritten :

$$dq = G_o\left(\frac{p'}{p_o'}\right)^{1/2} d\gamma^e \tag{15}$$

Now the functions $G_{\varepsilon\varepsilon}$, $G_{\gamma\gamma}$ and $G_{\varepsilon\gamma}$ can be related to the experimental parameters appearing in equations (11)-(15). By considering a q constant test, equation (2) together with equation (12) yields :

$$\frac{G_{\gamma\gamma}}{G_{\varepsilon\varepsilon}G_{\gamma\gamma}-G_{\varepsilon\gamma}^2} = (\frac{cm}{p_o^m})(p')^{m-1} \tag{16a}$$

Considering a p' constant test, equations (2) together with equation (15) yields:

$$\frac{G_{\varepsilon\varepsilon}G_{\gamma\gamma}-G_{\gamma\varepsilon}^2}{G_{\varepsilon\varepsilon}} = G_o (\frac{p'}{p_o'})^{1/2} \tag{16b}$$

From (16a) and (16b) :

$$G_{\gamma\gamma} = G_{\varepsilon\varepsilon} \cdot \frac{G_o \, c \, m}{(p_o')^{(m+1/2)}} (p')^{(m-1/2)} \tag{16c}$$

Returning to equations (2), they may be rewritten :

$$\frac{dp'}{G_{\varepsilon\gamma}} = \frac{G_{\varepsilon\varepsilon}}{G_{\varepsilon\gamma}} d\varepsilon^e + d\gamma^e$$

and $\frac{dq}{G_{\gamma\gamma}} = \frac{G_{\varepsilon\gamma}}{G_{\gamma\gamma}} d\varepsilon^e + d\gamma^e$

Solving these two equations for $d\varepsilon$, using equations (16a) and (6b) :

$$d\varepsilon^e = (\frac{cm}{p_o^m})(p')^{m-1} dp' + a \, dq \tag{17a}$$

In a similar fashion eqns. (2) can be solved for $d\gamma$, using equations (16b) and (6b) to give :

$$d\gamma^e = dq/G_o \ (\frac{p'}{p'_o})^{1/2} + a \ dp' \tag{17b}$$

From equation (17a), for the case of an undrained pure deviatoric test in which $d\epsilon$ is zero and $dp' = dp-du = du$;

$$du = \frac{a \ dq}{(\frac{cm}{p_o^m}) \ (p')^{m-1}} \tag{17c}$$

and $\quad d\gamma^e = \dfrac{dq}{G_o (\frac{p'}{p'_o})^{1/2}} - a \ du \tag{17d}$

Equations (17c) and (17d) enable calculation of the build up of pore pressure and shear strain in an elastic undrained specimen loaded under pure deviatoric conditions. The values of c, m, G_o and a can all be obtained from drained tests. On the basis of the observations of Finn et al., and the results shown in Figs. 1 and 2, it is clear that a is a negative quantity; it is obtained from the slope of the unload branch of Fig. 2. Consequently, during loading, the elastic portion of the model generates negative pore pressures or volume increases.

THE PLASTIC MODEL

There is experimental evidence (Poorooshasb (12), Frydman (4)) indicating that $\frac{q}{p'}$ serves as a yield function. However, this ratio cannot define the complete yield function, since it implies zero plastic strains along spherical loading paths. Consequently, it is assumed here that the overall plastic model consists of two systems - that due to spherical loading and that due to deviatoric loading, which are independent of each other. The total plastic strain increment is assumed equal to the sum of the increments generated by each system alone. A similar approach has been used by Lade (10).

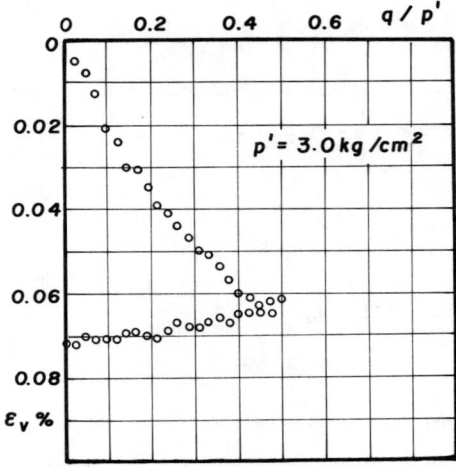

Figure 2

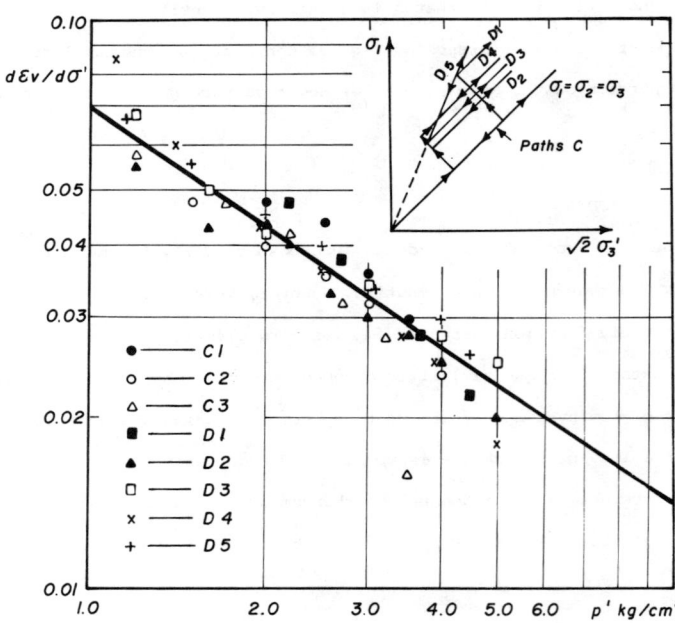

Figure 3

The deviatoric system used here was modified from the model developed by Frydman (4,5,6), and consists of the following components :

(a) A yield function $f^d = \frac{q}{p'}$, (18)

where the superscript d signifies the deviatric system.

(b) A plastic potential function

$$g^d = \frac{q}{p'} + K \ln \frac{p'}{p'_o}$$ (19)

where K is a soil parameter quantifying the departure from normality.

The value of K was found by Frydman (4) from drained pure deviatoric tests on specimens of glass microspheres to be approximately equal to $\tan \phi_\mu$ in both compression and extension, where ϕ_μ = true friction angle of the particles. However, in a similar series of tests performed as part of this study on relatively dense sand specimens, K was found to be equal to about 0.42 in compression (approximately $\tan \phi_\mu$), but to only about 0.28 in extension.

(c) A hardening function

$$H^d = A \bar{w}_p^B$$ (20a)

where A and B are soil parameters, and $\bar{w}_p$ is given by (Frydman (5)) :

$$\bar{w}_p = \int_o^\gamma p' \, d\gamma$$ (20b)

The original model developed by Frydman assumed the existence of an initial yield surface $(\frac{q}{p'})_y$ below which the material is elastic. This assumption is inacceptable in the present framework since it would imply the development of negative pore pressure during loading below $(\frac{q}{p'})_y$ (since the elastic cross-effect parameter a has been shown to be negative), contrary to observations (eq. Fig. 1) Consequently it is assumed that yielding occurs from the start of loading, and that the pore pressure build up observed during loading is due to the plastic effect overshadowing the elastic one.

Refer now to the spherical system. If yielding occurs from a stress state which is below the current deviatoric yield surface, but on the spherical surface, the volumetric plastic strain increment may be written as :

$$d\varepsilon^p = H^s \frac{\partial g^s}{\partial p'} \left(\frac{\partial f^s}{\partial p'} dp' + \frac{\partial f^s}{\partial q} dq \right)$$

where the superscript s refers to the spherical system For a spherical load increment (i.e. dq=0) :

$$d\varepsilon^p_{q=const} = H^s \frac{\partial g^s}{\partial p'} \frac{\partial f^s}{\partial p'} dp'$$

A series of tests was carried out in which $d\varepsilon^p$ was measured for various loading paths below the current deviatoric yield surface. The results of these tests are shown in Fig. 4, and it is seen that

$$\left(\frac{d\varepsilon^p}{dp'}\right)_{q=const} = \left(\frac{d\varepsilon^p}{dp'}\right)_{q \neq const}$$

This implies that $\frac{\partial f^s}{\partial q} dq = 0$, and hence that f^s is equal to p'.

It is further assumed that the spherical system is associative, so that

$$g^s = f^s = p' \tag{21}$$

PLASTICITY MODEL FOR SAND

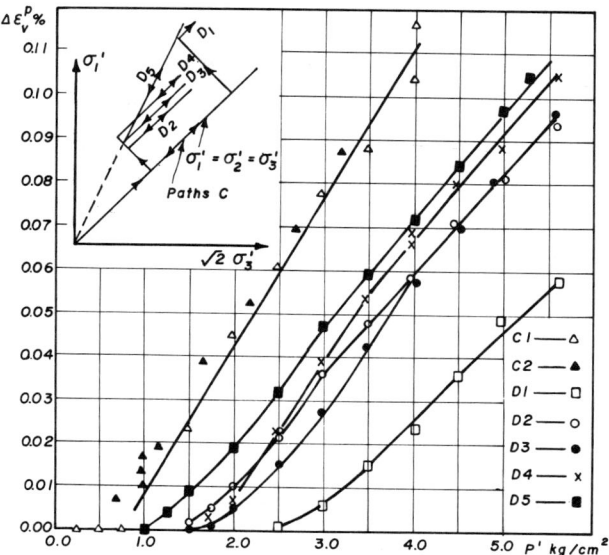

Figure 4

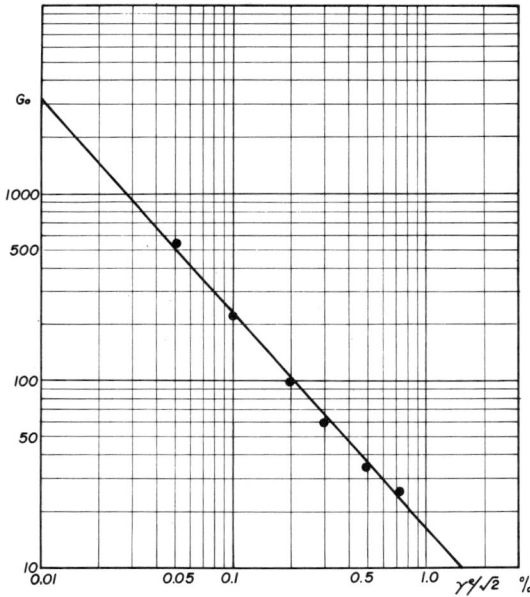

Figure 5

In order to complete the description of the spherical system, the hardening function H^s should be specified. This requirement is, however, avoided in the present study by making use of the observation (Elsohby (2)) that total volumetric strain resulting from spherical loading may be described by an equation similar to equation (11), as follows :

$$\varepsilon = \varepsilon^e + \varepsilon^p = c_2 \, (\frac{p'}{p_o})^{m_2}$$

leading to :

$$d\varepsilon = (\frac{c_2 m_2}{p_o^{m_2}}) \, (p')^{m_2 - 1} \, dp' \qquad (22)$$

where c_2 and m_2 are parameters similar to c and m in equation (11), but they refer to total rather than elastic volumetric strain. In principle, H^s can be obtained from equations (22) and (11); however this is not necessary for the present purpose, as will be seen below.

PREDICTION OF PORE PRESSURE AND SHEAR STRAIN DEVELOPMENT
DURING UNDRAINED TEST.

The Pore Pressure Scheme —— During the loading of a granular soil, the volumetric strain increment may be expressed by :

$$d\varepsilon = d\varepsilon^e + d\varepsilon^p \qquad (23)$$

Now
$$d\varepsilon^p = H \cdot \frac{\partial g}{\partial p'} \, df \qquad (24)$$

$$= H^d \frac{\partial g^d}{\partial p'} \, df^d + H^s \cdot \frac{\partial g^s}{\partial p'} \, df^s \qquad (25)$$

However $df^d = \frac{\partial f^d}{\partial p'} dp' + \frac{\partial f^d}{\partial q} dq$ (26a)

and $df^s = \frac{\partial f^s}{\partial p'} dp' + \frac{\partial f^s}{\partial q} dq$ (26b)

but $f^s = p'$, $df^s = dp'$ (27)

Substituting equations (26) and (27) into (25)

$d\varepsilon^p = H^d \frac{\partial g^d}{\partial p'} (\frac{\partial f^d}{\partial p'} dp' + \frac{\partial f^d}{\partial q} dq) + H^s dp' =$

$= (H^d \frac{\partial g^d}{\partial p'} \frac{\partial f^d}{\partial p'} + H^s) dp' + H^d \frac{\partial g^d}{\partial p'} \frac{\partial f^d}{\partial q} dq$ (28)

Now consider the loading system to be made up of a deviatoric and a spherical component. For a pure deviatoric loading component (i.e. dp'=0) :

$d\varepsilon^p_{p'=const} = H^d \frac{\partial g^d}{\partial p'} \frac{\partial f^d}{\partial q} dq$ (29)

From equation (17a), or as defined in equation (6b), the elastic increment of volumetric strain which develops during pure deviatoric loading is given by :

$d\varepsilon^e_{p'=const} = a \, dq$ (30)

and therefore $d\varepsilon_{p'=const} = H^d \frac{\partial g^d}{\partial p'} \frac{\partial f^d}{\partial q} dq + a \, dq$

For a spherical loading component, although use could once again be made of equation (28) it is more convenient to use the expression (22) :

$d\varepsilon_{q=const} = (\frac{c_2 m_2}{p_o^{m_2}}) (p')^{m_2-1} dp'$ (22)

Consequently, the total volumetric strain resulting from any load increment is given by :

$$d\varepsilon = (\frac{c_2 m_2}{p_o^{m_2}})(p')^{m_2-1} dp' + H^d \frac{\partial g^d}{\partial p'} \frac{\partial f^d}{\partial q} dq + a\, dq \qquad (31)$$

Now, for an undrained test, equation (31) must be equated to zero. Furthermore, the following expression may be written :

$$dp' = dp - du \; ; \quad du = \frac{\partial u}{\partial p} dp + \frac{\partial u}{\partial q} dq$$

For a saturated soil, $\frac{\partial u}{\partial p} = 1$

Therefore $\quad dp' = dp - dp - \frac{\partial u}{\partial q} dq = \frac{\partial u}{\partial q} dq \qquad (32)$

Substituting (32) into (31) and equating to zero :

$$\frac{\partial u}{\partial q} = \frac{H^d \frac{\partial g^d}{\partial p'} \frac{\partial f^d}{\partial q} dq + a\, dq}{(\frac{c_2 m_2}{p_o^{m_2}})(p')^{m_2-1}} \qquad (33)$$

But from equations (18a) and (19) :

$$\frac{\partial g^d}{\partial p'} = -\frac{q}{(p')^2} + K/p'$$

$$= \frac{1}{(p-u)} (K - \frac{q}{(p-u)}) \qquad (34)$$

and $\quad \dfrac{f^d}{q} = \dfrac{1}{p'} = \dfrac{1}{(p-u)} \qquad (35)$

Substituting (34) and (35) into (33) :

$$\frac{\partial u}{\partial q} = \frac{H^d [K-q/(p-u)] + a}{(\frac{c_2 m_2}{p_o^{m_2}})(p')^{m_2-1}} \qquad (36)$$

Equation (36) defines the pore pressure development to be expected due to increase of deviatoric stress q during an undrained test. If q is decreasing, then obviously no plastic effects will be taking place, and equation (36) must be replaced by

$$\frac{\partial u}{\partial q} = \frac{a}{(\frac{c_m}{p_o^m})(p')^{m-1}}$$

which is identical to equation (17c).

The Shear Strain Scheme — In a similar fashion to the above, we may write:

$$d\gamma = d\gamma^e + d\gamma^p \qquad (37)$$

Now, since the spherical portion of the plastic model results in spherical strains only, the total plastic strain increment $d\gamma^p$ may be expressed by :

$$d\gamma^p = 2 H^d \frac{\partial g^d}{\partial q} df^d \qquad (38)$$

From equation (19) :

$$\frac{\partial g^d}{\partial q} = \frac{1}{p'} \qquad (39a)$$

From equation (18):

$$df^d = \frac{\partial f^d}{\partial p'} dp' + \frac{\partial f^d}{\partial q} dq = -\frac{q}{(p')^2} dp' + \frac{dq}{p'} \tag{39b}$$

Consequently, substituting equations (39) into equation (38):

$$d\gamma^p = \frac{2H^d}{(p-u)^3} [(p-u)dq - q\,dp'] \tag{40}$$

Combining equation (40) with the elastic strain increment given by equation (17d), the following expression is obtained for the total shear strain increment:

$$d\gamma = \frac{dq}{G_o(\frac{p'}{p_o})^{1/2}} - a\,du + \frac{2H^d}{(p-u)^3}[(p-u)dq - qdp'] \tag{41}$$

Equation (41) defines the shear strain increment to be expected during loading; during unloading only elastic effects will occur, and the third term on the right hand side of equation (41) is ignored.

The two equations (36) and (41) are two coupled simultaneous differential equations, the solution of which enables the prediction of pore pressure and shear strain development during an undrained test. The parameters appearing in the equations may be obtained from a series of drained tests, and then the equations may be used for prediction of behaviour under any total stress path. For any small stress increment, equation (36) is first applied to obtain an estimate of du, and this is then substituted into equation (41) in order to obtain an estimate of $d\gamma$.
The equations may be solved iteratively in order to obtain improved estimates.

These equations have been used in order to predict the behaviour of a dense sand specimen during one cycle of undrained, pure deviatoric compression and extension. The specimen was consolidated under an all around pressure of 1 kg/cm^2 (100 KN/m^2), and then loaded to an octahedral shear stress in compression and then in extension of 0.49 kg/cm^2 (49 KN/m^2). The prediction was carried out for the consolidated specimen, using octahedral shear stress increments of 0.01 kg/cm^2 (1 KN/m^2).

The Model Parameters Used for Prediction ——

(a) The hardening parameters - on the basis of test results presented

by Frydman (6) for the Ezra Uvitzaron sand tested in this study, the

hardening parameters A and B appearing in equation (20a) range between 1.5-3

and 1.7-1.2 respectively. Predictions were made with several combinations

from these ranges, and the results presented below are for A=3 and B=1.2.

For the present prediction $\bar{W}_p$ was summed over eachloading increment both in compression and extension.

(b) The plastic parameter K - on the basis of the drained pure deveatoric

tests carried out in this study, K was taken as 0.42 when loading

in compression and 0.28 when loading in extension.

(c) The elastic parameters G_o - from available test data for Ezra Uvitzaron sand, Fig. 5 was developed relating G_o to γ^e. This relationship is fitted by the expression $G_o = 24(\gamma^e)^{-1.16}$, with a maximum value of $G_o = 700$.

(d) The elastic cross-effect factor a - this factor was assumed to be related to plastic shear strain. As insufficient data was available for sands, to develop this dependence, results obtained from tests on glass microspheres specimens were also incroporated, and the relationship is shown in Fig. 6. The relationship used for prediction was $a = 0.3\ (\gamma^p)^{1.5} \leq 0.1$

(e) The spherical loading parameters c, m, c_2, m_2 - from a series of spherical load-unload tests performed as part of this study, values of these parameters for dense specimens were found to be :

$c = 0.23, \quad m = 0.30\ ;\ c_2 = 0.18,\ m_2 = 0.58.$

Comparison Between Prediction and Measurement ——— Fig. 7 shows a comparison between predicted and measured behaviour for a cyclic octahedral shear stress of 0.49 kg/cm^2 (49 KN/m^2). The agrement is seen to be excellent for both pore pressure and shear strain development. Similar agreement was obtained for a test carried out with an octahedral shear stress amplitude of 0.28 kg/cm^2 (28 KN/m^2).

Attempts to extend the prediction procedure into the second cycle were unsuccessful, and liquefaction was predicted early in the cycle, while in fact no such phenomenon occurred within 10 cycles.

CONCLUSIONS

The non linear elastic-plastic model presented in this paper was used successfully to predict the behaviour of a sand specimen under one cycle of undrained compression - extension shear. Attempts to extend this prediction to further cycles have so far been unsuccessful , the prediction generally

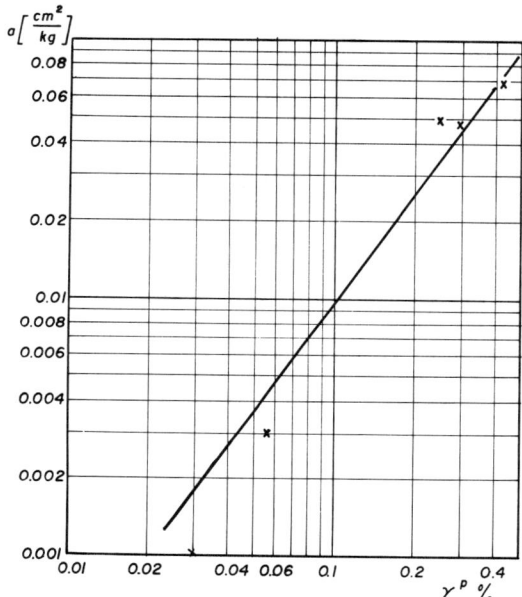

Figure 6

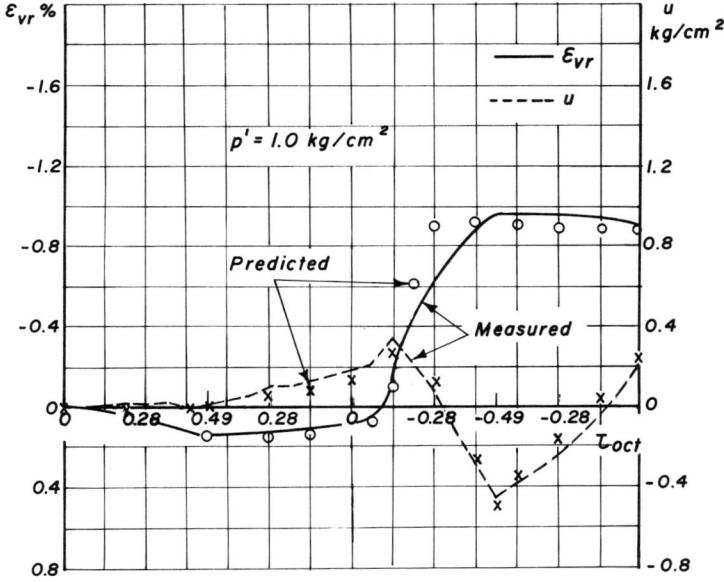

Figure 7

indicating a rapid development of pore water pressure and shear strain in the second cycle whereas the tests indicated no such phenomenon even after ten cycles. The difficulty in extension of the prediction is due mainly to insufficient information with regards the hardening nature of the model. The desire to apply a plastic model to the cyclic loading problem will require further study of this question. The present investigation has served the purpose, however, of indicating that the type of model necessary must include a non linear elastic component, in order to explain the pore pressure increases which develop during the unloading stages.

ACKNOWLEDGEMENT

The research described was partially based on the work carried out by the third author under the supervision of the other two towards the MSc degree.

APPENDIX 1 - REFERENCES

1. Desai, C.S., "State of the Art, Overview, Trends and Projections. Theory and Applications of the Finite Element Method in Geotechnical Engineering." Proc., Symposium on Application of the Finite Element Method in Geotechnical Engineering, Vicksburg. 1972, pp. 3-90. Edited by C.S. Desai.

2. El Sohby, M., "Elastic Behaviour of Sand", Journal of the Soil Mech. and Found. Eng. Div., ASCE, Vol. 95, No. SM6, 1969, pp. 1393-1409.

3. Finn, W.D.L., Lee, K.W. and Martin, G.R., "An Effective Stress Model for Liquefaction", Journal of the Geotechnical Engineering Division, ASCE, Vol. 103, No. GT6, 1977, pp. 517-532

4. Frydman, S., "An Inquiry into the Stress Strain Behaviour of Particulate Media", DSc. Thesis, Technion - Israel Institute of Technoloty, 1972.

5. Frydman, S., Zeitlen, J.G. and Alpan, I., "The Yielding Behaviour of Particulate Media", Canadian Geotech. J., Vol. 10, 1973, pp. 341-362.

6. Frydman, S., "The Strain Hardening Behaviour of Particulate Media", Canadian Geotech. J. Vol. 13, 1976, pp. 311-323.

7. Ishihara, K., Tatsuoka, F. and Yasuda, S., "Undrained Deformation and Liquefaction of Sand under Cyclic Stresses", Soils and Foundations, Vol. 15, No.1, 1975, pp. 29-44.

8. Ishihara, K., Lysmer, J., Yosuda, S. and Hirao, H., "Prediction of Liquefaction in Sand Deposits during Earthquakes", Soils and Foundations, Vol. 16, No.1, 1976, pp. 1-16.

9. Ishihara, K., and Okada, S., "Effects of Stress History on Cyclic Behaviour of Sand", Soils and Foundations, Vol. 18, No. 4, 1978, pp. 31-45.

10. Lade, P.V., "The Stress Strain and Strength Characteristics of Cohesionless Soils", Ph.D. Thesis, Univ. of California, Berkeley, 1972.

11. Martin, G.R., Finn, W.D.L. and Seed, H.B., "Fundamentals of Liquefaction under Cyclic Loading", Journal of the Geotechnical Engineering Division, ASCE, Vol. 101, No. GT5, 1975, pp. 423-438.

12. Poorooshasb, H.B., Holubec, I. and Sherbourne, A.N., "Yielding and Flow of Sand in Triaxial Compression", Canadian Geotech. J., Vol.3, No.4, 1966, pp. 179-190.

13. Schnabel, B., Lysmer, J. and Seed, H.B., "SHAKE - A Computer Program for Earthquake Response Analysis of Horizontally Layered Site". Report EERC 72-12, Earthquake Engineering Research Center, Berkeley, California.

14. Tatsuoka, F. and Ishihara, K., "Drained Deformation of Sand under Cyclic Stresses Reversing Direction", Soils and Foundations Vol. 14, No.3, 1974, pp. 51-65.

15. Wong, R.T., Seed, H.B. and Chan, C.K. "Cyclic Loading Liquefaction of Gravelly Soil", Journal of Geotechnical Engineering Div., ASCE, Vol. 101, No. GT6, 1975, pp. 571-583.

A REASSESSMENT OF LIMIT EQUILIBRIUM CONCEPTS IN GEOTECHNIQUE

Robin N. Chowdhury[1] M.ASCE

INTRODUCTION

The concept of limit equilibrium is rooted in classical soil mechanics and has had a considerable influence on the whole field of geotechnique incorporating related disciplines such as rock mechanics and engineering geology. It has made possible the use of simple theoretical approaches in traditional geotechnical practice thus placing the latter on a sound footing. It is an ingenious concept which helped surmount many obstacles to problem-solving at a time when sophisticated approaches had yet to be developed and relatively little was known about the mechanical behaviour of earth masses. Both before and after the era of modern computers it has provided geotechnical engineers and theoreticians with flexible and simple analytical tools.

Many researches and advances have been made in the field of geotechnical engineering in recent years. Interrelationships between soil mechanics, rock mechanics and engineering geology have been explored and their significance in practice emphasised. At the same time remarkable progress has been made in the area of stress analysis of continua and discontinua. Development of sophisticated numerical techniques (e.g. finite difference and finite element methods) and fast computers has facilitated this progress. Concentrated research efforts have been made to develop reliable stress-strain theories. Against this background it is remarkable that the limit equilibrium concept has survived the arrival and acceptance of sophisticated approaches and computations. Most practitioners still consider limit equilibrium analyses and concepts to be valid and reliable. Research papers which deal with such procedures also continue to be published (24,34) and even the latest scholarly or reference books emphasise the value of the basic concepts (12,40).

To some extent, this situation is a reflection of the imbalance between research and application or between theory and practice. Benefits from the vast research efforts of recent years have not been such as to warrant that limit equilibrium procedures be abandoned. It is certainly not the case that these procedures are always correct or successful. In fact, experienced practitioners use such calculations mainly as aids to judgement. Judicious and careful use of such tools will continue to be necessary. It is, therefore, most desirable to reassess their attributes and limitations and to consider possibilities of their further development.

<u>Different Limit Equilibrium Methods</u>.-A variety of methods and procedures based on the limit equilibrium concept, have been developed over several decades. Three broad groups of stability problems can be tackled by one or more of these methods viz. (1) earth retaining structures (active and passive pressures) (2) foundations (bearing capacity) and (3) soil and rock slopes (factor of safety). Almost all the interesting and controver-

---

[1] Reader, Department of Civil Engineering University of Wollongong, NSW, Australia, 2500.

sial aspects of limit equilibrium procedures are clearly brought out in the class of problems related to slopes. Accordingly this paper is devoted primarily to a consideration of the concept of limit equilibrium in relation to the stability of slopes or sloping earth masses.

Significant or commonly used methods include: Fellenius (Swedish) or ordinary method of slices, Taylor's modified Swedish method, Friction-circle method and Taylor's modified friction circle method, Bishop's method, Bishop's simplified method, Wedge or sliding block method, Morgenstern and Price method, Janbu's method, simplified Janbu method, Sarma's method, Spencer's method using parallel interslice forces, and Sarma's generalisation of the wedge method (34). Some of the methods were developed initially for failure surfaces which are circular in cross-section (e.g. friction-circle, Swedish, Bishop and Spencer's method). Spencer (38) developed a generalisation of his method for slip surfaces of arbitrary shape. Sometimes practitioners find it convenient to approximate real slip surfaces by corresponding circular shapes in order to use stability charts or simple methods of analysis. It may be noted that nearly all slope stability charts are based on the assumption of circular slip surfaces (a plane slip surface may be accommodated as a circle of infinite radius).

THE CONCEPT AND ITS ATTRIBUTES

Free Body Concept and Factor of Safety. -An earth mass is identified and regarded as a free body with known external slope boundaries and an internal boundary along a real or assumed discontinuity generally known as the slip surface. The latter may consist of one or more planes or it may be curved. The free body is under the action of disturbing and resisting forces. The former are predominantly body forces and the latter reflect the shearing resistance along the assumed slip surface. Equilibrium between forces requires that the real shear strength be divided by a factor F better known as the factor of safety. Thus the mobilised shear strength is given by:

$$\tau_m = \frac{\tau_f}{F} = \frac{c'}{F} + \sigma' \frac{\tan \phi'}{F} \qquad \ldots \ldots \ldots \ldots \ldots \ldots (1)$$

In this equation F is essentially a local safety factor at any point on a slip surface. In most limit equilibrium methods the overall factor of safety is considered identical to the local factor of safety. Further, the factor of safety with respect to the cohesive component of shear strength F_c is often considered identical to that with respect to the frictional component F_ϕ i.e.

$$F_c = \frac{c'}{c_m'} = F_\phi = \frac{\tan \phi'}{\tan \phi_m'} = F \qquad \ldots \ldots \ldots \ldots \ldots \ldots (2)$$

The value of F is the chief unknown in a limit equilibrium analysis. Simplified or rigorous methods may be used. In the former simplifying assumptions make calculations easy while force and moment equilibrium is not strictly satisfied. Often the overall factor of safety is obtained directly as the ratio

$$F = \frac{\text{Sum of Resisting Forces (or Moments)}}{\text{Sum of Disturbing Force (or Moments)}} \qquad \ldots \ldots \ldots \ldots \ldots (3)$$

For example such a ratio of forces is considered adequate for hard rock slopes with planar discontinuities (Fig.1) whereas a ratio of moments is appropriate for soil slopes where slip surfaces are usually curved (Figs.2 and 3). In some methods of analysis, e.g. Fellenius or Ordinary method of slices, definition for local safety factor as in Eq.1 is avoided. However in others, e.g. Bishop simplified method, Eq.1 forms the basis of calculation along with other assumptions (including constant F) before Eq.3 is used to obtain the expression for overall safety factor. As a third alternative, e.g. in the wedge method, Eq.1 is used as a basis of calculation considering successive wedges within the whole mass (free body) and Eq.3 is not used at all.

For slip surfaces which are single planes simplified and rigorous methods lead to identical results. Rigorous methods result in complex expressions for calculating F where slip surfaces are curved or consist of several planes. Simplified methods lead to convenient expressions even in these cases. Some simple formulae are given in table 1 and essential features of some rigorous methods are summarised in table 2.

The concept of limit equilibrium should not be confused with the condition of critical (or limiting) equilibrium. The latter is automatically incorporated as one of a variety of possible solutions in an analysis. This becomes clear in the case of slopes since values of the factor of safety F above one indicate stability and values below one indicate instability. Only a value F=1 corresponds to the theoretical condition of critical equilibrium.

<u>Bounds to Safety Factor F.</u>-The general slope stability problem is highly statically indeterminate and this is the reason for the cumbersome nature of rigorous methods of analysis e.g. Morgenstern and Price, Sarma etc. (see table 2). Considering a soil mass with a curved slip surface to be divided into n vertical slices (Fig.3) there are (6n-2) unknowns and only 4n equations. Hence (2n-2) assumptions are required to solve the problem even though F is considered constant along the slip surface.

It is important to note that a unique solution may not exist since a variety of stress distributions along the slip surface may be consistent with statical equilibrium. Each set of assumptions appropriate to a given rigorous method of analysis leads to a corresponding stress distribution along the assumed slip surface. Moreover, the value of F obtained by different methods (or by the same method using different assumptions) may not be the same.

This situation has suggested the useful concept of upper and lower bounds to the magnitude of F. For example, in the friction circle method the resultant normal reaction may be assumed to be concentrated at one point on the slip surface and this leads to a lower bound F(LB) for all values of F that satisfy statics. On the other hand if the resultant normal reaction is assumed to be concentrated at the two ends of the slip surface, the value of F is an upper bound F(UB). For a specific problem it has been shown by Lambe and Whitman (23) that F(UB)=1.62 and F(LB)=1.27. Intuitively reasonable stress distributions corresponded to the range F=1.30-1.36. It is conservative and hence popular to use the lower bound regardless of what stress distribution it implies. Moreover simplified methods, which are convenient to use, lead to conservative, lower bound results.

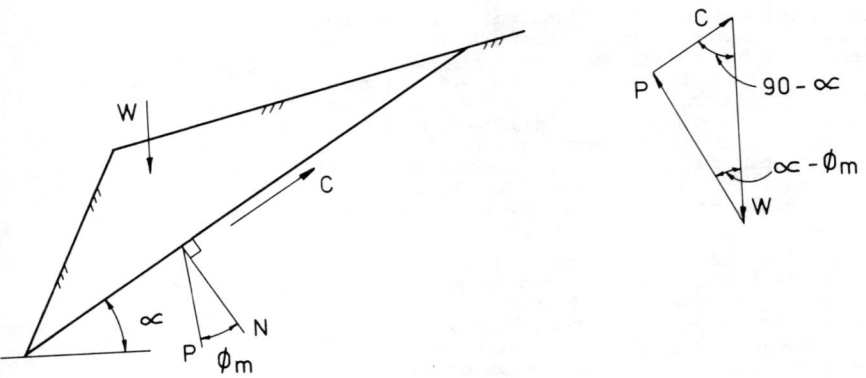

Fig.1 Plane Slip Surface (e.g. a discontinuity in a rock mass) along with force polygon when there is no water pressure.

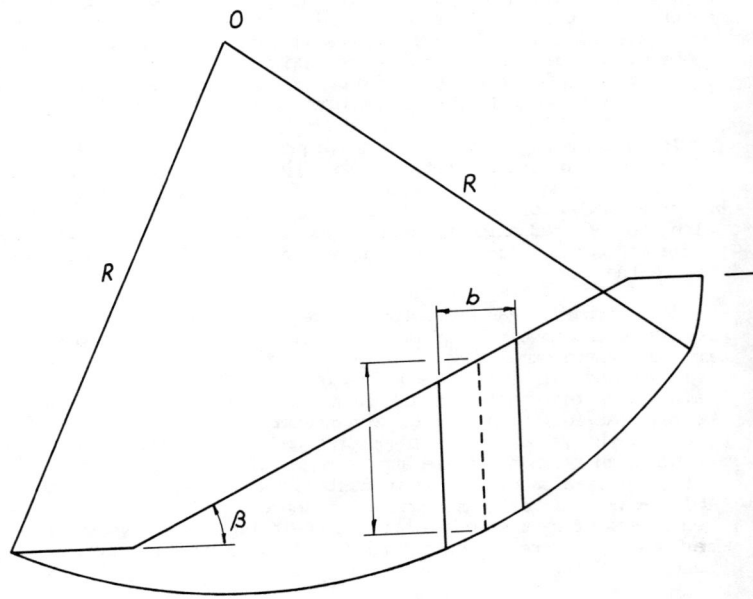

Fig.2 Slip surface circular in cross section and a typical vertical slice for use in any 'method of slices'.

LIMIT EQUILIBRIUM CONCEPTS

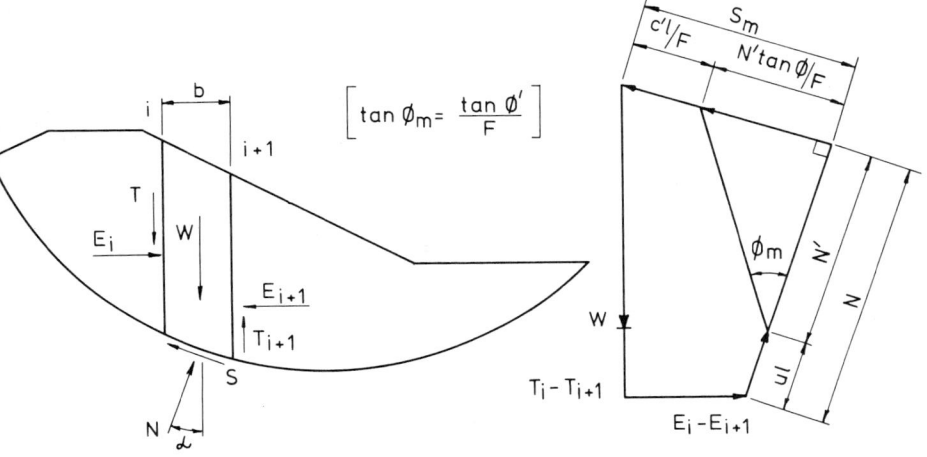

Fig. 3 Slip surface of arbitrary curved shape and forces acting on a typical vertical slice with force polygon.

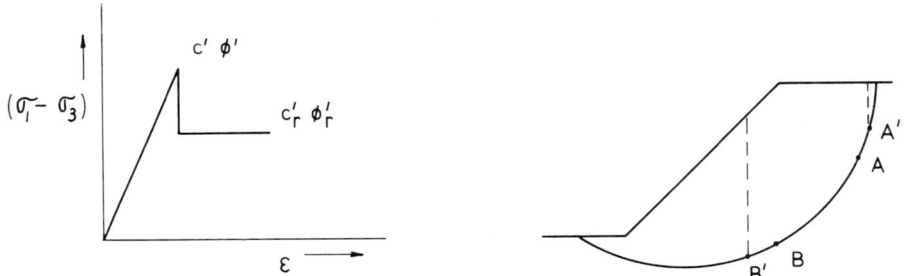

Fig. 4 (a) Idealised deviator stress-strain curve for strain softening soil (b) Softened part of slip surface; length AB before redistribution of excess shear stress and $\overline{AB}'$ after redistribution.

TABLE 1. SOME SIMPLE FORMULAE - CURVED SLIP SURFACES

PROBLEM AND METHOD	FORMULAE AND REMARKS
Undrained failure of clay slope $\theta_u = 0$ analysis	$F = \dfrac{c_u R^2 \theta}{W a}$ θ is central angle of circular slip surface, a is horizontal distance from c.g. of potential sliding mass to the centre of circle of radius R, W is weight of soil above slip surface
Fellenius or Swedish Method or Ordinary Method of Slices	$F = \dfrac{\Sigma\{c \ell + \tan \phi (W \cos \alpha - u\ell)\}}{\Sigma W \sin \alpha}$ where ℓ is the length of base of a slice, W the weight and α the inclination of the base of any vertical slice and Σ is summation over all slices.
Bishop Simplified Method	$F = \dfrac{\Sigma\{cb + (W - ub) \tan \phi\}/m_\alpha}{\Sigma W \sin \alpha}$ $m_\alpha = (1 + \dfrac{\tan \alpha \tan \phi}{F}) \cos \alpha$, $b = \ell \cos \alpha$
Janbu's Simplified Method	$F = \dfrac{f \Sigma b s \sec^2 \alpha}{\Sigma W \tan \alpha}$ correction factor f increases from 1.00 to 1.12 as ratio d/L increases from 0 to 0.4. Note d is max. depth of slip surface normal to line joining crest and toe of length L. $s = \dfrac{c = (\dfrac{W}{b} - u) \tan \phi}{1 + \tan \alpha \tan \phi / F}$ This method is applicable to slip surfaces of arbitrary shape. The three methods mentioned above are strictly correct only for circular slip surfaces.

TABLE 2. ESSENTIAL FEATURES OF SOME 'RIGOROUS' LIMIT EQUILIBRIUM METHODS

Morgenstern & Price (Slip surface of arbitrary shape)	Assume $T = \lambda f(x) E$, where $f(x)$ is an arbitrary function and λ is an unknown found along with F from the solution. This relationship means that for n slices (n-1) relationships, are assumed between T and E forces. (The solution method assumes that n tends to infinity i.e. infinitesimal slices). The total assumptions made are thus (2n-1) which is one more than (2n-2) assumptions required*. For this extra assumption, the extra unknown λ is introduced. Statics is thus satisfied rigorously i.e. (6n-1) unknowns and (6n-1) equations.
Spencer - 1973 (Slip surface of arbitrary shape)	The same type of assumption as Morgenstern and Price in the following form $\tan \delta_i = k_i \tan \theta$ k_i corresponds to $f(x)$ $k_i=1$ when Spencer assumes parallel interslice forces $\tan \theta$ corresponds to λ $\tan \delta_i$ corresponds to $\frac{T}{E}$ Relates interslice force on the right of a slice to one on the left of a slice and iterates on this relationship starting from one boundary condition and moving to the other end of the slip surface, slice by slice. This satisfies force equilibrium. Moment of pairs of interslice forces are also taken about the middle of the base of each slice.
Sarma - 1979	A modified wedge solution; slices do not necessarily have vertical sides and assumed inclination of sides adjusted during solution for optimum results. Shear strength is considered to be mobilised on the sides of any slice or wedge and by iteration the most critical inclination of each wedge side is determined. The solution is again obtained in terms of a critical acceleration factor K_c. Considered suitable for back-analysis of actual slips because shear failure in the sliding mass invoked.

NOTE:- Total unknowns (6n-2) consist of n normal and n shear forces on base of slice, n point of application of normal forces, (n-1) forces E with (n-1) points of application and (n-1) forces T and one factor of safety F. Available equations $\Sigma M=0$, $\Sigma H=0$ for $\Sigma V=0$ for each slice. Mohr-Coulomb Criterion each slice i.e. 4n equations. Hence $\{(6n-2) - 4n\} = (2n-2)$ assumptions required.

Stability Charts.-The development of stability charts in terms of dimensionless parameters has proved to be one of the major assets of the limit equilibrium approach. Taylor's (39) original charts are valid for analysis in terms of total stresses and incorporate the dimensionless stability number or factor $N = \frac{\gamma H}{c_m} = \frac{F\gamma H}{c}$. Many types of charts have been developed in terms of effective stresses and Janbu (22) introduced the additional dimensionless parameter $\lambda_{c\phi} = \frac{\gamma H \tan \phi'}{c'}$ in order to reduce the number of charts required. The most comprehensive set of charts has recently been developed by Cousins (15). Different types of charts for the evaluation of F and the location of critical slip surfaces have been presented and compared in Ref.(12).

It is necessary to point out that correspondence between predicted and observed slip surfaces is rare. However, theoretical solutions help in visualising possible failure mechanisms in different situations and in developing sound judgement concerning stability problems. Care should be exercised in the use of charts for back-analysis of any failure since the actual shape and location of slip surface must be taken into consideration. Appendix 3 shows how stability charts may be misused and an example of misuse as recently as 1978 is given. In the same appendix reference is made to theoretical shape of critical slip surfaces based on variational calculus (see also Ref.12).

Comparison with Limit Analysis Solutions.-It is a common feature of all limit equilibrium methods that material behaviour is assumed to be rigid-plastic. Strains and deformations are ignored. No consideration is given to dilatancy either on the slip surface or within the earth mass. However, it is significant that limit equilibrium solutions agree very closely to plasticity solutions obtained by the upper bound technique of limit analysis. In the latter a kinematically admissible mechanism is first assumed and the external rate of work is then equated to the internal rate of energy dissipation along the discontinuity or slip surface. In this upper bound limit analysis technique the stresses need not be in equilibrium. Alternative mechanisms, each of which is kinematically admissible, may lead to different results. For example the assumption of plane failure and logarithmic spiral mechanisms for the problem of a vertical cut in cohesive soil lead respectively to the following expressions for critical height H_c using limit analysis (upper bound)

$$H = \frac{4c}{\gamma} \tan (45 + \phi/2) \ldots \text{Plane failure} \ldots \ldots \ldots \ldots \ldots (4a)$$

$$H_c = \frac{3.83c}{\gamma} \tan (45 + \phi/2) \ldots \text{Log Spiral Surface} \ldots \ldots \ldots (4b)$$

These solutions are identical to corresponding limit equilibrium solutions: Note that lower bound limit analysis gives a different result:

$$H_c = \frac{2c}{\gamma} \tan (45 + \phi/2) \ldots \ldots \ldots \ldots \ldots \ldots \ldots \ldots (4c)$$

It is interesting in contrast to 4(a) and 4(b) that limit equilibrium solutions using either circular or log spiral slip surfaces agree very closely. These results are almost identical to the upper bound limit analysis solution using log spiral mechanism. (Note that circular failure mechanism is only admissible in limit analysis when $\phi=0$). Many such comparisons have been made and tabulated by Chen (8).

Upper and lower bounds to the value of F should not be confused with upper and lower bound solutions of limit analysis. Conventional F is conservative and close to the lower bound of possible limit equilibrium solutions. However it corresponds to the upper bound of limit analysis. The lower bound of limit analysis is associated with any statically admissible stress distribution but not with a kinematically admissible velocity field. It does not correspond to conventional F unless it coincides with the upper bound of limit analysis (being, in that case, a unique or exact solution).

BOUNDARY CONDITIONS AND GENERAL STABILITY ANALYSIS

Any limit equilibrium method is required to satisfy given boundary conditions. Slope surfaces (external boundaries) may be free or loaded, there may be parts of the slope which are submerged, and tension crack at the crest may be full of water. In any one of the methods which involve subdivision of the slope mass into vertical slices, the condition of free slope boundaries may be expressed as follows:

$$\Sigma d \ T = 0 \ , \ \Sigma d \ E = 0 \quad\quad\quad\quad\quad\quad\quad\quad\quad\quad\quad\quad (5)$$

in which $dT = (T_{i+1} - T_i)$ is the difference of tangential forces acting on the two sides of the ith slice and $dE = (E_{i+1}-E_i)$ is the difference of normal forces acting on the two sides of the ith slice. Eq.5 may be modified in an appropriate manner when there is a net boundary force acting on a slope. Loads on the crest of a slope can be included by increasing the weight W of each slice to include the surcharge.

With these modifications it is thus possible to tackle problems of bearing capacity and earth pressure in addition to those of slope stability using the same computational technique. Morgenstern (27) has emphasised this versatile role of limit equilibrium techniques. It is necessary to note that such solutions rarely give results identical with those given by classical plasticity theory. Sarma (34) has claimed that his generalised wedge method can obtain the classical solution for the Prandtl bearing capacity problem provided the same mechanism of failure is assumed. This is certainly an exception. However, failure of the results to correspond to exact plasticity solutions does not detract from the practical value of limit equilibrium techniques in all kinds of stability problems.

In this connection it is useful to note that Hamel et al (19) successfully used a rigorous slope stability limit equilibrium program to analyse the sliding stability of gravity dam monoliths. Factors of safety obtained in this way are typically 30% greater than those obtained by the traditional method i.e. ratio of resisting forces to disturbing forces along the dam-rock interface or any plane below it. Hamel (20) wondered whether a limit analysis solution corrresponded to the rigorous or the traditional solution. It is interesting that the writer found upper bound limit analysis to give exactly the same solution as the simple traditional method. In fact even when the slip surface was considered at any arbitrary inclinations (other than horizontal) the same correspondence was achieved. It is well to remember that the gravity dam sliding stability problem is similar to the problem of a rock wedge with a deep, water-filled tension crack at the rear. Thus one finds here a significant problem in which a 'rigorous' solution is not necessarily the best one. However, it helps one to understand that the simple solution is conservative and theoretically sound.

Admissibility Criteria.-It is common practice in traditional limit equilibrium to specify criteria for the acceptability of a solution. Since many stress distributions can satisfy equilibrium conditions in these statically indeterminate problems, such criteria serve as a useful guide when rigorous methods are used. (1) Many sets of interslice forces may satisfy equilibrium and it is considered wise to accept only those which do not imply tension within the free body. (2) The criterion of failure should not be violated along any of the interslice boundaries. (3) The effective normal stresses along the slip surface are required to be always compressive. In practice these criteria may be modified depending upon the type of slope, the nature of soil or rock and the reliability of soil parameters used in the analysis. All the criteria are seldom satisfied throughout the soil or rock mass under consideration. Sometimes several repeated trials with different assumptions may fail to produce an admissible solution.

LOCAL SAFETY FACTORS

Consider a vertical slice within a failure mass. Let it have a width b, weight W, base length ℓ and inclination α. There are net interslice forces dT and dE acting as defined before, and also shear and normal stresses act on the base of the slice. Considering force equilibrium in directions normal and tangential to the base of the slice, the normal and shear stresses on the base are:-

$$p = \frac{W+dT}{b} \cos^2\alpha - \frac{dE}{b} \sin\alpha \cos\alpha$$
$$\tau = \frac{W+dT}{b} \sin\alpha \cos\alpha + \frac{dE}{b} \cos^2\alpha \qquad \ldots \ldots \ldots (6)$$

Using Coulomb's shear strength equation modified for effective stress and assuming pore pressure u at the base of a slice, the local factor of safety F_ℓ may be obtained as a ratio of available shear strength τ_f to shear stress as follows:

$$F_\ell = \frac{\tau_f \ell}{\tau \ell} = \frac{c'\ell + (W\cos\alpha + dT\cos\alpha - dE\sin\alpha - u\ell)\tan\phi'}{W\sin\alpha + dE\cos\alpha + dT\sin\alpha} \quad \ldots (7)$$

It is interesting to note that the same equation is obtained even if forces equilibrium is considered in vertical and horizontal directions after assuming the shear stress at is mobilised value τ_f/F_ℓ (as in the well known Bishop method). On the contrary the expression for p thus obtained is different from the one in Eq.6, as it includes the unknown F_ℓ.

From Eq.7 it is obvious that F_ℓ can vary widely along a slip surface depending on factors such as geometry, strength parameters, pore pressure distribution, net interslice forces, all of which may vary from slice to slice. Assumption of a constant F is, therefore, a gross oversimplification in traditional methods which is hardly compensated by adopting a 'rigorous' approach to force or moment equilibrium. Experience with finite element analyses has shown that invariably some regions of a slope fail before others (16,17,44). Again local failure may not develop in certain regions right until the conditions of critical equilibrium for the whole mass are satisfied; yet failure at other points may have developed when the slope was in stable equilibrium with F>1 (17,44). Direct observations of natural slopes, excavations and embankments have confirmed the fact that local overstressing and local failure can occur even when a slope is stable. Variation in local factors of safety has important significance for progressive failure of earth materials especially when strain-softening is considered and this is discussed in subsequent sections.

THE STRESS FIELD AND ITS IMPLICATIONS

Consider a long natural slope of inclination β. A simple gravitational stress field implies the following expressions for normal and shear stresses on any plane parallel to the slope

$$p = \gamma Z \cos^2 \beta \, , \quad \tau = \gamma Z \sin \beta \cos \beta \quad \ldots \ldots \ldots \ldots \ldots \ldots (8)$$

It is pertinent to ask how one may account for lateral stresses of arbitrary magnitude that may exist in the ground. In this connection the concept of a conjugate stress field (39) is both simple and appealing. An element with two boundaries parallel to the ground surface and vertical sides is considered under the action of resultant stresses σ_V and σ_β so that $\sigma_\beta = K \sigma_V$ in which K is the lateral stress ratio (see Fig.5); σ_V is vertical and σ_β is parallel to the ground surface.

A conjugate stress field is well rooted in traditional soil mechanics and widely accepted even today. Moreover Taylor (39) proved that it represents equilibrium stress distributions as long as the conjugate stress ratio K is not high or low enough to violate the criterion of failure. Thus it can not be brushed aside in limit equilibrium terms where any stress distribution which satisfies equilibrium is acceptable. The following equations have been derived (9) for normal and shear stresses on a plane of inclination α in terms of K and β:

$$p = \gamma Z \{\cos^2 \alpha (1+K \sin^2 \beta) + K \sin^2 \alpha \cos^2 \beta - K \sin 2\alpha \sin \beta \cos \beta\}$$
$$\ldots \ldots (9)$$
$$\tau = \gamma Z \{\sin \alpha \cos \alpha (1+K \sin^2 \beta - K \cos^2 \beta) - (\sin^2 \alpha - \cos^2 \alpha) K \sin \beta \text{ as } \beta\}$$

These equations reduce to Eqs.8 only when K=0. Thus a traditional limit equilibrium stress field can not represent situations in which the lateral stress parameter K is important. (Only infinite slope analyses are not affected whatever the value of K because when $\alpha=\beta$, Eqs.8 and 9 are identical.) It may be argued that those methods which include interslice forces implicitly consider lateral stresses in the ground. However, the interslice forces are not based on real in situ stresses and are simply adjusted according to the need of a given technique of solution to obtain a condition of equilibrium. Values of K can be very high even near the surface of natural soil and rock masses. Evidence in support of high K values (of the order of 2.5 or 3) continues to accumulate for overconsolidated soils. However values of K for rocks much higher than these have been measured or inferred.

In finite element analyses of excavations and natural slopes the validity of a conjugate stress field is widely accepted. In fact the analysis of an excavation can not even proceed unless an appropriate value of K is chosen. In general, excavations in horizontal ground are

Fig.5 Simulation of stress release in natural slope assuming $c'=0$ showing effect of K on variation in $F_p/\tan \phi$ from $F_i/\tan \phi$ to $F/\tan \phi$ as failure propagation from the crest of the slope is considered.
(Note F is the conventional safety factor and F_i the one based on initial stresses).

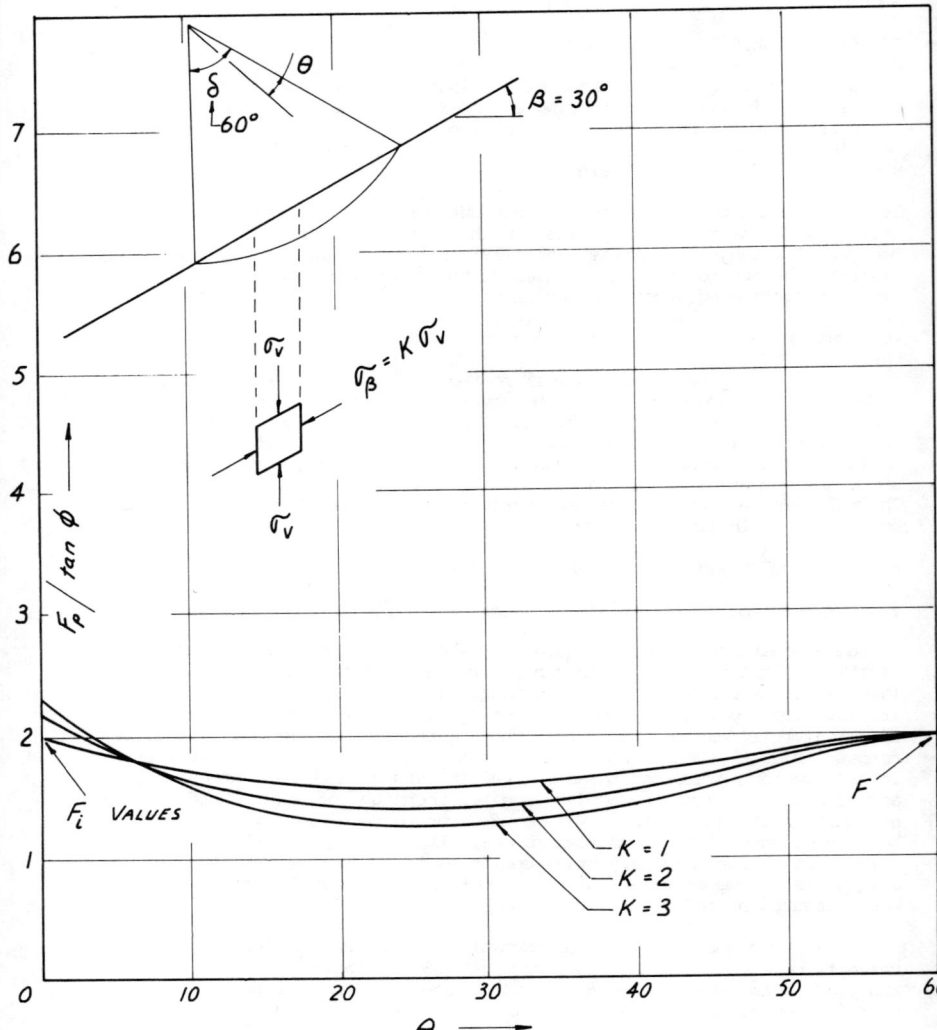

analysed on the basis of known or assumed values of K_o (16,17) which is related to as follows:

$$K_o = \frac{K - r_u}{1 - r_u} \qquad \qquad (10)$$

in which r_u is the pore pressure ratio $\frac{u}{\gamma z}$. In general when the ground is sloping:

$$K = \frac{K \cos^2 \beta - r_u}{1+K \sin^2 \beta - r_u} \qquad \qquad (11)$$

Two significant limitations of traditional limit equilibrium methods, which are widely recognised in principle but not given sufficient emphasis in routine practice, may now be mentioned viz:

(1) Whereas stress analysis techniques can simulate construction history and slope formation realistically, limit equilibrium techniques deal only with the final completed shape of a slope or soil structure. Consequently significant differences between natural, excavated and man-made slopes may be overlooked when these techniques are used in isolation.

(2) Non-uniform stress distribution and non-uniform strain distribution imply progressive failure which does not receive adequate attention in traditional limit equilibrium procedures.

BACK ANALYSES USING LIMIT EQUILIBRIUM

Introduction.-It is useful to analyse a failed slope to understand why failure occurred. An overall factor of safety is calculated with data relevant to slope conditions just before failure. A value of calculated F=1 is often considered to vindicate the correctness of the assumed failure mechanism and in particular that of the limit equilibrium analytical technique. It should be noted that there are often many uncertainties in the data that are used in such analyses. Shear strength parameters obtained from laboratory tests, however elaborate, may not represent the correct in situ bulk shear strength. Pore water pressures at the time of failure may not have been monitored and reliance on measurements made later could be very misleading. More importantly, minor geological details are often very significant. The presence of thin seams of silt or sand or clay within an otherwise homogeneous soil mass may remain undetected. Also perched water tables, often undetected, may have a dramatic effect on stability.

Back analyses may show that calculated F is not close to unity. In such cases, reference is made to the above uncertainties and explanations are offered in terms of changed strength parameters and pore water pressures which do produce the result F=1. The limit equilibrium concept is rarely called into question. It is obvious that even in case histories with calculated values F=1, that result may itself be fortuitous due to self-compensating errors in strength parameters, pore water pressures and the analytical procedure. It is well to remember that, in effective stress analyses of clay slopes, apparently minor changes in c' may have a significant influence on F. Again, a failure plane in a rock mass may predominantly consist of discontinuities which it intersects. A small proportion may, however, pass through intact rock. Minor changes in the proportion of the intact material often have a dramatic influence on the value of F.

First-Time Slides in Overconsolidated Clay.-Considerable difficulties have been experienced in properly explaining delayed, first-time slides in overconsolidated clays. The best example of changing explanations is offered by case histories of London Clay. These failures have been studied with thoroughness and with concern for detail over several decades. By 1964 the concept of residual shear strength was invoked to explain the failures. Peak strength values gave factors of safety well above one and residual shear strength parameters gave values below one. Skempton (35) proposed that a part of the slip surface mobilised peak and the other part residual shear strength. The proportion which fell to the residual (the residual factor) was considered to increase with time. By 1970, this explanation was discarded and Skempton (36) suggested that the whole clay mass softened after excavation with loss of almost all the cohesion parameter and no significant change in the friction parameter. The fully softened strength was, therefore, much higher than the residual shear strength but lower than the peak strength and appeared to produce values of F=1. However, delays of 50 years or more in failures could not be explained. Skempton (37) attributed the delays to the delay in attainment of pore pressure equilibrium. The earlier view that in this fissurred clay pore pressure equilibration should be very rapid (due to high mass permeability) was discarded due to new field evidence. In all these papers the limit equilibrium concept has not been questioned.

Three of the case histories were analysed by the finite element method (26). High values of the lateral stress ratio K which has been measured or inferred in London clay were used in the analysis. The failures were explained primarily on the basis of the 1964 residual strength concept. However, Lo and Lee (26) did not simulate the excavations incrementally and it would be unwise to consider their conclusion more reliable than that of Skempton (37) based on the limit equilibrium approach. Recently (25) it has been suggested that progressive structural breakdown continually generates negative excess pore pressures so that the time necessary to reach equilibrium with the local ground water regime primarily reflects the time-rate of structural breakdown. Thus different explanations could be offered for the delay in failures without the necessity of questioning the use of limit equilibrium concepts.

On the other hand, it should also be possible to simulate critical equilibrium by means of a stress analysis approach. Such an approach can not only account for the history of slope formation but also the delay in pore pressure equilibration. A fair comparison between limit equilibrium and stress analysis approaches (as regards their reliability for back-analysis of actual failures) can only be made if and when comprehensive stress analysis results for these case histories become available. Work along these lines is at present in progress (18) and ideally, both approaches should lead to the same conclusion as regards the average shear strength mobilised at the time of failure.

Embankments - Some Unusual Evidence.-Short-term stability of embankments (and even of excavated slopes in soft clay) has not presented any significant difficulties of analysis by limit equilibrium techniques. In fact, these methods have been considered most successful for such problems. However, redistribution of stresses cannot be considered and it is of interest to refer to a case history presented by Al-Dhahir et al (1) in which pore pressures were monitored during construction. At one stage before completion of the embankment, a sudden rise in pore pressure was monitored without placement of any additional fill. This was then followed by movement and slipping. Neither the sudden rise in pore pressure nor the failure could be properly explained by analysis and it is relevant to quote the authors' conclusions: "The actual behaviour illustrates some limitations of the limit equilibrium analysis of earth structures and the control of construction by observations of displacements and pore pressures" and again "An investigation of the effective stresses using elastic theory showed that the material was not overstressed in terms of effective strength parameters prior to structural collapse. Had the bank been completed at that time the overall factor of safety would have been greater than unity and conventional design considerations would have expected it to increase with time".

In a discussion of the above paper, Rowe (32) stated that he had knowledge of several case histories in which pore pressure rises and failure could not be explained in conventional terms. He cited one case where the piezometer rise was three times the height of the fill. He suggested that internal shear and total stress redistributions occurred in such cases: "As failure was approached suctions began in the fill near the surface, progressed in size at a given point and at the same time moved along a given path. This is consistent with failure starting in the active region and developing progressively towards the passive zone. This mechanism, which would be accentuated by a stiff fill on a soft foundation, is one which is also consistent with observations on uniform dense sand".

Parry (30) made a thorough study of several case records involving $\phi_u=0$ limit equilibrium analyses and concluded that strain rate effects may result in an overestimate of F by up to 50%. This is because the conventional quick shear test shows significant strain-softening behaviour with respect to undrained shear strength c_u. (This behaviour is not noted with slow, undrained shear tests). Thus conventional analyses, which ignore reduction of c_u with strain, must be called into question. Even if test rates in the laboratory or field are carefully selected to eliminate peaking, F is over-estimated by 15%. Similarly ignoring stress redistribution can cause an error of up to 20% in F, whereas pore pressure effects can cause an error of up to 30%.

Vaughan (42) suggested that the brittleness of the embankment material should be carefully assessed and commented that: "Existing limit analysis methods are unsatisfactory for till type materials with non-brittle shear behaviour and improved methods of determining pre-failure deformations, which are the limiting factors in these materials, should be sought. These methods should include the prediction of pore pressures and the influence of partial drainage. Limit analysis methods .... are apparently satisfactory for brittle materials if the behaviour of a laboratory test specimen is considered but unfortunately, the effect of progressive failure in reducing average shear stress at failure in the field below the peak failure stresses measured in the laboratory can not be analytically assessed".

Rockslides and Conventional Limit Equilibrium.—There are many examples of rockslides which have not been adequately explained by limit equilibrium methods. One example is the Vaiont slide which has been studied by numerous investigators. Most conventional limit equilibrium analyses gave factors of safety above one even with most pessimistic assumptions with regard to shear strength and pore pressure. Even those analyses which produced a value of F=1 on the basis of assumed artesian pore pressures or unrealistically low value of ϕ, could not explain the tremendous speed and violence with which the slide occurred. A new approach based on an appreciation of high in situ stresses and appropriate adaptation of analyses has provided satisfactory explanation of this rockslide (10,29).

The Vaiont slide is not an isolated case. Many rockslides and avalanches have not been properly explained on the basis of conventional limit equilibrium. A volume edited by Voight (43) provides comprehensive information on several important natural slide events which are not fully explained on the basis of traditional concepts. All these cases indicate the need for analyses to simulate the release of energy and the conversion of potential energy and internal energy to kinetic energy once the value of F drops below one. The question of dynamics of slides and energy release is outside the scope of this paper and is considered elsewhere (14).

SOME DEVELOPMENTS FOR SPECIAL CASES

Strain-Softening and Limit Equilibrium Analysis.—One of the important factors which may lead to stress redistribution is strain-softening. This may be discussed with reference to Eqs.1 and 6. Consider a brittle soil (Fig.4) in which the strength drops from a peak value τ_f (parameter c' and ϕ') to a residual or reduced value τ_r (parameters c_r' and ϕ_r') as soon as shear stress τ exceeds τ_f. It is well known that large relative deformations are required for shear strength to drop to the absolute residual value on any surface. In connection with slopes where large deformations or pre-formed shear surfaces are not involved, τ_r may be considered as a reduced value mobilised in the field after peak strength has been exceeded (for example the softened shear strength of London clay, assuming it to be perfectly brittle).

Local failure at the base of a slice will occur if the shear stress $\tau > \tau_f$. From Eqs. 6 and 1 this leads to the condition

$$\{\frac{W+dt}{b} \sin\alpha\cos\alpha + \frac{dE}{b} \cos^2\alpha\} > \{c' + (\frac{W+dT}{b} \cos^2\alpha - \frac{dE}{b} \sin\alpha\cos\alpha - u)\tan\phi'\}. . (12)$$

When overall factor of safety F>1, Eq.(12) will not be satisfied if dT and dE are from any type of rigorous limit equilibrium solution in which the local factor of safety F_ℓ=F and therefore F_ℓ>1 , $\tau < \tau_f$.

Accordingly an independent definition of local failure becomes necessary. Law and Lumb (24) assumed dT and dE to be zero initially and then used the simpler form of Eq.12 to identify failed slices. For these slices the strength on their base falls to its reduced value based on c_r' and ϕ_r'. Failure implies that excess shear force $(\tau-\tau_r)\ell$ over the base of relevant slice can not be carried by that slice. This excess must be redistributed in an appropriate manner and special iterative procedures are necessary (24,2).

LIMIT EQUILIBRIUM CONCEPTS

During iteration the number of failed slices increases due to stress redistribution. The final overall factor of safety F_f reduces in proportion to the length of the slip surfaces over which strength has dropped to the residual value and may be written as follows:

$$F_f = \frac{\Sigma_{n-m} (\tau_f \ell) + \Sigma_m \tau_r \ell}{\Sigma_n \tau \ell} \quad \quad \quad (13)$$

in which m is the number of failed slices. It may be noted that use of Eq.13 is not required if a rigorous limit equilibrium method is used during iterations. The local factor of safety anywhere on the slip surface also reduces to the final overall factor of safety F_f.

Stress redistribution for soils which are not perfectly brittle can not be studied accurately unless the level of strain is considered in relation to the real stress strain curves. Further, pore pressure effects require a great deal of attention in further developments of the limit equilibrium method. Not only should the rate and manner of pore pressure changes be considered but also the possibility of pore pressure redistribution and partial drainage (e.g. in embankment stability problems). Attention also requires to be given to the development of failure from the crest or toe or both in many slopes. Simulation of these phenomena in analysis could lead to a fuller understanding of progressive failure effects.

<u>Initial Stress Factor F_i.</u> Assuming in situ stresses can be measured or inferred reliably, an initial stress slope factor F_i may be defined by intergrating the shear strength (based on in situ effective normal stress $(p_i - \mu)$ and the in situ shear stress τ_i and taking the ratio of corresponding forces or moments. For a slip surface approximating to circular shape of radius R and total length L.

$$F_i = \frac{R \int_0^L [c´ + (P_i - \mu) \tan\phi´] \, dL}{R \int_0^L \tau_i \, dL} \quad \quad \quad (14)$$

The magnitude of F_i is found to be different from the conventional limit equilibrium value F depending on the particular initial stress field (10, 12). Assuming a conjugate stress field in a uniform natural slope, increase in the value of K increases the value of F_i for a given circular slip surface. For small values of K approaching zero, the value of F_i is found to be close to the conventional F value. In many situations the stress field is neither a simple gravitational one nor a conjugate one. Stress analyses would be required to determine the influence of excavation, alteration, embankment construction etc. on the stresses. Thus the initial stresses for use in Eq.14 may be obtained by superimposing the changes in stress due to construction on the natural stresses p_n, τ_n thus:

$$p_i = p_n + \Delta p, \quad \tau_i = \tau_n + \Delta \tau \quad \quad \quad (15)$$

The question now arises as to what use one can make of the new quantity F_i. If the slope is relieved of initial stresses completely and suddenly,

F_i changes to conventional F. However stresses may be relieved progressively. There is a great deal of evidence, for instance, that failure in natural slopes progresses from the crest or the toe depending on stress concentrations, tension cracks, initial conditions etc. (6,31). Progressive relief of stresses starting from the crest or the toe was simulated in several studies assuming a conjugate initial stress field. It was found that, for assumed circular slip surfaces, progression from the toe was unrealistic (12). In other words the slope would have a tendency to be relieved of stresses gradually starting from the crest. This release of stresses or energy can be analysed by an appropriate modification of the limit equilibrium technique.

   Relief of Initial Stresses - Simulation in Analysis.-Let the factor of safety of a slope at any stage during the progressive stress relief process be F_p when an upper part A of the potential sliding mass has a simple gravitational stress field and the lower part B the original stress field. The two parts have separate factors of safety:

$$F_A = \frac{R_A}{D_A} \text{ and } F_B = \frac{R_B}{D_B} \quad \ldots \ldots \ldots \ldots \ldots \ldots \ldots \ldots \ldots (16)$$

in which R_A and D_A are given by conventional limit equilibrium and R_B and D_B by the procedure discussed above. The overall factor of safety is

$$F(\text{Progressive}) = F_p = \frac{R_A + R_B}{D_A + D_B} = \frac{F_A D_A + F_B D_B}{D_A + D_B} \quad \ldots \ldots \ldots (17)$$

F_p has an initial value F_i and final value identical with the limit equilibrium value F. It is found, however, that the variation from F_i to F is not linear during the progressive process. For any curved or biplanar slip surface it is found that F_p first decreases to a minimum value (lower than F_i or F) and then increases to the value F (Fig.5). Moreover, it is found that as K increases (indicating increase in initial stresses) the minimum decreases. Thus, while the limit equilibrium factor of safety is independent of K, the minimum value of factor F_p decreases with increase K when progressive stress relief is simulated. This approach may explain failure of natural soil or rock masses with high initial stresses better then the conventional limit equilibrium approach (10). Just as the actual magnitude of F is not as important as its sensitivity to different parameters, the value of F_p should serve as a guide concerning the influence of K (and, in general, of the initial stress field) on the potential stability of a sliding mass.

DISCUSSION OF SOME RECENT TRENDS

   Limit Equilibrium vs. Stress Analysis.-Stress analysis procedures based on the finite element method have been increasingly used in the last decade to supplement traditional methods and to strengthen the basis of engineering judgement. Sophisticated methods of analysis enable the simulation of construction history and the study of how stresses, strains and deformation may be developed in a soil or rock mass. However, there are several difficulties of a practical nature to be overcome before such methods can be used to advantage e.g. (1) it is often difficult to obtain the necessary input data on deformation parameters of a soil or rock mass and of joints or discontinuities within it (2) it is also difficult to obtain data on initial ground stresses (3) stress-strain or stress-deformation theories for real soils are still in the process of development and idealisations are often necessary (4) while zones of failure may be identified, it has not yet been possible to simulate the propagation or development of slip surfaces by these methods. Thus interpretation of results in simple terms (e.g. a value of the factor of safety) is not possible unless the concept of trial slip surfaces is used and some additional limit equilibrium calculation made. These factors explain why the concept of limit equilibrium has not been replaced by stress analysis procedures.

Field instrumentation and performance monitoring are receiving increasing attention in geotechnical practice. In order to take full advantage from their use, both limit equilibrium and stress analysis procedure should be used where possible. Again there may be situations in which the estimation of deformations is of primary importance. In such cases stress-deformation studies based on the finite element method must be made using realistic data and with due consideration to stress paths and construction history.

Reliability Analyses based on Probability Theory.-A recent trend in geotechnical engineering is the use of reliability analyses based on the theory of probability. The capacity R (resisting force or moment) and the demand D (disturbing force or moment) are treated as random variables rather than single-valued quantities. Therefore, the factor of safety $F=(R/D)$ and the safety margin $SM=(R-D)$ are also random variables. Failure is regarded as the even:

$$\text{Failure} \equiv [F \leq 1] \equiv [SM \leq 0] \quad \ldots\ldots\ldots\ldots\ldots\ldots\ldots\ldots (18)$$

the probability of failure p_f is given by

$$p_f = P[F \leq 1] = P[SM \leq 0] \quad \ldots\ldots\ldots\ldots\ldots\ldots\ldots (19)$$

The probability of failure depends on the probability density functions (PDF) or the distributions of capacity and demand. Reliability or the probability of success p_s is given by:

$$\text{Reliability} = p_s = (1-p_f) \quad \ldots\ldots\ldots\ldots\ldots\ldots\ldots\ldots (20)$$

Uncertainties in capacity and demand can be analysed logically; consequently the ability to exercise good judgement is enhanced.

The capacity-demand model is the simplest one that can be used. In general, significant parameters (e.g. cohesion, friction, pore water pressure) may be treated as random variables with given probability distributions. The statistical parameters of capacity and demand (the mean, variance etc.) can then be evaluated on the basis of statistical parameters of the basic random variables. Consequently the influence of these parameters on the reliability can be analysed (21). It is also possible to determine the most probable extent of failure of an embankment (41) or a natural slope (13).

Reliability approaches enable problems to be considered in a broader framework whereby a so-called homogeneous medium is visualised as one which is, at best, only statistically homogeneous. Consequently one may look at the available data for significant information (e.g. means and variances of parameters and statistical correlations between them) and obtain information which would be neglected within a narrow deterministic framework.

It is significant that limit equilibrium concepts often form the basis for reliability analyses. This is due, in large measure, to their simplicity and versatility. Just as traditional procedures require adoption of allowable values of the factor of safety, reliability analyses are useful when the acceptable values of the probability of failure can be specified. This is where the experience gained from limit equilibrium methods is invaluable and will continue to be so. Generalised relationships between the probability of failure and conventional F (which are suitable for a whole class of problems) can not be developed. However it is possible to develop such relationships for a given problem so that decisions about optimum design can be made. It becomes obvious that there is no one-to-one relationship between conventional F and p_f (e.g. doubling the factor of safety F does not mean that safety or reliability is also doubled).

Given the present state of development of probabilistic approaches, the choice of a correct deterministic model is essential for their success. Therefore, it is important that basic assumptions underlying limit equilibrium concepts are not forgotten and that there is an appreciation of both their capabilities and limitations. Independent probabilistic models may, of course, be developed in the near future. Again the use of such models in practice will be facilitated by continual reference to traditional procedures and experience gained in their use.

CONCLUSIONS

Limit equilibrium concepts have proved to be of immense value in dealing with geotechnical stability problems and especially the stability of slopes. Considering the nature of assumptions on which traditional methods are based, the precise magnitude of the factor of safety may not be important. Results of limit equilibrium calculations should be evaluated on the basis of experience and should serve to strengthen engineering judgement. Moreover, the concept of limit equilibrium is a simple tool with which one can study the sensitivity of a soil or rock mass to variations in the values of significant parameters.

It can not be claimed that accurate or reliable predictions of stability can always be made on the basis of limit equilibrium studies. However this is not necessarily a reflection on the basic concepts since there are often many uncertainties concerning strength parameters and pore water pressures and since minor geological details may remain undetected during investigation. Present knowledge about mechanisms of failure is not complete although considerable progress has been made in recent decades; this also contributes to difficulties in making successful predictions.

The importance of progressive failure has been known for decades (e.g. 7,39) and mechanisms of progressive failure have been emphasised continually (4,5,6,31). Yet traditional analysis procedures have not been strengthened to enable studies to be made of progressive action within geological media. Accordingly some suggestions have been made in this paper for extending the scope of limit equilibrium studies. Before proposed developments or modifications can be accepted in principle, it is necessary to recall the basic assumptions of the traditional approaches. There should be a recognition of the fact that the concept of limit equilibrium is not fundamental to phenomena concerning stability. It is only a device which enables the determination of a 'factor of safety' for a soil or rock mass. The state of critical or limiting equilibrium should not be confused with the concept of limit equilibrium.

There can be little progress if the limit equilibrium concept is treated as a principle which can not be violated e.g. the principle of effective stress. To illustrate the point, stresses computed on the basis of sophisticated procedures are, in general, not in agreement with those based on the limit equilibrium concept and no one expects such agreement. The degree of disagreement increases as the magnitude of K_o increases and depends on the type of slope, construction history and other factors. On the contrary pore pressure or effective stress estimates are expected always to be consistent with the principle of effective stress, whatever method is used to estimate pore pressures.

It should be noted that accurate predictions of stability can sometimes not be made even with the most sophisticated procedures. The use of such analyses is dependent on a great deal of accurate and comprehensive data which is often difficult to obtain. Moreover, there is not enough evidence to support the view that back-analyses of failures by sophisticated methods are more successful than back-analyses based on limit equilibrium. Again reliability analyses based on probability theory are of immense value in enhancing engineering judgement. Yet such analyses require simple deterministic models such as those based on the limit equilibrium concept.

It is obvious that there are some situations in which traditional limit equilibrium concepts may not be applicable. Natural slopes with high initial stresses and exceptional rockslides belong to this category of problems but this represents an abuse of the traditional methods. The reason for such misuse is obvious; even relatively simple methods of dealing with release of energy, post-failure behaviour and the dynamics of slides remain to be developed (14). Again, limit equilibrium calculations are not sufficient in problems where deformations are of primary significance. As long as traditional methods are used within the range of their capabilities and there are no unrealistic expectations from their use, the concept of limit equilibrium will continue to have immense practical value.

ACKNOWLEDGEMENTS

The writer would like to acknowledge the support of the Australian Research Grants Committee, the Water Research Foundation of Australia and the University of Wollongong for research concerning the stability of slopes.

APPENDIX 1.-REFERENCES

1. Al-Dhahir, Z.A., Kennard, M.F. and Morgenstern, N.R., "Observations on pore pressures beneath the ash lagoon embankments at Fiddler's Ferry power station, In-Situ Investigations in Soils and Rocks, B.G.S., London, Paper 20, 1970, pp.265-277.
2. Bertoldi, C., and Chowdhury, R.N., "Simulation of Strain-softening in Limit Equilibrium Analysis", unpublished notes, University of Wollongong, 1980.
3. Bishop, A.W., "The Use of the Slip Circle in the Stability Analysis of Slopes", Geotechnique, Vol.5, 1955, pp.7-17.
4. Bishop, A.W., "Progressive Failure with Special Reference to the Mechanism Causing It", Proc.Geotechnical Conference, Oslo, Vol.2, 1967, pp.142-154.
5. Bishop, A.W., "The Influence of Progressive Failure on the Method of Stability Analysis", Geotechnique, Vol.21, 1971, pp.168-172.
6. Bjerrum, L., "Progressive Failure in Slopes in Overconsolidated Plastic Clays and Clay-Shale", Journal of the Soil Mechanics and Foundations Division, ASCE, Vol.93, SM5, 1967, pp.3-49.
7. Casagrande, A., "Notes on the Design of Earth Dams", Journal, Boston Soc.Civil Engrs., Vol.34, No.4, (reprinted in Contributions to Soil Mechanics, 1941-1953, p.237.
8. Chen, W.F., "Limit Analysis and Soil Plasticity, Elsevier, Amsterdam, 1975, pp.638.
9. Chowdhury, R.N., "Initial Stresses in Natural Slopes Analysis", Rock Engineering for Foundations and Slopes, ASCE, Proc.Geotech.Eng. Specialty Conf., Vol.1, 1976, pp.404-415.
10. Chowdhury, R.N. "Analysis of the Vaiont Slide - New Approach", Rock Mechanics, Vol.11, 1978, pp.29-38.
11. Chowdhury, R.N., "Propagation of Failure Surfaces in Natural Slopes", Journal of Geophysical Research, Vol.83, No.B12, 1978, pp.5983-88.
12. Chowdhury, R.N. "Slope Analysis", Elsevier, Amsterdam, 1978, 424pp.
13. Chowdhury, R.N., "Probabilistic Evaluation of Natural Slope Failures", Proc.Int.Conf. on Engineering for Protection from Natural Disasters, Bangkok, 1980, pp.605-15.
14. Chowdhury, R.N., "Landslides as Natural Hazards - Mechanisms and Uncertainties", Keynote Special Lecture, Int.Conf. on Engineering for Prevention from Natural Disasters, Bangkok, 1980, (University of Wollongong Report, 1980,pp.1-68).
15. Cousins, B.F., "Stability Charts for Simple Earth Slopes, Journal of Geotechnical Engineering Division, ASCE, Vol.104, GT2, 1978, pp.267-279.
16. Duncan, J.M., and Dunlop, P., "Slopes in Stiff-Fissured Clays and Shales", Journal Soil Mech. and Foundations Division, ASCE, Vol.95, SM2, 1969, pp.467-49) (See also Vol.96, SM1, pp.336-338, 1970 and Vol.97, SM4, 1971 for discussion).
17. Dunlop, P., and Duncan, J.M., "Development of Failure in Excavated Slopes", Jour.Soil Mech. and Foundations Division, ASCE, Vol.96, SM2, 1970, pp.471-95.
18. Gray, P.A., and Chowdhury, R.N., "Stress Analysis of Excavations", unpublished notes, University of Wollongong, 1979.
19. Hamel, J.V., Long, S.B., and Ferguson, H.F., "Mahoning Dam Foundation Re-evaluation", Rock Engineering for Foundations and Slopes, Proc. Geotech.Eng.Specialty Conf., Vol.1, 1976, pp.217-43.
20. Hamel, J.V., Personal Communication, 1980.

21. Harr, M.E., Mechanics of Particulate Media - A Probabilistic Approach, McGraw Hill, New York, 1977, 543pp.
22. Janbu, N., "Stability Analysis of Slopes with Dimensionless Parameters", Harvard Soil Mechanics Series, No.46, 1954, 81pp.
23. Lambe, T.W., and Whitman, R.V., Soil Mechanics, John Wiley, New York, 1969, 553pp.
24. Law, K.T., and Lumb, P., "A Limit Equilibrium Analysis of Progressive Failure in the Stability of Slopes", Canadian Geotechnical Journal, Vol.15, 1978, pp.113-122.
25. Leonards, G.A., "Stability of Slopes in Soft Clays", Special Lecture, VI Panamerican Conference, SMFE, Lima, Peru, 1979, pp.1-50.
26. Lo, K.Y., and Lee, C.F., "Stress Analysis and Slope Stability in Strain-Softening Materials", Geotechnique, Vol.23, 1, 1973, pp.1-11.
27. Morgenstern, N.R., "Ultimate Bheaviour of Rock Structures", Rock Mechanics in Engineering Practice, Ed. K.C. Stagg and O.C. Zienkiewicz, Wiley, 1968, pp.321-346.
28. Muller, L., "New Considerations on the Vaiont Slide", Rock Mechanics Eng.Geology, Vol.2, 1968, pp.1-91.
29. Muller, L., Personal communication 1978.
30. Parry, R.H.G., "Stability Analysis for Low Embanbkments on Soft Clays", Stress Strain Behaviour of Soils. Foulis & Co., 1972, London, England, pp.643-668.
31. Peck, R.B., "Stability of Natural Slopes", Journ.Soil Mech. and Foundations Division, ASCE, Vol.93, SM4, 1967, pp.403-417.
32. Rowe, P.W., "Discussion on Paper 20", In-Situ Investigations in Soils and Rocks, British Geotechnical Society, London, 1970, pp.310-311.
33. Sarma, S.K., "Stability Analysis of Embankments and Slopes", Geotechnique, Vol.23, 1973, pp.423-433.
34. Sarma, S.K., "Stability Analysis of Embankments and Slopes", Journ. Geotechnical Engineering Division, ASCE, Vol.105, GT12, 1979, pp.1511-1524.
35. Skempton, A.W., "Long-Term Stability of Clay Slopes, Rankine Lecture, Geotechnique, 14, 1964, pp.77-101.
36. Skempton, A.W., "First-Time Slides in Overconsolidated Clays, Geotechnique, 20, 1970, pp.320-324.
37. Skempton, A.W., "Slope Stability of Cuttings in Brown London Clay", Special Lectures Volume, 9th Int.Conf.Soil Mechanics and Foundations Engineering, Tokyo, 1977, pp.25-33.
38. Spencer, E., "The Thrust Line Criterion in Embankment Stability Analysis", Geotechnique, Vol.23, 1973, pp.85-101.
39. Taylor, D.W. "Fundamentals of Soil Mechanics", Wiley, New York, 1948, 700pp.
40. Transportation Research Board "Landslides-Analysis and Control", TRB Special Report 176, Washington, D.C.1978.
41. Vanmarcke, E.H., "Reliability of Earth Slopes", Journ.Geotech.Eng. Division, ASCE, Vol.103, GT11, 1977, pp.1247-1265.
42. Vaughan, P.R., "Undrained Failure of Clay Embankments", Stress Strain Behaviour of Soils, Foulis & Co., London, 1972, pp.683-691.
43. Voight, B., Editor, "Rockslides and Avalanches", Elsevier, Amsterdam, 1977.
44. Wright, S.G., Kulhawy, E.D., and Duncan, J.M., "Accuracy Equilibrium Slope Stability Analysis", Journ.Soil Mechanics Foundations Division, ASCE, Vol.99, SM10, 1973, pp.783-793.

APPENDIX II.-NOTATION

APPENDIX II.–NOTATION

b = width of a vertical slice
$c, c´, c´_m$ = cohesion, effective cohesion and its mobilised value
c_u = undrained cohesion
$c´_r$ = residual effective cohesion
dE = difference of normal forces on the two vertical sides of a slice
dT = difference of tangential forces on the two vertical sides of a slice
F, F_c, F_ϕ, F_ℓ = factor of safety; F w.r.t. cohesion; F w.r.t. friction; local value of F.
F_f = factor of safety after simulation of strain-softening and stress redistribution
F_i = An initial stress slope factor
F_p = factor of safety during simulation of progressive stress release
H = slope height
K = conjugate stress ratio or initial stress ratio
K_o = ratio of in-situ normal effective horizontal stress to in-situ normal effective vertical stress
ℓ = length of base of a slice
L = total length of slip surface
m = number of failed slices
n = total number of slices
N = stability factor, a dimensionless parameter
p = normal stress
p_f = probability of failure
p_i, p_n = initial normal stress, existing normal stress before slope formation or alteration
p_s = reliability
r_u = pore pressure ratio
u = pore water pressure
W = weight of slice
z = depth below ground surface, depth of slice
α = inclination at base of slice
β = inclination of slope
γ = unit weight of slope material
$\sigma´$ = normal effective stress
σ_v, σ_β = conjugate stresses in vertical direction and parallel to slope
τ, τ_m, τ_f = shear stress, its mobilised value and its value at failure (i.e. shear strength)
τ_i, τ_n = initial shear stress, shear stress before slope formation or alteration
$\lambda_{c\phi}$ = dimensionless stability parameter in terms of c and ϕ
$\phi, \phi´, \phi´_m$ = friction angle, effective friction angle and its mobilised value
$\phi´_r$ = residual friction angle

LIMIT EQUILIBRIUM CONCEPTS

APPENDIX III.-CRITICAL FAILURE SURFACE AND STABILITY CHARTS

A. <u>Misuse of Stability Charts</u> - Back Analyses of Failure.-Several trial failure surfaces are usually assumed and the critical surface is one which gives the minimum factor of safety. Often the shape is assumed in advance and most charts assume a surface circular in cross section. Misleading conclusions may result from these charts because (a) the actual surface of sliding is non-circular and (b) the actual surface does not correspond to the theoretical critical location.

Consider a recent example (24) where charts of Bishop and Morgenstern were used in connection with three case histories of long-term slope failure in London clay. The factors of safety from charts were significantly lower than those given by simple calculations based on actual slip surfaces (37). It so happens that when $c'=0$ for any soil, minimum theoretical F corresponds to a plane slip surface parallel to the slope. (For all the three case histories a value of c' approaching zero is considered appropriate due to softening of clay mass.) The theoretical and actual slip surfaces being different in shape and location, the results from charts in this case are totally misleading since these are compared by the authors with other results based on actual slip surfaces.

B. <u>Prediction from Charts when $c'=0$</u>.-The limit equilibrium approach predicts a plane critical failure surface when $c'=0$ for any soil. Yet, slip surfaces in homogeneous clays are invariably curved whether failure takes place in the short-term under $\phi_u=0$ conditions or in the long term with $c'=0$. This is an important point which is likely to be overlooked in practice when assessing the stability of slopes.

C. <u>Variational Calculus and Theoretical Shape of Critical Failure Surfaces</u>.-Use of variational calculus has shown that the critical slip surface in a homogeneous clay soil may be markedly non-circular even under undrained $\phi_u=0$ conditions. In studies based on variational calculus, the shape of the potential failure surface need not be assumed in advance. Minimisation of a functional leads to the critical shape and location. Moreover, such a critical surface gives a significantly lower value of F in comparison to the conventional value based on a surface approximating to circular shape (for further discussion see Ref.12).

A GENERALIZED BOUNDING SURFACE CONSTITUTIVE MODEL FOR CLAYS

by

Yannis F. Dafalias[1], A.M. ASCE
and
Leonard R. Herrmann[2], M. ASCE

ABSTRACT

The concept of the bounding surface in stress space is used to formulate a generalized constitutive model for cohesive soils within the framework of rate independent plasticity and critical state soil mechanics. The salient feature of the bounding surface concept is that plastic deformation can occur for stress points within the surface at a pace which depends on the proximity of the actual stress point from a properly defined "image" point on the bounding surface. The state is expressed in terms of the effective stress and the plastic change of the void ratio, and the bounding surface is defined in the space of three properly chosen stress invariants. General incremental constitutive relations are given for three dimensional analysis, and the corresponding expressions for triaxial conditions can be obtained as a particular case. Introducing a small number of material constants, the model is shown to realistically predict the soil response under monotonic, cyclic, drained and undrained loading conditions at any overconsolidation ratio for triaxial experiments in both compression and extension.

KEY WORDS: Bounding surface; Clays; Cohesive soils; Constitutive relations; Critical state; Cyclic loading; Monotonic loading; Overconsolidation; Plasticity; Plastic properties; Pore-water pressure; Soil mechanics.

---

[1] Associate Professor, Department of Civil Engineering, University of California, Davis.

[2] Professor, Department of Civil Engineering, University of California, Davis.

INTRODUCTION CONSTITUTIVE MODEL FOR CLAY

The purpose of this paper is to present a general constitutive model for clays within the framework of critical state soil mechanics, employing the new concept of the bounding surface in plasticity theory. The primary motivation, which led to the bounding surface concept, is the desire to obtain a unique set of constitutive relations, which can realistically describe the material response under both cyclic and monotonic loading. A serious shortcoming of classical yield surface plasticity in relation to soils, is that plastic deformation cannot occur by definition within the purely elastic range enclosed by the yield surface. An important consequence is that overconsolidated states are treated as purely elastic states, contrary to the observed plastic deformation which begins far before the stress point reaches the existing yield surface from within. For the same reason, during an undrained cyclic deviatoric loading after normal consolidation, classical plasticity predicts pore-water pressure build up only during the first half cycle, and subsequently, the stress oscillates within an expanded yield surface, being unable to cause any additional plastic volumetric strain which is the underlying cause for the observed continuing pore-water pressure development and decrease of the effective stress.

The general framework of the bounding surface plasticity, which encompasses past and current versions of the different models, has been briefly presented by Dafalias [5]. The main idea can be described as follows. Instead of viewing the yield surface as a locus which separates distinctly the elastic and elastoplastic ranges in stress space, one may view it as a bounding envelope of all stress states within, such that the value of the plastic modulus is a function of the proximity of a given stress point from this envelope. The envelope is called the "Bounding Surface". More specifically, to each stress point within, a properly defined mapping rule associates one only corresponding "image" stress point on the bounding surface with two consequences. First, the normal to the bounding surface at the "image" point defines the loading-unloading direction at the actual stress point. Second, the plastic modulus is a function of a "bounding" plastic modulus associated with the "image" point and obtained by the consistency condition for the bounding surface, and the Euclidean distance δ in stress space between actual and "image" stress points. Thus, the salient feature of the bounding surface concept is that plastic deformation can occur at a progressive pace, depending on this distance. The actual stress may eventually reach the hardening or softening bounding surface, where it becomes identical with its "image" and moves with it until unloading and inward motion occurs and so forth.

The bounding surface was originally introduced for metal plasticity in conjunction with an enclosed yield surface of finite or zero size [6,7,8,12]. This formulation was adopted in [3,13] for soil plasticity. Here, a much simpler formulation is presented without an explicitly introduced yield surface following the basic idea in [4], and which has been applied to soil plasticity in the triaxial and two stress invariants space for isotropic soils [9,10]. The bounding surface encloses always the origin (a realistic assumption for soil plasticity) and is origin-convex, i.e. any radius emanating from the origin intersects the surface at one only point. This does not preclude a general non-convexity. For any stress point inside the bounding surface, the "image" stress point is obtained as the intersection of the surface with the extension of the radius connecting the origin with the stress point. This simple "radial" mapping rule has the attractive feature of simplicity and has proved to be capable of predicting realistically the soil response under many different loading conditions. Here, the above formulation is generalized to include all three stress invariants, thus making possible to take into account the different values that some material parameters assume in triaxial compression and extension. The introduction of the third stress invariant is straightforward, but not trivial. It is first introduced without any reference to the particular shape of the bounding surface, and its effect on the constitutive relations is expressed in general. Subsequently, a specific form of the shape of the bounding surface is adopted in the two stress invariants space (meridional section), which depends on certain parameters. The dependence of the bounding surface on the third stress invariant is now obtained by rendering these parameters functions of it. This is achieved by a proper interpolation between the values of the parameters in triaxial compression and extension, where the third stress invariant assumes specific values. The above third stress invariant dependence can be extended to some or all the parameters, which are either associated with the shape of the bounding surface or enter the rate constitutive relations directly. In this presentation a rather restricted dependence is introduced which, however, has been proved to be simple and efficient enough to depict satisfactorily the compression-extension differences. Only two material parameters have been made functions of the third invariant: the one which corresponds to the slope of the critical state line projection on the two stress invariants space, and the hardening shape parameter which enters the relation between actual and bounding plastic moduli. Few additional parameters are introduced which are easily determined by conventional triaxial experiments. The model response is presented qualitatively and successfully compared quantitatively with experimental data for clays. Consolidation, dilatation, hardening, softening, cyclic pore-water pressure build-up, etc., at different OCR, can be described realistically by a single set of material constants within the formulation of the present model.

BOUNDING SURFACE FORMULATION FOR ISOTROPIC SOILS

Subsequently effective stresses σ_{ij} are considered, which are taken positive if compressive. A bar over a stress quantity implies a state on the bounding surface. The strain ε_{ij} is decomposed into an elastic and a plastic part, indicated by one and two primes, respectively. A dot indicates rate and the summation convention over repeated indices is employed. The deviatoric stress and strain components are denoted by s_{ij} and e_{ij}, respectively. For isotropic material, the elastic stress-strain rate relations become

$$\dot{\sigma}_{kk} = 3 K \dot{\varepsilon}'_{ii}, \qquad \dot{s}_{ij} = 2 G \dot{e}'_{ij} \qquad (1)$$

with K and G, the bulk and shear modulus respectively, possibly proper functions of the stress invariants.

It is further assumed that the equation of the bounding surface, $F = 0$, depends on three properly chosen stress invariants and one single scalar plastic internal variable, controlling hardening/softening and measuring the plastic volumetric strain. Such a variable can be defined as the plastic rate of the total void ratio rate e given by

$$\dot{e}'' = -(1 + e_o) \dot{\varepsilon}''_{kk} \qquad (2)$$

with e_o the initial void ratio corresponding to the reference configuration with respect to which strains are measured. For natural strains, it follows that $e_o = e$. The stress invariants are defined by

$$I = \sigma_{kk}, \quad J = (\tfrac{1}{2} s_{ij} s_{ij})^{1/2}, \quad S = (\tfrac{1}{3} s_{ij} s_{jk} s_{ki})^{1/3}, \tag{3}$$

Instead of S, it is convenient to introduce the "Lode" angle α:

$$-\frac{\pi}{6} \leq \alpha = \tfrac{1}{3} \sin^{-1}\left[\frac{3\sqrt{3}}{2}\left(\frac{S}{J}\right)^3\right] \leq \frac{\pi}{6} \tag{4}$$

where $\alpha = \pm \pi/6$ corresponds to triaxial compression and extension respectively.

The "radial" mapping rule mentioned in the introduction can now be written as

$$\bar{\sigma}_{ij} = \beta\,(\sigma_{kl}, e^{\ast\ast})\,\sigma_{ij} \tag{5}$$

with the radial factor $1 \leq \beta < \infty$ determined from the equation of the bounding surface:

$$F(\bar{I}, \bar{J}, \bar{\alpha}, e^{\ast\ast}) = 0 \tag{6}$$

where according to Eqs. (3), (4), (5), we have $\bar{I} = \beta I$, $\bar{J} = \beta J$, $\bar{S} = \beta S$ and $\bar{\alpha} = \alpha$. An important property following from Eq. (5) is that the derivatives of $\bar{I}, \bar{J}, \bar{S}$ with respect to $\bar{\sigma}_{ij}$ are identical to the derivatives of I, J and S with respect to σ_{ij}. Assuming the associated flow rule, the plastic strain rate is given by

$$\dot{\varepsilon}_{ij}^{\ast\ast} = <L>\frac{\partial F}{\partial \bar{\sigma}_{ij}}, \quad L = \frac{1}{K_p}\frac{\partial F}{\partial \bar{\sigma}_{kl}}\dot{\sigma}_{kl} = \frac{1}{\bar{K}_p}\frac{\partial F}{\partial \bar{\sigma}_{kl}}\dot{\bar{\sigma}}_{kl} \tag{7}$$

where K_p is the plastic modulus associated with the actual stress rate and $\bar{K}_p$ the "bounding" plastic modulus associated with the "image" stress rate. The $<\,>$ denotes the operation $<\ast> = \ast\,\bar{H}(\ast)$ with $\bar{H}$ the heavyside step function. Plastic loading, unloading and neutral loading is defined by $L > 0$, $L < 0$ and $L = 0$ respectively, with L called the loading function. The inclusion of K_p (or $\bar{K}_p$) in L treats simultaneously stable (hardening) response when $K_p > 0$ and unstable (softening) response when $K_p \leq 0$. On the basis of Eqs. (3), (4), (5), and (6) it follows that

$$\frac{\partial F}{\partial \bar{\sigma}_{ij}} = F_{,\bar{I}}\,\delta_{ij} + \frac{F_{,\bar{J}}}{2J}s_{ij} + \frac{\sqrt{3}\,F_{,\alpha}}{2\beta J\cos 3\alpha}\left[\frac{s_{ik}s_{kj}}{J^2} - \frac{3}{2}\frac{S^3 s_{ij}}{J^4} - \frac{2}{3}\delta_{ij}\right] \tag{8}$$

$$L = \frac{1}{K_p}\left(F_{,\bar{I}}\,\dot{I} + F_{,\bar{J}}\,\dot{J} + \tfrac{1}{\beta}F_{,\alpha}\,\dot{\alpha}\right) = \frac{1}{\bar{K}_p}\left(F_{,\bar{I}}\,\dot{\bar{I}} + F_{,\bar{J}}\,\dot{\bar{J}} + F_{,\alpha}\,\dot{\alpha}\right) \tag{9}$$

where a comma followed by the symbol of an invariant as a subscript indicates partial differentiation with respect to that invariant. From the consistency condition $\dot{F} = 0$ and Eqs. (5), (6), (7) and (8) it follows that

$$\bar{K}_p = 3(1 + e_o) \; (\partial F/\partial e^{\prime\prime}) \; F_{,\bar{I}} \tag{10}$$

Observe from Eq. (10) that with $\partial F/\partial e^{\prime\prime} > 0$, the bounding plastic modulus $\bar{K}_p$ is positive (consolidation), negative (dilatation), or zero (unrestricted shear flow), according to the value of $F_{,\bar{I}}$. Correspondingly, the bounding surface expands, contracts, or does not harden.

The functional dependence of K_p on $\bar{K}_p$ and the distance δ between σ_{ij} and $\bar{\sigma}_{ij}$ in invariant stress space is assumed to be given by

$$K_p = \bar{K}_p + H \; \frac{\delta}{r - \delta} = \bar{K}_p + H (\beta - 1) \tag{11}$$

where H is a proper hardening shape function of I, J, a, $e^{\prime\prime}$ and certain material constants, r is a reference stress which here has been chosen to be the distance between $\bar{\sigma}_{ij}$ and the origin and where Eq. (5) was used to obtain the second part of Eq. (11). A different choice of r is also possible [10], but the present choice appears to be simpler, more efficient, and preserves the invariance, with respect to different stress spaces, of the right hand side of Eq. (11). The H and the associated constants are intimately related to the soil response for states within $F = 0$ (overconsolidation). For $H \to \infty$, observe that $K_p \to \infty$ and $K_p = \bar{K}_p$ only for $\delta = 0$. In this case, the bounding surface behaves as a yield surface. It is possible to have $\bar{K}_p < 0$ and $K_p > 0$ if δ is large enough, which allows the description of an initially rising stress-strain curve as the stress point approaches the contracting bounding surface ($\bar{K}_p < 0$) during dilatation, and the subsequent unstable falling curve behavior when eventually δ becomes small enough to have both $\bar{K}_p$, $K_p < 0$ as in the case of heavily overconsolidated clays.

If the superscript t denotes total stresses and u the pore water pressure one has

$$\sigma_{ij}^t = \sigma_{ij} + \delta_{ij} u \tag{12}$$

Imposing the undrained condition $\dot{\epsilon}_{kk} = 0$, it follows that

$$\dot{u} = 3 K F_{,\bar{I}} <L> + \frac{1}{3} \dot{I}^t \tag{13}$$

$$\left[F_{,\bar{I}} + \frac{K_p}{9 K F_{,\bar{I}}} \right] \dot{I} + F_{,\bar{J}} \dot{J} + \frac{1}{\beta} F_{,\alpha} \dot{\alpha} = 0 \tag{14}$$

where Eq. (14) is the differential equation of the effective undrained stress path in the three stress invariants space.

SPECIFIC FORM OF THE BOUNDING SURFACE

Specific analytical expressions for the bounding surface are presented which allow the computation of the quantities $\partial F/\partial \bar{I}$, $\partial F/\partial \bar{J}$, $\partial F/\partial \alpha$ and $\partial F/\partial e^{\prime\prime}$ entering the constitutive relations (7), (8), (9) and (10).

Extending the ideas of the critical state soil mechanics [14] from the triaxial to the invariant space, a meridional section of the bounding surface is eloquently shown in Fig. 1 as a combination of two ellipses and one hyperbola. The quantity N can be identified as the slope of the projection CSL of the critical state line in invariant stress space which intersects the bounding surface at C where $\partial F/\partial \bar{I} = 0$. For triaxial conditions the N is related to the triaxial CSL slope M by $M = 3\sqrt{3}\,N$. The radial rule associating the "image" stress point $\bar{I}, \bar{J}$ to the stress point I, J is illustrated. The projection of C on the I axis is denoted by I_1 and is related to I_o by $I_1 = I_o/R$, with I_o the intersection of $F = 0$ with the hydrostatic axis I. The dependence of $F = 0$ on the third stress invariant is introduced through N, which is assumed to vary according to [11]:

$$N(\alpha) = \frac{2n}{1 + n - (1 - n)\sin 3\alpha} N_c \qquad (15)$$

where $N_c = N(\pi/6)$ is the value of N for triaxial compression and $n = N_e/N_c$ with $N_e = N(-\pi/6)$ the value of N for triaxial extension. It is interesting to note that the trace of the bounding surface on the π-plane may be non-convex depending on the value of n, but still the surface is origin-convex as commented in the introduction.

The dependence of $F = 0$ on $e^{\prime\prime}$ is introduced through I_o, thus $\partial F/\partial e^{\prime\prime} = (\partial F/\partial I_o)(\partial I_o/\partial e^{\prime\prime})$. The expression $dI_o/de^{\prime\prime}$ necessary for the hardening behavior is sought subsequently. Let κ and λ denote the slopes of the rebound and consolidation lines in the $e - \ell n I$ plot (just as in the $e - \ell n p$ plot). In order to prevent excessive softening of the elastic stiffness around $I = 0$ for cohesive soils, we assume that there is a limit value $I_\ell > 0$ such that for $I \le I_\ell$, the relation between I and the elastic part $e^\prime$ of the void ratio changes continuously from logarithmic to linear. Then, one has

$$\frac{dI}{de^\prime} = -\frac{\langle I - I_\ell\rangle + I_\ell}{\kappa}, \qquad \frac{dI_o}{de} = -\frac{I_o}{\lambda} \qquad (16)$$

where the first part of Eq. (16) applies also for $I = I_o$. It follows immediately from $e^{\prime\prime} = \dot{e} - \dot{e}^\prime$ and Eq. (16) that

$$\frac{dI_o}{de^{\prime\prime}} = -\frac{I_o}{\lambda - \omega\kappa} \quad \text{with} \quad \omega = \frac{I_o}{\langle I_o - I_\ell\rangle + I_\ell} \qquad (17)$$

The first part of Eq. (16) in combination with Eq. (1) and $\dot{e}^\prime = -(1 + e_o)\dot{\epsilon}_{kk}^\prime$ yields an expression for the bulk modulus K in terms of κ as

$$K = \frac{(1+e_o)(<I-I_\ell> + I_\ell)}{3\kappa} \qquad (18)$$

For further reference, let us introduce the quantities

$$\theta = J/I, \qquad x = \theta/N \qquad (19)$$

For $0 \leq \theta \leq N$ the bounding surface is the ellipse 1, Fig. 1, given by the equation

$$F = \bar{I}^2 + (R-1)^2 \left(\frac{\bar{J}}{N}\right)^2 - \frac{2 I_o}{R} \bar{I} + \frac{2-R}{R} I_o^2 = 0 \qquad (20)$$

Eq. (20) needs one only material parameter, the value of R.

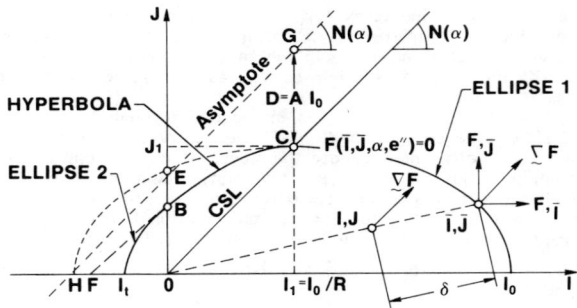

FIG. 1. - Schematic illustration of the bounding surface and the "radial" mapping rule in the invariant stress space.

Extension of the ellipse in the range $N \leq \theta \leq \infty$ is possible with $R \geq 2$ in order to include the origin, but it was proved to be unsatisfactory for the prediction of heavily overconsolidated states. Instead, a hyperbola is proposed, as shown in Fig. 1, defined by means of the parameter A, which positions the hyperbolic asymptote with respect to CSL by means of the distance $D = AI_o$. A varies with α as N does according to Eq. (15), i.e., $n = N_e/N_c = A_e/A_c$ with A_e and A_c the values of A in triaxial extension and compression, respectively. The equation of the hyperbola is now given by

$$F = \bar{I}^2 - \left(\frac{\bar{J}}{N}\right)^2 - \frac{2 I_o}{R}\bar{I} + 2 I_o\left[\frac{1}{R} + \frac{A_c}{N_c}\right]\frac{\bar{J}}{N} - \frac{2 A_c}{R N_c}I_o^2 = 0 \qquad (21)$$

Finally, extension in the tension regime for $-\infty < \theta \leq 0$ is obtained by a second ellipse, Fig. 1, as discussed in [9]. The intersection I_t of ellipse 2 with the I axis measures the tensile strength of the soil.

On the basis of Eqs. (10), (17), (20) and (21), one can easily compute $\bar{K}_p$, $\partial F/\partial \bar{I}$, and $\partial F/\partial \bar{J}$. In addition, based on Eq. (15) and the dependence of Eqs. (20) and (21) on the ratio $\bar{J}/N$, it follows that

$$\frac{F_{,\alpha}}{\beta \bar{J}} = \frac{3(n-1)\cos 3\alpha}{1 + n - (1-n)\sin 3\alpha} F_{,\bar{J}} \qquad (22)$$

Finally, for triaxial undrained deviatoric loading of a normally consolidated sample at $I_o = 3p_o$, Eq. (14) can be integrated in closed form to yield the undrained stress path in terms of the triaxial variables $q = \sigma_1 - \sigma_3 = \pm\sqrt{3}J$, $p = (1/3)(\sigma_1 + 2\sigma_3) = I/3$, $M = \sqrt{3}/N$ as:

$$\frac{q}{p_o} = \frac{M}{R-1}\left[\frac{2}{R}\left(\frac{p}{p_o}\right)^{\frac{\lambda-2\kappa}{\lambda-\kappa}} + \left(1 - \frac{2}{R}\right)\left(\frac{p}{p_o}\right)^{\frac{-2\kappa}{\lambda-\kappa}} - \left(\frac{p}{p_o}\right)^2\right]^{1/2} \qquad (23)$$

where M assumes different values, M_c and M_e, in triaxial compression and extension.

THE SHAPE HARDENING FUNCTION, IDENTIFICATION AND DETERMINATION OF MATERIAL CONSTANTS

The following form for H entering the key equation (11) is assumed

$$H = p_a h(\alpha)\left[1 + |x|^{-m}\right]\left[9 F_{,\bar{I}}^2 + \frac{1}{3} F_{,\bar{J}}^2\right] \qquad (24)$$

where p_a is the atmospheric pressure providing the proper stress units. The introduction of the second bracket is associated with the corresponding "unit normal" triaxial formulation [10] and the continuity of the constitutive relations at point C, Fig. 1. The introduction of $|x|^{-m}$ with $m > 0$, does not allow plastic deformation to occur within the bounding surface for $\theta = 0$ (zero deviatoric stress) rendering K_p infinite except when $\delta = 0$ for which $K_p = \bar{K}_p$. The absolute value has been introduced for use in tensile response where $x < 0$. A small value of m suppresses the influence of $|x|^{-m}$ for $\theta > 0$. The shape hardening parameter h bears mainly the responsibility for the description of the material response for states within the surface. The h is considered a function of the "Lode" angle α varying as N does, according to an equation similar to Eq. (15), with n being substituted by $\mu = h_e/h_c$, with h_e and h_c the values of h at triaxial extension and compression, respectively.

It is now possible to identify and suggest calibration procedures for the material constants entering the formulation. The set of these constants will be divided in two groups, the old and the new.

(i) Old material constants

This group includes the elastic constants κ (or K) and G (or Poisson's ratio ν), the slope of the consolidation lines λ, the slopes of the CSL N_c, N_e or equivalently N_c and $n = N_e/N_c$, and I_ϱ, Eq. (16). Their determination follows well known methods. The I_ϱ although introduced as a new parameter, Eqs. (16)-(18), is not related to the bounding surface concept and refers to a better description of the elastic response near the origin. It can be taken equal to 1 psi or about 10 kPa.

(ii) New material constants

The new constants are R, A_c, m, h_c and μ. The first two refer to the determination of the shape of the bounding surface (R for ellipse 1 and A_c for the hyperbola) while the last three are associated with the change of the plastic modulus for states within the surface. The ellipse 2 is determined by a parameter T such that $I_t = TI_o$ [9], but we refrain from further discussion on T since no sufficient information exists at this stage on the tensile strength of the soil. The R, A_c can be considered also as appropriate constants for a classical yield surface formulation, improving the shape of the surface. In the past, R has been assigned a fixed value for all clays [14]. The role of m has been explained earlier in connection with the use of $|x|^{-m}$ and does not effect the material response considerably except near x = 0. It was found that m = 0.2 can be used for most clays. For this value, the first bracket of Eq. (24) varies between 2.58 and 1 as x changes from 0.10 to ∞, a variation which is negligible compared to the values of $\bar{K}_p$ and the other quantities entering Eqs. (11) and (24). The h_c, μ are the most important constants.

The following concrete steps are now suggested for the calibration of R, A_c, h_c and μ from triaxial experiments once the old material constants have been defined and m = 0.2.

Step 1: Determination of R

After normal consolidation, one must obtain the undrained stress path in triaxial compression and determine experimentally the projection p on the p-axis of the point of intersection of the path with the CSL. For this point $q = M_c p$ and substitution of this value in Eq. (23) eliminates $M = M_c$ and yields an equation for R in terms of p/p_o. The same procedure in extension yields, in general, a slightly different value of p, thus of R. An average value of R is chosen. In this case, none of the other new constants appears.

Step 2: Determination of h_c, μ

Repeat the same undrained loading as in step 1 but for an overconsolidated sample at an overconsolidation ratio between 1 and R, preferably at OCR = 3R/(2R + 1). Obtain the experimental curves q-p, q-ε_1, u-ε_1, for both compression and extension. Determine h_c, h_e and $\mu = h_e/h_c$ by curve fitting the experimental data using the developed incremental relations. This can be done by a trial-and-error process observing that increasing h_c or h_e implies a stiffer response. For this range of OCR, the A does not appear in the constitutive equations because it is associated with the hyperbola for heavy overconsolidation, which is not active.

Step 3: Determination of A_c

Repeat the same experiment as in step 2 for compression and for OCR $\geq$ 5 at least. With h_c known from step 2, A_c is determined by a similar trial-and-error numerical process. Increasing A_c implies a "flatter" hyperbola with a stiffer response and reduced dilatation, while a very small A_c brings the hyperbola down close to the CSL for materials with small cohesion. Recall that A_e is determined from $A_e/A_c = n = N_e/N_c$.

The great advantage of the above steps is that the parameters can be determined by a sequence of triaxial tests, where each parameter becomes active at a time. At the end, an overall slight rearrangement may be necessasry. The sets of material constants used subsequently are tabulated in Table 1.

It is important to emphasize that the response to cyclic loading is obtained on the basis of the above state dependent formulation as a sequence of monotonic loading/unloading incremental events without introducing any additional cyclic empirical parameters except one scalar material constant necessary for cyclic stabilization as it will be discussed subsequently.

TABLE 1 - Two Sets of Material Constants

Old Constants	Sets 1	2	New Constants	Sets 1	2
κ	0.05	0.05	R	2.60	2.72
ν	0.20	0.20	A_c	0.06	0.06
λ	0.14	0.14	m	0.20	0.20
N_c	0.20	0.20	h_c	11	16
n	0.81	1.00	μ	2	1

QUALITATIVE BEHAVIOR OF THE MODEL

Monotonic Loading

Using the set 1, Table 1, of material constants, the drained and undrained behavior of the model at OCR = 1, 2 and 10 for compression and OCR = 1, 2 and 8 for extension with initial void ratios e = 0.94, 0.97 and 1.00 correspondingly is shown in Figures 2a, b, c and 3a, b, c, where also the initial position of the bounding surface is marked. Many interesting features can be observed, such as the failure envelope between the initial position of the bounding surface and the CSL, consolidation for small OCR, initial small consolidation and subsequent dilatation for large OCR, unstable softening response showing a falling stress-strain curve for the drained loading at large OCR, a more stable response for the corresponding undrained loading with simultaneous negative pore water-pressure build up, which has a well known stabilizing effect, the characteristic "hook" shape of the undrained stress path for large OCR, etc.

Cyclic Loading

Using the set 2 of soil parameters with e_o = 0.95 the undrained cyclic deviatoric loading for 6 cycles and for amplitude $q/p_o = 0.35$ for compression and $q/p_o = 0.25$

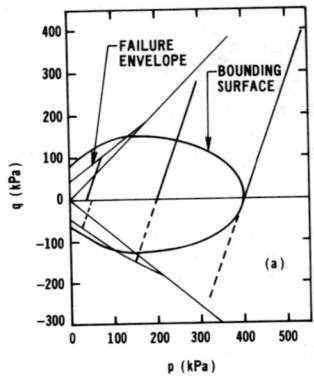
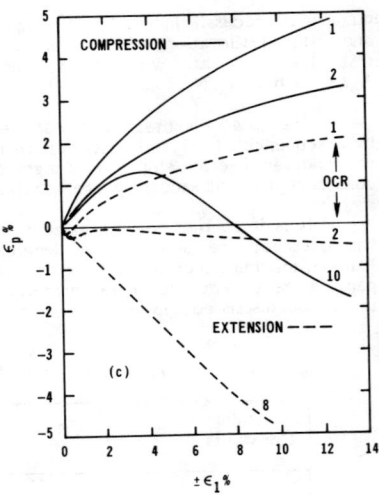
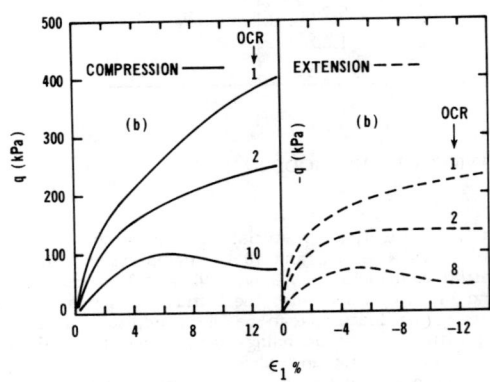

FIG. 2a,b,c. - Drained behavior of the model at different OCR for monotonic loading.

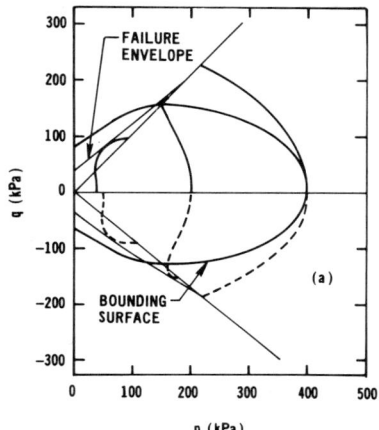
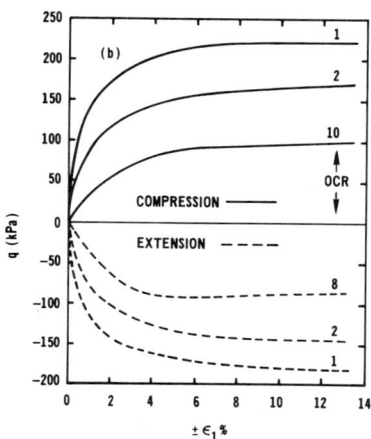
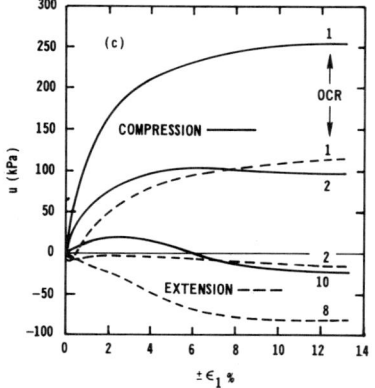

FIG. 3a,b,c. - Undrained behavior of the model at different OCR for monotonic loading.

90 SOIL STRESS STRAIN APPLICATIONS

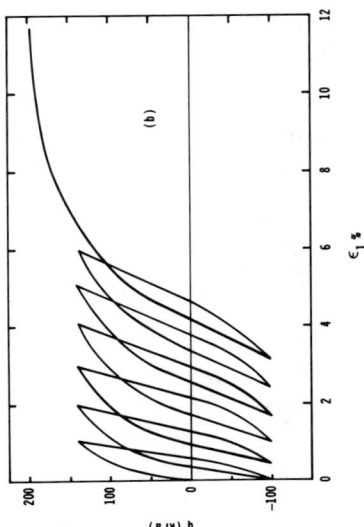

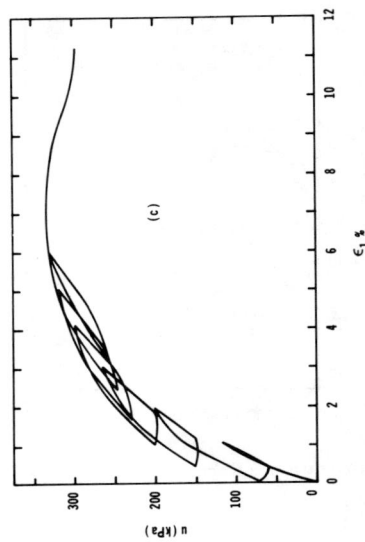

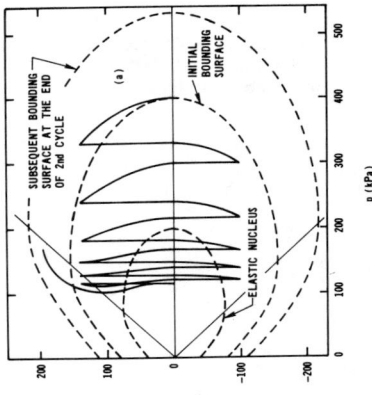

FIG. 4a,b,c. - Undrained behavior of the model for cyclic loading.

for extension yields the soil response as shown in Fig. 4a, b, c. Observe the progressive motion of the effective stress loops towards the critical state line and the simultaneous expansion of the bounding surface shown by discontinuous lines in Fig. 4a. Corresponding loops are observed in the q-ε_1 and u-ε_1 graphs, Figs. 4b and 4c. Unloading is followed by loading when q changes sign. For failure, a final q increase is necessary at the end of the 6th cycle. The realistic model description of the cyclic loading response is intimately related to the fact that cyclic loading brings a normally consolidated sample to states of increasing overconsolidation for which the present constitutive law has been proved successful.

If it becomes necessary to have full stabilization of the cyclic loops before they reach the CSL, it is suggested to substitute the quantity $r - \delta$ in Eq. (11) by $<r-s\delta>$ with s being a stabilization factor called also the elastic factor. Then, whenever $\delta > r/s$ the brackets yield a zero value and $K_p = \infty$. Observe that $K_p \to \infty$ continuously as $\delta \to r/s$. This modification introduces indirectly the concept of the "elastic nucleus" as suggested in [5], Fig. 4a, within which a purely elastic response occurs for any stress rate direction. The boundary of the elastic nucleus is equivalent but not identical to the concept of the yield surface, since a stress point can cross and move outside the elastic nucleus with a smooth elastoplastic transition. When a cyclic loop enters the elastic nucleus, depending on the q amplitude, full stabilization occurs.

COMPARISON WITH EXPERIMENTS

Monotonic Loading

The experimental results of soft clay response under undrained monotonic deviatoric loading in compression and extension are shown by corresponding symbols in Fig. 5a, b, c and 6a, b for OCR = 1, 1.2, 2, 5, 8 and 12 for compression with initial void ratios e_o = 0.94, 0.95, 0.97, 0.95, and 0.95, respectively, and for OCR = 1, 1.2, 2, 6 and 10 for extension with initial void ratios e_o = 0.93, 0.94, 0.96, 0.95 and 0.95, respectively, as reported by Banerjee and Stipho [1,2]. Using the set 1 of material constants, Table 1, the predictions of the developed model are superimposed in the same figures and shown by continuous and dashed lines for compression and extension, respectively. The old constants are taken from the above references. On the top right corner of Fig. 6b the prediction of the model for OCR = 5 is shown if the ellipse 1 is used instead of the hyperbola for the range $\theta > N$. The observed deviation of the predicted results from the experimental data justifies the necessity to introduce the hyperbola instead of the ellipse 1 for that range of θ.

Cyclic Loading

The undrained stress path under cyclic deviatoric loading of kaolin samples in compression, as obtained experimentally by Wroth and Loudon [15] and predicted by the model, are shown for comparison on the same Fig. 7. The associated material constants are also shown on Fig. 7. The old material constants were taken from [15]. The experimental data are not complete for the calibration of all the new constants described before, but still, it was possible to obtain most of them. The R and h_c (referred to as h on the figure) were determined from two successive cycles, m = 0, and the value 1 was assigned to n and μ for simplicity, although not necessary, since cyclic deviatoric loading occurs only in compression. Hence, $h_c = h_e = h$ and $M_c = M_e = M$. Except for the last the other cycles are not associated with the hyperbola, thus A_c cannot be specified accurately. For this last cycle which exhibits the "hook" behavior of the stress path, it was taken A = 0.06. A final observation is that for the cyclic prediction of Fig. 7, the reference stress r in Eq. (11) was chosen to be equal to p_o for the triaxial space as in Ref. [10].

92 SOIL STRESS STRAIN APPLICATIONS

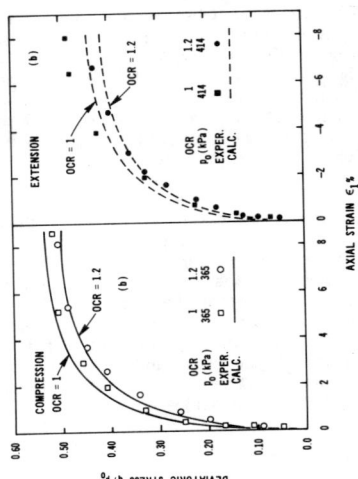
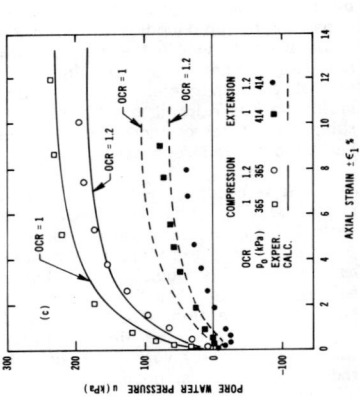

FIG. 5a,b,c. - Theory versus experiments for lightly overconsolidated clay. Experimental data after Banerjee and Stipho [1].

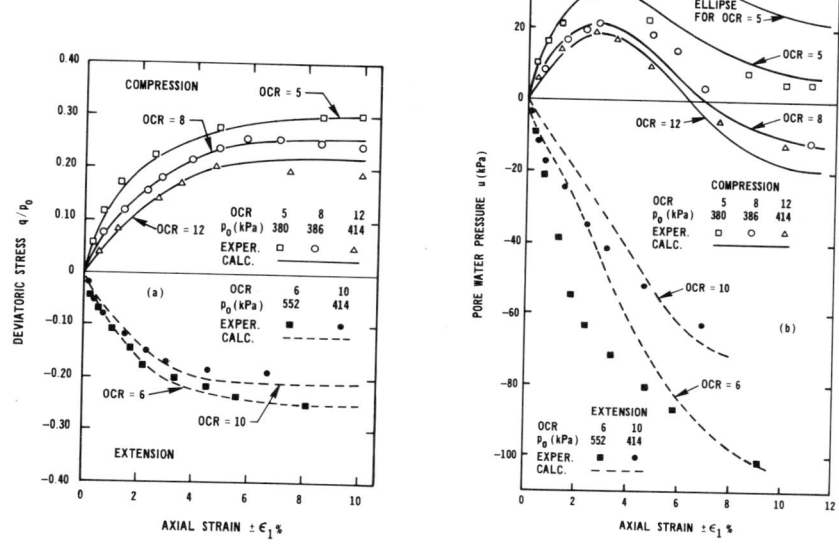

FIG. 6a,b. - Theory versus experiments for heavily overconsolidated clay. Experimental data after Banerjee and Stipho [2].

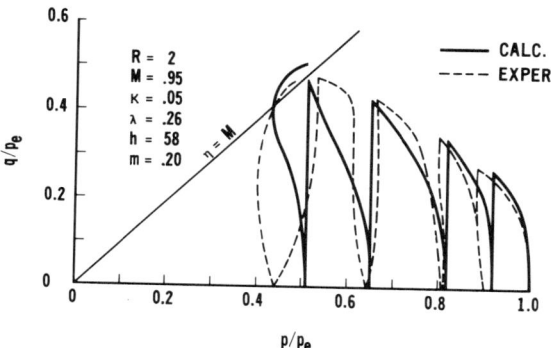

FIG. 7. - Theory versus experiment for undrained cyclic deviatoric loading. Experimental data after Wroth and Loudon [15].

CONCLUSION

Despite its simplicity, the present bounding surface model can describe realistically the soil response under different monotonic and cyclic, drained and undrained loading conditions at any OCR, including unstable behavior and cyclic mobility with further potential to include tension response. The constitutive relations are general and tensorial in character and applicable to any loading stress path, although their applicability was demonstrated here only for triaxial loading. The inclusion of the third stress invariant makes possible to take into account the different values that some important material parameters assume in the particular cases of triaxial compression and extension. This, of course, will also yield a more accurate material description for any loading path for which the three principal stresses vary independently. Comparison with experiments demonstrates these properties. The present formulation introduces only three new parameters, m, h_c, and μ, associated with the bounding surface concept. Two other parameters, R and A_c, aim at improving the shape of the used bounding surface, but they are not essential to the general concept (other shapes can be used). Detailed procedures for the calibration of R, A_c, m, h_c, and μ are proposed and applied.

Future improvement must include initial and developing anisotropy and a more complete dependence on the third stress invariant than the one introduced here by means of N and h. For example, the important parameter R should be made a function of α rather than being determined by its average value between compression and extension as done in Step 1 of the calibration.

As a final conclusion, perhaps the value of this work can be embodied in the demonstrated simple idea that any sound classical yield surface soil plasticity model can be easily transformed into a corresponding and more flexible bounding surface model on the basis of the general rules expounded in [5] and briefly described earlier in this work.

Acknowledgment

The research reported in this paper was conducted, in part, under National Science Foundation Grant NSF-CME-79-10835. The authors would also like to acknowledge the assistance in the numerical applications provided by Mr. J. S. DeNatale, Graduate Student in the Department of Civil Engineering at U.C. Davis.

REFERENCES

1. Banerjee, P.K., and Stipho, A.S., "Associated and Non-Associated Constitutive Relations for Undrained Behavior of Isotropic Soft Clays", **International Journal for Numerical and Analytical Methods of Geomechanics**, Vol. 2, 1978, pp. 35-56.

2. Banerjee, P.K., and Stipho, A.S., "An Elastoplastic Model for Undrained Behavior of Heavily Overconsolidated Clays", **International Journal for Numerical and Analytical Methods in Geomechanics** (Short Communication), Vol. 3, 1979, pp. 97-103.

3. Dafalias, Y.F., "A Model for Soil Behavior under Monotonic and Cyclic Loading Conditions", **Transactions of the 5th International Conference on SMiRT,** Berlin, Germany, Vol. K, No. 1/8, 1979.

4. Dafalias, Y.F., "A Bounding Surface Plasticity Model", **Proceedings of the 7th Canadian Congress of Applied Mechanics,** Sherbrooke, Canada, 1979, pp. 89-90.

5. Dafalias, Y.F., "The Concept and Application of the Bounding Surface in Plasticity Theory", **Symposium IUTAM on Physical Non-Linearities in Structural Analysis,** CETIM, Senlis, France, 1980, Springer Verlag publs., in press.

6. Dafalias, Y.F., and Popov, E.P., "A Model of Nonlinearly Hardening Materials for Complex Loadings", **Proceedings of the 7th U.S. National Congress of Applied Mechanics,** Boulder, USA, 1974, p.149 (Abstract), and **Acta Mechanica,** Vol. 21, 1975, pp. 173-192.

7. Dafalias, Y.F., and Popov, E.P., "Plastic Internal Variables Formalism of Cyclic Plasticity", **Journal of Applied Mechanics,** Vol. 98, No. 4., 1976, pp. 645-650.

8. Dafalias, Y.F., and Popov, E.P., "Cyclic Loading for Materials with a Vanishing Elastic Region", **Nuclear Engineering and Design,** Vol. 41, No. 2, 1977, pp. 293-302.

9. Dafalias, Y.F., and Herrmann, L.R., "A Bounding Surface Soil Plasticity Model", **International Symposium on Soils under Cyclic and Transient Loading,** Swansea, U.K., Vol. 1, 1980, pp. 335-345.

10. Dafalias, Y.F., and Herrmann, L.R., "Bounding Surface Formulation of Soil Plasticity", **Soils under Cyclic and Transient Load,** G.N. Pande and O.C. Zienkiewicz eds., John Wiley and Sons, Inc., New York, in press.

11. Gudehus, G., "Elastoplastische Stoffleichungen fur trockenen Sand", **Ingenieur-Archiv,** Vol. 42, 1973.

12. Krieg, R.D., "A Practical Two-Surface Plasticity Theory", **Journal of Applied Mechanics,** Vol. 42, 1975, pp. 641-646.

13. Mroz, Z., Norris, V.A., and Zienkiewicz, O.C., "Application of an Anisotropic Hardening Model in the Analysis of Elastoplastic Deformation of Soils", **Geotechnique,** Vol. 29, No. 1, 1979, pp. 1-34.

14. Schofield, A.N. and Wroth, C.P., **Critical State Soil Mechanics,** McGraw-Hill, London, 1968.

15. Wroth, C.P. and Loudon, P.A., "The Correlation of Strains with a Family of Triaxial Tests on Overconsolidated Samples of Kaolin", **Procedings, Geotechnical Conference,** Oslo, Vol. 1, 1967, pp. 159-163.

SOIL AS AN ANISOTROPIC KINEMATIC HARDENING SOLID

W.D. Liam Finn[1], M.ASCE and G.R. Martin, M.ASCE[2]

INTRODUCTION

Recently, the authors have been assessing the merits of a variety of constitutive relations proposed for modelling soil behaviour under both static and cyclic loading conditions. The immediate objective of this assessment is the selection of a constitutive model that might be used in parametric studies of certain effects of cyclic loading on proposed offshore installations. One of the models studied extensively was the anisotropic theory of plasticity as formulated for soils by Prevost (14,15,16) and Mroz, Norris and Zienkiewicz (10,11). The evaluation of the assumptions, procedures and potential of these different formulations of anisotropic plasticity theory is the subject of this paper.

Anisotropic plasticity theory is a relatively new development and its concepts are not widely known. To facilitate the presentation of the model and access to the original papers for engineers not familiar with the theory some of the more important concepts of the model are discussed in the next section.

BASIC ELEMENTS OF ANISOTROPIC THEORY

On the basis of observed behaviour in cyclic loading tests, soil is a non-linear, hysteretic, strain-hardening material over the range of strains usually considered tolerable for engineering design. The behaviour in extension and compression may be quite different and the Bauschinger effect (different yield stresses in extension and compression) is usually pronounced. Furthermore, because of their geological history soils are always anisotropic to some degree. Although soils can exhibit strain-softening, this behaviour is excluded from discussion. Therefore, only strain-hardening theories of plasticity are considered.

For strain-hardening materials, represented for instance by the piecewise linear stress-strain curve in Fig. 1a, a yield surface exists in stress space separating regions of elastic and plastic response. The yield surface corresponding to the initial elastic range defined by the stress σ_0 is given by the inner ellipse in the stress space σ_1, σ_2 shown in Fig. 1b. The equation, f=0, defining the yield surface, shapes the mathematical forms of the equations of the computational model. Two different yield surfaces will be discussed, the von Mises yield criterion and the Roscoe-Burland surface (17).

If loading is continued past σ_0 in Fig. 1a to σ_r, what happens to the yield surface? The assumptions made at this point introduce a major distinction between plasticity theories and a radical difference in the predictive capabilities of the resulting plasticity models. It is frequently assumed that unloading from F is elastic and continues until G is reached in the opposite loading sense at a similar stress magnitude. It

---

[1] Professor, Soil Dynamics Group, Faculty of Graduate Studies and Dept. of Civil Engineering, University of British Columbia, Vancouver, B.C., Canada, V6T 1W5.

[2] Associate, Fugro Incorporated, Long Beach, California 90807, U.S.A.

SOIL AS HARDENING SOLID   97

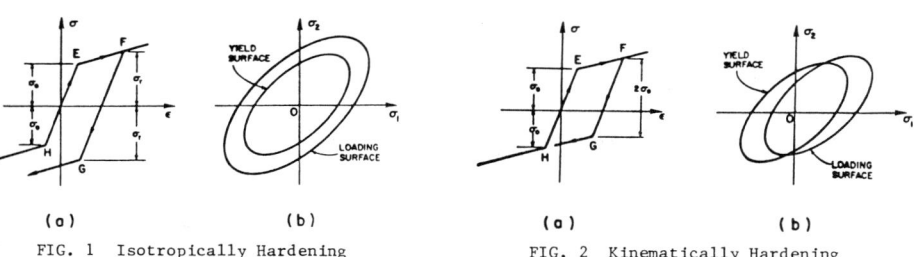

FIG. 1 Isotropically Hardening Material

FIG. 2 Kinematically Hardening Material

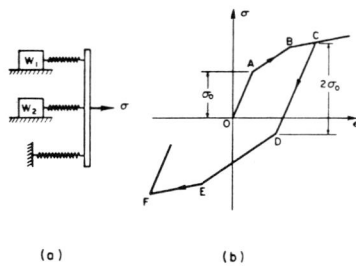

FIG. 3 Piece-wise Linear Material Idealization-Parallel or Sublayer Model   (After Popov (12))

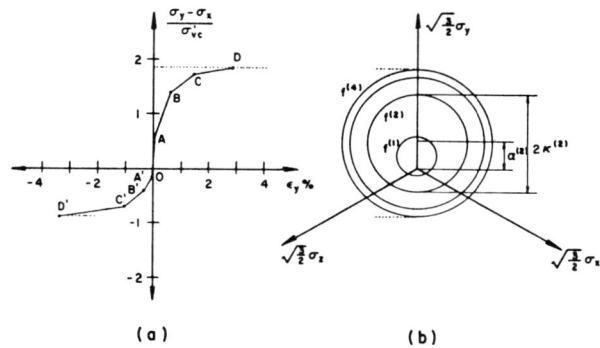

FIG. 4 Relationship between Uniaxial Stress-Strain Curve and Initial Yield Surfaces

is clear that the elastic region is now extended from $2\sigma_0$ to $2\sigma_r$. The new yield surface is given by the outer ellipse in Fig. 1b. All paths within this enlarged surface are elastic. The yield surface has been expanded uniformly by the loading causing plastic deformation. A material with this type of response is called an isotropically hardening material (6). Clearly, this alone is not a good model for soil under cyclic loading. Nevertheless, as will be seen later, it is useful in modelling some aspects of soil behaviour and will be retained as a component of the more general anisotropic model.

The yield surface is also used to define loading states. For strain-hardening materials, stress-increments directed outward from the yield surface are termed loading states, those directed tangential to it are neutral and those directed inwards are termed unloading states. Because of this close connection with states of loading, the yield surface is also called a loading surface. If the yield surface is defined by $F(\sigma_{ij}) = 0$ where σ_{ij} are the stresses, then

for loading
$$\frac{\partial f}{\partial \sigma_{ij}} d\sigma_{ij} > 0 \tag{1}$$

for neutral loading
$$\frac{\partial f}{\partial \sigma_{ij}} d\sigma_{ij} = 0 \tag{2}$$

for unloading
$$\frac{\partial f}{\partial \sigma_{ij}} d\sigma_{ij} < 0 \tag{3}$$

The vector with components $\partial f/\partial \sigma_{ij}$ is in the direction of the normal to the yield surface and the vector dot product of $\partial f/\partial \sigma_{ij}$ with the stress increment vector $d\sigma_{ij}$ defines the nature of the loading increment and hence the loading state. This precise definition of loading or unloading is one of the advantages of plasticity theory for the analysis of multi-axial stress-states. Generalizations of successful one-dimensional methods not based on plasticity theory to multi-axial stress-states often have difficulty in defining the loading states precisely.

Prager (13) made a different assumption about the behaviour of the yield surface as loading proceeded from E to F in Fig. 2a. He assumed that the yield surface did not change in size or shape but translated without rotation in stress-space as in Fig. 2b. In effect, he assumed the yield surface to be represented by a frictionless rigid ring which was pushed around the stress-space σ_1, σ_2 (Fig. 2b) by the stress point. As the stress increased from E to F, the stress point pushed the yield surface to the stress-state associated with F. On unloading from this state it is clear that the maximum elastic response range is $2\sigma_0$ since the yield surface has not changed in size and yielding will occur again when loading in the opposite sense touches the boundary of the yield surface. From Fig. 2a, it can be seen that the yield stress in extension is different from that in compression, thus modelling the Bauschinger effect. This type of hardening is called kinematic hardening as it is specified by the motion of the yield surface. Kinematic hardening is one of the key elements of the anisotropic theory of plasticity which makes it a useful model for soils.

The hardening that occurs beyond initial yield in Figs. 1a and 2a is of the simplest kind - linear. In general, the hardening is non-linear but can be represented by a piecewise linear approximation. The slope of the line EF (Fig. 2a) is used to define a work-hardening or plastic modulus. In the linear hardening case, a single modulus associated with a single yield surface is sufficient to define the plastic deformations. For the piecewise linear representation a nest of yield surfaces and associated plastic moduli are required, each associated with a particular linear segment of the strain-hardening stress-strain curve. This generalization of the basic model is due to Mroz (9). When more than one yield surface is used, the translation of the yield surfaces in stress-space must be controlled by some criterion to ensure that the yield surfaces do not intersect. A rule which has been developed by Mroz (9) is used in the models described later. The Mroz rule states that if the stress point P is on yield surface $f^{(m)}$ then on loading $f^{(m)}$ translates towards the next yield surface $f^{(m+1)}$ along the line PR where R is the point on $f^{(m+1)}$ with outward normal in the same direction as the normal at P. P and R are called conjugate points. (See points P_3 and R_3 in Fig. 9b).

For one-dimensional problems mechanical sub-layer models (Fig. 3a) are often used in soil mechanics. Taylor and Larkin (18) have used such models for non-linear dynamic analysis. Finn, Lee and Martin (4), for dynamic effective stress analysis, use an analytic curve representing the limiting case of the piecewise linear approximation as the number of linear segments becomes large. These models give the unloading-reloading pattern shown in Fig. 3b. Note that each segment of the curve is twice as long in unloading as in loading. This is the Masing criterion (8) for establishing the unloading and reloading curves from an initial loading curve. Note that the Masing effect is also produced by the kinematic hardening model in Fig. 2a. As will be seen shortly, pure Masing response is limited to isotropic materials.

A crucial element of an incremental theory of plasticity is the flow rule which defines the directions of the plastic strain increments during loading. The strain increments are assumed to be normal to a surface in stress-space called the potential surface, g=0. If the potential surface coincides with the yield surface the flow rule is called an associative flow rule, otherwise it is called a non-associative flow rule. Most formulations of the anisotropic theory of plasticity assume an associative flow rule, although Prevost (16) has recently introduced a model with a restricted form of a non-associative flow rule which will be discussed later.

The associative flow rule is a consequence of Drucker's definition of a work-hardening or strain-hardening material (5);

$$d\sigma_{ij} \, d\varepsilon_{ij} > 0 \qquad \text{on loading} \qquad (4)$$

and over a complete cycle

$$d\sigma_{ij} \, (d\varepsilon_{ij} - d\varepsilon_{ij}^e) \geq 0 \qquad (5)$$

in which $d\varepsilon_{ij}$ are the total strain increments due to the application of the self-equilibrating stress increments $d\sigma_{ij}$. The repeated subscripts indicate summation in all cases over the number of dimensions of the problem under review. The elastic strains are given by $d\varepsilon_{ij}^e$ and $(d\varepsilon_{ij} - d\varepsilon_{ij}^e) = d\varepsilon_{ij}^p$ is the plastic strain which is not recovered during the cycle. The latter relationship is a common assumption of plasticity and not necessarily true in all cases (7). Drucker's definition may have important consequences for the analysis of liquefaction problems by the anisotropic theory of plasticity which will be analysed in a later section.

The incremental relationship between stress increments and strain-increments is the corner-stone of any computational model. The relationship for elastic response is given by one of the usual formulations of the theory of elasticity. The relationship between stress increments and plastic strain increments depends on the flow rule. For materials satisfying Drucker's definition of work-hardening it can be shown that the relationship is linear. A simple heuristic development of the relationship for this case is given below.

Only stress increments normal to the yield surface cause plastic strain increments. The components of the unit normal, n_{ij}, are defined by

$$n_{ij} = \frac{\partial f/\partial \sigma_{ij}}{\{\partial f/\partial \sigma_{mn} \cdot \partial f/\partial \sigma_{mn}\}^{\frac{1}{2}}} \tag{6}$$

where the quantity below the line is the magnitude of the vector with components $\partial f/\partial \sigma_{ij}$. The normal stress increment components $d\sigma_{ij}$ are given by

$$d\sigma_{ij} = n_{ij} (n_{mn} \cdot d\sigma_{mn}) \tag{7}$$

Assuming a linear relationship between $d\varepsilon^p_{ij}$ and $d\sigma_{ij}$ the following relationship is obtained

$$d\varepsilon^p_{ij} = \frac{1}{H} \cdot d\sigma_{ij} \tag{8}$$

where H is the plastic modulus. Written in tensor form Eqn. (8) appears as

$$d\varepsilon_{ij} = \frac{1}{H} \cdot \frac{\partial f}{\partial \sigma_{ij}} \cdot \frac{\partial f/\partial \sigma_{rs} \cdot d\sigma_{rs}}{\partial f/\partial \sigma_{mn} \cdot \partial f/\partial \sigma_{mn}} \tag{9}$$

These are some of the more important elements of anisotropic plasticity theory. Two specific formulations of the theory will now be discussed. These were selected for the following reasons: the theories are relatively well developed, experimental validation is offered, indications are given on how to measure the parameters of the models. These two models are the multi-yield surface model developed expressly for soils by Prevost (15,16) and the two-surface model of Mroz, Norris and Zienkiewicz (10,11). The theoretical potential of these models are assessed.

MULTI-YIELD SURFACE MODEL

As mentioned before, the formulation of the anisotropic plasticity model for computational purposes depends on the surface or function chosen to define yielding. The Prevost formulation for undrained stress-strain behaviour for clay (15) will be presented and some of the changes proposed to include drained behaviour will be discussed.

Prevost assumes that yielding is defined by deviatoric stresses and occurs at constant plastic volume. He uses the von Mises yield criterion which may be represented by a circle in the deviatoric plane, the plane normal to the hydrostatic line in stress space. In the isotropic theory of plasticity the von Mises yield criterion for a single yield surface may be defined by

$$f = \frac{1}{2} S_{ij} S_{ij} - K_0^2 = 0 \tag{10}$$

in which S_{ij} are the deviatoric stresses and K_0 is a material constant. For the form of f given above, K_0 is the yield stress in pure shear. For anisotropic materials with a single yield surface the yield function becomes

$$f = \frac{1}{2} (S_{ij} - \alpha_{ij})(S_{ij} - \alpha_{ij}) - K_0^2 = 0 \tag{11}$$

This is the equation of a circle in the deviatoric plane with its centre defined by the deviatoric stress co-ordinates α_{ij}. The circle has been displaced from the origin. Recalling that anisotropy can be modelled by displacing the yield surface, Eqn. (11) defines the yield surface for a pure kinematically hardening material. The generality of the model and, hence, its predictive capability, is increased by allowing the constant K_0 to change. This implies that some isotropic hardening may also occur. Thus, the multi-surface anisotropic plasticity model for undrained shear is given by the yield function

$$f^{(m)} = \frac{1}{2} (S_{ij} - \alpha_{ij}^{(m)})(S_{ij} - \alpha_{ij}^{(m)}) - K_m^2 = 0 \tag{12}$$

where $f^{(m)}$ is the yield function for the mth yield surface, K_m is the radius of the yield circle in deviatoric stress-space and $\alpha_{ij}^{(m)}$ its centre.

All the equations needed for a computational model follow from this definition of yield. The stress-strain relations may be determined in a routine way using Eqn. (9) if, as Prevost does, an associative flow rule is assumed.

The determination of the parameters of the model will be explained with the aid of Fig. 4. A piecewise linear representation of a stress-strain curve obtained in a triaxial test is shown in Fig. 4a, non-dimensionalized with respect to the initial confining pressure σ'_{vc}. The behaviour in extension and compression is different but segments of similar slope exist in both the compression and extension regions. This need not be so as is shown by Prevost (Fig. 2a, Ref. 15). The yield circles delimiting regions of similar mechanical properties are shown in Fig. 4b. The scale has been increased by the factor $\sqrt{3/2}$ so that the scales on the stress-axes in Fig. 4a and 4b are directly comparable. This factor does not result from a co-ordinate transformation but from the stress conditions in the triaxial test and the yield function. Note that the centres are displaced with respect to the origin reflecting the anisotropy induced by previous stress history and are centred on the σ_y axis. This implies that the σ_y axis is an axis of symmetry. Thus, the yield curves are indicative of a K_0-consolidated sample loaded along the axis of consolidation in a triaxial test.

The parameters $\alpha_{ij}^{(m)}$ and K_m shown in Fig. 4b, are easily obtained from a triaxial test. Let $\sigma_y = \sigma_{11}, \sigma_x = \sigma_z = \sigma_{33}$. It follows that $S_{11} = 2/3(\sigma_{11}-\sigma_{33})$, $S_{22} = S_{33} = -1/3(\sigma_{11}-\sigma_{33})$ and $S_{ij} = 0$, $i \neq j$. Since the y-axis is an axis of symmetry in the case under consideration,

$$\alpha_{xx}^{(m)} = \alpha_{zz}^{(m)} \quad \text{and} \quad \alpha_{ij}^{(m)} = 0 \quad \text{for} \quad i \neq j.$$

Furthermore,

$$\alpha_{xx} + \alpha_{yy} + \alpha_{zz} = 0$$

since α_{ij} are deviatoric components. Therefore,

$$\alpha_{xx} = \alpha_{zz} = -\frac{\alpha_{yy}}{2} = -\frac{\alpha_y}{2}.$$

Then, defining $(\sigma_{11}-\sigma_{33}) = q$ the yield function $f=0$ gives

$$\frac{1}{2}[(\frac{2}{3}q - \alpha_y^{(m)})^2 + 2(-\frac{q}{3} + \alpha_y^{(m)})^2] = K_m^2 \tag{13}$$

$$q - \frac{3}{2}\alpha_y^{(m)} = \pm\sqrt{3}\,K_m \tag{14}$$

Now consider the yield surface associated with the points BB' in Fig. 4a. Then

$$q_B = \frac{3}{2}\alpha_y^{(m)} + \sqrt{3}\,K_m \tag{15}$$

$$q_{B'} = \frac{3}{2}\alpha_y^m - \sqrt{3}\,K_m \tag{16}$$

or

$$K_m = \frac{1}{2\sqrt{3}} \cdot (q_B - q_{B'}) \tag{17}$$

$$\alpha_y^{(m)} = \frac{q_B + q_{B'}}{3} \tag{18}$$

and

$$\alpha_x^{(m)} = \alpha_z^{(m)} = -\alpha_y^m/2 \tag{19}$$

Knowing the location of the centres $\alpha_{ij}^{(m)}$ and the radii K_m, the circles in Fig. 4b are easily drawn. The yield circles and stress-strain curve in Fig. 4 are rough approximations to the behaviour of Drammen Clay, OCR=4 and are given only as an example. The data for Fig. 4, given in Tables 1 and 2 was supplied by Bardet (1).

During cyclic loading the diameters of the circles may change in size. This will be reflected in a difference in the range of elastic behaviour or in the range of applicability of a given plastic modulus from the predicted value based on the initial sizes of the circles. The new diameters are easily determined from an examination of the unloading stress-strain curve as shown in Fig. 5.

The remaining elements needed to specify the initial state of the soil are the plastic moduli associated with the yield surfaces. They may be determined from the plastic strains using Eqn. (9). A more

TABLE 1: Points on the Approximated Uniaxial Response in Fig. 4a

	q/σ'_{vc}	$\varepsilon_y \%$		q/σ'_{vc}	$\|\varepsilon_y\|\%$
A	0.59	0.05	A'	−0.110	0.0093
B	1.41	0.64	B'	−0.390	0.210
C	1.75	1.45	C'	−0.70	0.950
D	1.85	3.08	D'	−0.88	3.884

Shear Modulus $G = \frac{1}{3}$ slope $A'A = 382\ \sigma'_{vc}$.

TABLE 2: Non-dimensional Model Parameters for the 4 Yield Surfaces in Fig. 4b

m	$\alpha^{(m)}/\sigma'_{vc}$	K_m/σ'_{vc}	H'_m/σ'_{vc}
2	0.24	0.350	106
3	0.51	0.90	28
4	0.525	1.23	4
5	0.485	1.36	0

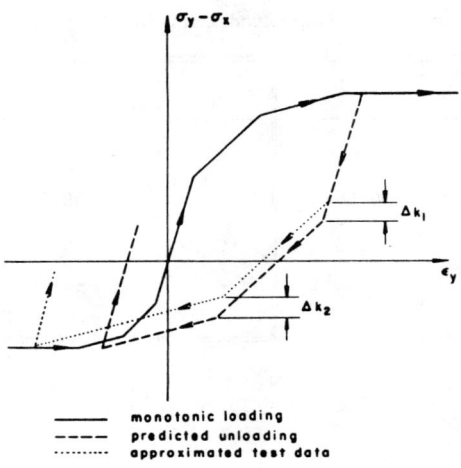

FIG. 5 Changes in Isotropic Hardening During Cyclic Loading (After Bardet (1))

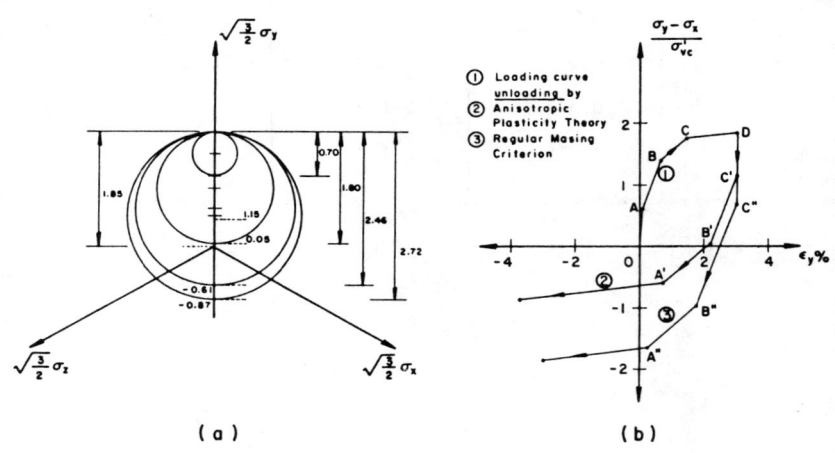

FIG. 6 Unloading Curves by Anisotropic Plasticity Theory and the Masing Criterion

convenient form of Eqn. (9) for use with the von Mises criterion is

$$d\varepsilon_{ij}^p = \frac{1}{H} \cdot \frac{\partial f}{\partial S_{ij}} \cdot \frac{\partial f/\partial S_{rs} \cdot dS_{rs}}{\partial f/\partial S_{mn} \cdot df/dS_{mn}} \qquad (20)$$

For the cross-anisotropic triaxial specimen loaded vertically along the axis of symmetry this equation gives

$$d\varepsilon_{11}^p = \frac{2}{3H_m'} \cdot d(\sigma_{11}-\sigma_{33}) \qquad (21)$$

where $d\varepsilon_{11}^p$ is the principal deviatoric strain and H_m' is the plastic modulus associated with the mth yield surface.

Since a von Mises yield surface is used, plastic strains occur at constant volume and deviatoric plastic strains and total plastic strains are identical. The major elastic deviatoric principal strain ε_{ij}^e is given by

$$d\varepsilon_{11}^e = \frac{1}{3G} \cdot d(\sigma_{11}-\sigma_{33}) \qquad (22)$$

in which G is the shear modulus and is assumed to remain constant. $G = S_{ij}/\gamma_{ij}$ where γ_{ij} = engineering shear strain = $2\varepsilon_{ij}^e$. The relationship between total deviatoric strain $d\varepsilon_{11}$ and the applied stress increments is then given by

$$d\varepsilon_{11} = d\varepsilon_{11}^e + d\varepsilon_{11}^p = (\frac{1}{3G} + \frac{2}{3H_m'}) d(\sigma_{11}-\sigma_{33}) \qquad (23)$$

or more simply by

$$d\varepsilon_{11} = \frac{2}{3\overline{H}_m} d(\sigma_{11}-\sigma_{33}) \qquad (24)$$

in which $3\overline{H}_m/2$ is the slope of the triaxial stress-strain curve at the stress level in question. The plastic modulus H_m' is given by

$$\frac{1}{H_m'} = \frac{1}{\overline{H}_m} - \frac{1}{2G} \qquad (25)$$

Values of the plastic modulus H_m' computed using the data in Table 1 and Eqn. (25) are given in Table 2 by Bardet (1).

The stress-strain curve in Fig. 4a is generated by loading from 0 to D along the $\sigma_y = \sigma_{11}$ axis. During loading, the stress point hits yield surface $f^{(1)}$ at the stress corresponding to pt. A. Plastic strains are now generated by further loading and these strains may be computed using Eqn. (23) with $H_m' = H_1'$ given in Table 2. On further loading, yield surface $f^{(1)}$ touches yield surface $f^{(2)}$ and the stress point then translates both surfaces together along the σ_y axis. Plastic strains are computed using H_2'. Finally, when the stress point D is reached all circles have been translated as in Fig. 6a.

On unloading from D the response is <u>elastic</u> until the diameter of $f^{(1)}$ has been traversed. Thus, the range of initial elastic response on unloading is $2K_1$. The initial elastic range in compression is

$\alpha^{(1)} + K_1$. In general, $2K_1 \neq \alpha^{(1)} + K_1$. The pure Masing response in which the elastic range on unloading is double the elastic range on loading is true only when $\alpha^{(1)} = 0$, that is, when the material is isotropic. The complete unloading curve according to anisotropic theory is given by curve 2 in Fig. 6b. The unloading curve by the Masing criterion is given by curve 3. The use of the Masing criterion for a fairly strong anisotropic material apparently can lead to an overestimation of hysteretic damping.

Prevost (16) has extended the basic model to drained conditions by including a hydrostatic pressure term in the yield function, using

$$f^{(m)} = \frac{3}{2}(S_{ij}-\alpha_{ij}^{(m)})(S_{ij}-\alpha_{ij}^{(m)}) + \frac{9}{2}(p-\beta^{(m)})^2 - K_m^2 = 0 \tag{26}$$

in which p is the hydrostatic effective stress and $\alpha_{ij}^{(m)}$, β^m are the coordinates of the mth yield surface in the deviatoric stress subspace and along the hydrostatic pressure axis, respectively. A non-associative flow rule is used with $f^{(m)}$. Thus, a potential surface must be selected to define the directions of plastic strain. This selection is made subject to the condition that the plastic deviatoric strain increment vector is normal to the projection of the yield surface on the deviatoric plane.

The drained and undrained models have been verified by comparing model predictions with laboratory test data. Predicted and measured stress-strain behaviour is shown in Fig. 7a for a drained triaxial test on sand and predicted and measured volumetric stress-strain behaviour of the same sand is given in Fig. 7b. In Fig. 8, predicted and measured stress-strain behaviour for Drammen Clay, OCR=4, are shown. The predictions of the models for drained and undrained behaviour are remarkably good and give confidence in the use of the method for static loading. The verification of the model for cyclic loading so far appears to be limited to undrained cyclic loading tests on Drammen Clay OCR=4. Comparisons between predicted and measured response of this clay to cyclic loading are given in Ref. (16). In this case also, the predictions are quite good although, as might be expected they are not as good as for static loading conditions.

THE TWO-SURFACE PLASTICITY MODEL

There are three basic concepts underlying the two-surface anisotropic theory of plasticity developed by Mroz, Norris and Zienkiewicz (10):

(i) the consolidation history of the soil is represented by a bounding surface F=0, which herein is assumed to be the Roscoe-Burland surface (Fig. 9a);

(ii) a yield surface f=0 defines an elastic domain within the bounding surface. This surface may undergo contraction or expansion or translation in the stress space but cannot intersect the existing bounding surface. Its translation is in accordance with the Mroz kinematic constraint (9);

(iii) the plastic modulus varies from a value H_0 near the yield surface to a value H_b on the bounding surface.

(The notation for plastic modulus is not standard. In referring to Mroz et al (10,11) and to Figs. 9-12 by them, $H_0 = K_{po}$ and $H_b = K_{PR}$.)

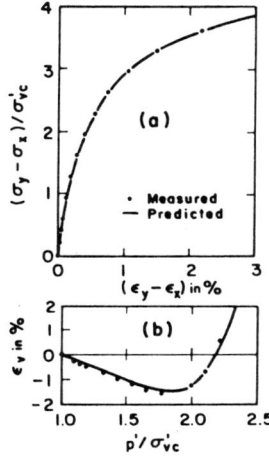

FIG. 7 Calculated and Measured Stress-Strain Behaviour for Cook's Bayou Sand (16)

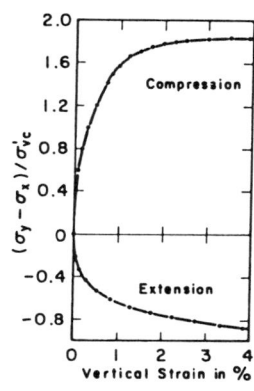

FIG. 8 Calculated and Measured Stress-Strain Behaviour for Drammen Clay OCR=4 (16)

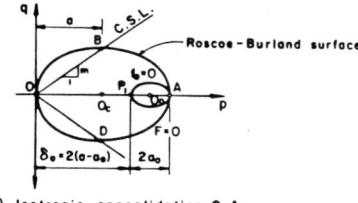

(a) Isotropic consolidation O,A.

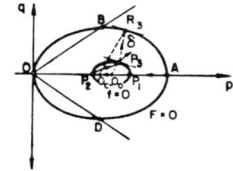

(b) Stress history O, P_1, P_2, P_3.

FIG. 9 Elements of the Two-Surface Model

The third concept is what sets this formulation apart from the multi-surface model. Instead of using a nest of yield surfaces with associated plastic moduli between the yield and bounding surfaces, the two-surface model uses an interpolation rule. The interpolation rule used in the model follows the work of Dafalias and Popov (2), in which the hardening is expressed as a function of the distance δ from the bounding surface. The greater the distance, the greater the plastic modulus. The maximum distance δ_0 after isotropic consolidation is shown in Fig. 9a. As a loading stress point traverses stress space, it translates the yield surface. At the stress point P_3 on the stress path $OP_1P_2P_3$ the distance from the conjugate point R_3 is δ. Mroz et al (11) have suggested the following interpolation formula to determine H at distance δ.

$$H = H_0 - (H_0 - H_b)(1 - \frac{\delta}{\delta_0})^\gamma \tag{27}$$

in which γ is an experimental constant. The theory of the two-surface model has been very lucidly presented in considerable detail by Mroz, Norris and Zienkiewicz (11) for triaxial test conditions and need not be repeated here.

It is suggested by Mroz et al (11) that the simplest way to determine the model parameters and the interpolation rule is by isotropic consolidation - swelling tests. The assumed elastic path shown in Fig. 10a is the exponential swelling line and deviations from it are assumed to be due to plastic strains. The region over which the total unloading path and the assumed elastic unloading path coincides defines the major axis of the yield surface $2a_0$ (Fig. 10a). The plastic component $(e-e^0)^p$ is used to determine the interpolation rule for H. A comparison of predicted and measured unloading-reloading paths is shown in Fig. 10b and appears satisfactory. A comparison of predicted and experimentally determined stress paths shown in Fig. 11 is very good. Note, however, that now although the same clay is consolidated to the same isotropic pressure the value of $H_0(=K_{p0})$ for a good data fit is 19 times less (11). This is somewhat disturbing as H_0 might be expected to be a constant for a given soil at a given isotropic consolidation stress. Computed results for a cyclic loading test are shown in Fig. 12.

EVALUATION OF THE MODELS

These models are being evaluated for possible use in practice. It is important to keep this objective in mind because it means that the discussion will not be confined to the strict limits within which the models have been developed and described. Some speculation and intuitive judgements have a valid role in this kind of process.

Both models are solidly based on the elements of plasticity theory so an evaluation is confined to assessing their compatibility with the real world and the practical aspects of using them as computational tools

First, some of the assumptions made will be discussed. Both models assume that the shear modulus remains constant. Since the modulus is a function of mean effective pressure it will not remain constant if pore-water pressures are developed during loading. When large increases in pore-water occur, for example, when sands approach liquefaction or the structure of sensitive clays breaks down during cyclic loading, then a considerable reduction will occur in the value of G. For these cases maintaining the initial stiffness may be inappropriate. However, if G is allowed to

SOIL AS HARDENING SOLID   109

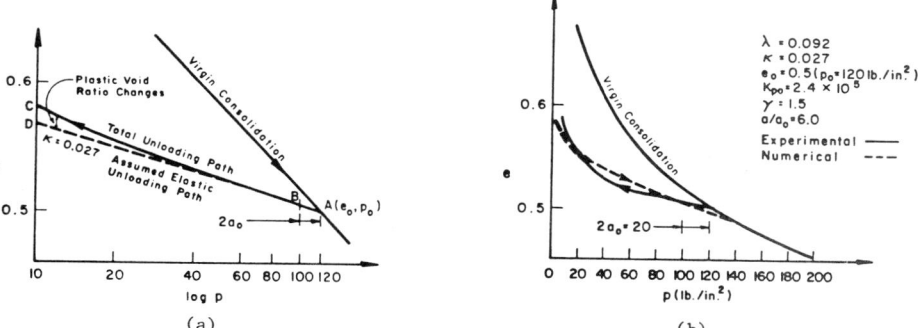

FIG. 10 Parameter Determination (a) and Model Verification (b)

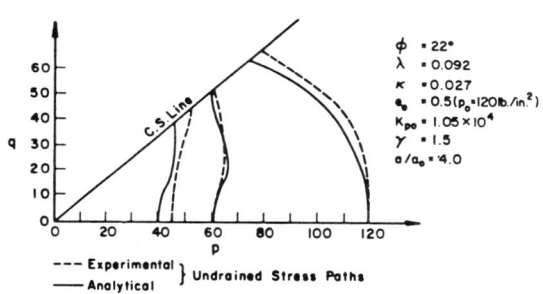

FIG. 11 Predicted and Measured Undrained Stress Paths

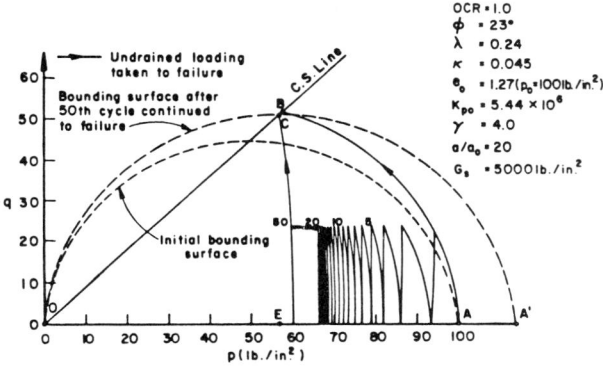

FIG. 12 Computed Cyclic Loading Response

vary some theoretical problems may be introduced. There is a possibility that Drucker's work-hardening criterion may be violated unless the energy consequences of changing G can be properly taken into account. Thus, the possibility exists that an associative flow rule may be incompatible with a changing shear modulus.

In the application of the model to the undrained behaviour of clay Prevost (15) assumes that plastic deformations occur at constant volume. Mroz et al (11) make the more usual assumption that the total straining process occurs at constant volume. This distinction may not be very important in some applications but it appears to be crucial for the successful predictions of pore-water pressure. The Mroz assumption is in line with the assumptions of critical state soil mechanics that during undrained deformation elastic and compressive plastic strains are occurring to maintain constant volume and the plastic volumetric strains induce some isotropic hardening. Pore-water pressure generation models such as used by Finn et al (4) are based on the same assumption of constant total volume. The influence of the assumption of plastic incompressibility on static stress-strain behaviour would be clarified by comparing computed and experimental effective stress-paths for undrained loading of clays such as is shown in Fig. 11. Verification by undrained stress-paths alone as in Fig. 8 does not resolve this point as the parameters of the model are determined by fitting such a stress-strain curve and, therefore, predictions of undrained response will not be affected by the assumptions of plastic compressibility.

It should be pointed out that Prevost (15) has constructed his undrained model only to predict undrained stress-strain response. This has resulted in a very simple model. However, since these models are being reviewed with a view to possible adoption the authors are concerned about all the implications of the assumption of plastic incompressibility.

The Prevost model for drained behaviour does allow plastic volume changes but it has not been used yet to predict total and effective stress-paths in undrained shear by imposing the constant total volume condition.

In attempting to form conclusions about the compatibility of the model and the real world it is useful to consider three separate cases: (i) static loading; (ii) cyclic loading of stable materials; (iii) cyclic loading of unstable materials such as strongly degrading clays and readily liquefiable sands. In static loading the available verification such as that discussed earlier indicates that either formulation of the anisotropic theory of plasticity predicts soil response very well. This is not unexpected. After all, parameters of the models are determined by obtaining as close an approximation as desired to a static stress strain curve determined by a triaxial test.

If, as in the case of stable materials, the model parameters remain essentially constant or undergo modest changes then one would expect the predictions of cyclic loading response to be good. This is the case for the lightly overconsolidated Drammen Clay (16). However, for material inherently unstable under cyclic loading the practical capabilities of the models need verification. Comparisons of predictions with experimental data on the response of strongly degrading clays or of liquefiable sands to cyclic loading have not yet been made. In this type of application the parameters of the system will be changing and probably rapidly. In these circumstances, it will be necessary to measure the initial values of the parameters and to forecast the changes in plastic moduli and circle radii with continued loading or deformation by appropriate functional relationships of, probably, strain. In theory this seems possible but it has not yet been done for multi-axial analyses. Finn, Lee and Martin (4) have incorporated such a parameter forecasting feature in their one-dimensional non-linear model.

The multi-surface model makes the determination of the initial parameters of the system very easy and also allows one to approximate the static stress-strain curve to any degree of accuracy without difficulty. It is possible that the use of an interpolation rule in the two-surface model simplifies the programming of the anisotropic theory for relatively stable materials at some cost in the facility to match the stress-strain curves to any desired accuracy. Although not sure about it, one suspects that the interpolation rule may prove unwieldy if soil properties are changing rapidly.

Although both formulations of anisotropic plasticity discussed herein suggest that the elastic domain may be ignored and totally plastic behaviour assumed (e.g., the vanishing elastic region, no yield surface case considered in (11)) it seems that the retention of a real or assumed elastic domain is essential for the controlled operation of the model under cyclic loading conditions.

Dafalias et al (3) have shown that an expanding elastic nucleus corresponding to the yield surface of the two-surface model is necessary to control behaviour during cyclic loading. Without an appropriate elastic nucleus N cycles of any deviatoric stress q will cause liquefaction or a critical state if N is sufficiently large, a result contrary to experimental experience.

There may be some practical difficulties associated with the use of the Roscoe-Burland surface for sands in the two-surface model. As may be seen in Fig. 10 or in (11), the properties of the normal consolidation line figure prominently in this model. It is difficult to establish these properties reliably for sand and it requires higher than usual pressures in testing for all but the loosest sands.

The anisotropy of soils demands that the analysis for general loadings be carried out in a 9-dimensional stress space. This fact is easily forgotten as the published verifications of the models are concerned with K_0 or isotropic consolidation-swelling response or stress-strain response in triaxial tests. Therefore, for example, the computational formulation in (11) for the analysis of these loading situations would not be applicable to loading in the field unless the effects of anisotropy were ignored. Even initially isotropic materials become anisotropic under general loading so, in theory at least, stress and strain invariants should not be used as variables. The exceptions are for analysis of tests like triaxial tests where the axes of material symmetry, principal stress and strain axes can all be maintained coincident during loading.

Soil is a very complex material and, in spite of the undoubtedly formidable potential of the anisotropic theory of plasticity, one is still left with the impression that no one theory can model all aspects of its behaviour. There is still room for special models for particular kinds of problems. However, one can readily agree that the anisotropic theory of plasticity is one of the more powerful models available at present for the analysis of the response of stable soils to static and cyclic loading. However, considerable research and verification remains to be done before the capability of the model for analysing the response of strongly degrading clays and liquefiable sands can be properly assessed.

The formulations of anisotropic theory for cyclic loading are for the pseudo-static case only. The models have not yet been extended to include earthquake loading which requires consideration of inertia effects. Thus, the practical applications of the model at present have been limited to the behaviour of offshore structures under long-period ocean wave loading (16).

CONCLUSIONS

Two different formulations of anisotropic plasticity theory for soils have been evaluated critically; the multi-surface model of Prevost (15,16) and two-surface model of Mroz, Norris and Zienkiewicz (10,11). The models are very new and are still in the process of development but progress to-date has been impressive.

Both models represent the response of soils to static loading very well. Both have demonstrated the potential to model the phenomenological aspects of cyclic loading but verification has been limited. The two-surface model has not yet been verified for cyclic loading; the multi-surface model has been verified for the case of undrained cyclic loading of a lightly overconsolidated clay only - a fairly stable material. None of the models has yet been applied to the analysis of strongly degrading clays or readily liquefiable sands. Nor have the models been expanded at this time to include loading such as earthquake loading which requires consideration of inertia effects.

The search for a good constitutive model for soil is not yet over but the anisotropic theory of plasticity is probably the best general model available at the moment for the analysis of the pseudo-static response of relatively stable soils to static and cyclic loading. There are reservations about the capability of the method at its present stage of development to analyse the response of strongly degrading clays or liquefiable sands efficiently.

ACKNOWLEDGEMENTS

The authors are grateful to J.P. Bardet and Ron Scott, California Institute of Technology; J.H. Prevost, Princeton University; and O.C. Zienkiewicz, University College, Swansea for helpful discussions. The paper was part of a study financed by Fugro Incorporated, Long Beach, California. The assessment of the two-surface model was part of a study by the senior author at University College, Swansea, Wales supported by a Senior Visiting Fellowship from the Science Research Council of Great Britain. Appreciation is expressed to Professor O.C. Zienkiewicz and the Science Council.

APPENDIX I - REFERENCES

1. Bardet, J.P. (1980), Private Communication.

2. Dafalias, Y.F. and Popov, E.P. (1976), "Plastic Internal Variables Formalism of Cyclic Plasticity," Jour. Applied Mechanics, 98(4) : 645-650.

3. Dafalias, Yannis F. and Hermann, Leonard R. (1980), "A Boundary Surface Soil Plasticity Model," International Symposium on Soils Under Cyclic and Transient Loading, Swansea, Vol. 1, 7-11 Jan., 1980, pp. 335-345.

4. Finn, W.D. Liam, Lee, Kwok W. and Martin, G.R. (1977), "An Effective Stress Model for Liquefaction," Jour. of the Geotechnical Engineering Division, ASCE, Vol. 103, No. GT6, Proc. Paper 13008, June, pp. 517-533.

5. Fung, Y.C. (1965), Foundations of Solid Mechanics, Prentice-Hall, N.J., 525 pp.

6. Hill, R. (1950), The Mathematical Theory of Plasticity, Oxford University Press, London, 356 pp.

7. Lee, E.H. (1980), Lecture, Simon Fraser University, Burnaby, B.C.

8. Masing, G. (1926), "Eigenspannungen und Verfestigung beim Messing," Proceedings, 2nd International Congress for Applied Mechanics, Zurich, Sept. 1926.

9. Mroz, Z. (1967), "On the Description of Anisotropic Hardening," Jour. Mech. Phys. Solids, Vol. 15, pp. 163-175.

10. Mroz, Z., Norris, V.A. and Zienkiewicz, O.C. (1978), "An Anisotropic Hardening Model for Soils and Its Application to Cyclic Loading," Int. J. Numerical and Analytical Methods in Geomechanics, 2:203-221.

11. Mroz, Z., Norris, V.A. and Zienkiewicz, O.C. (1979), "Application of an Anisotropic Hardening Model in the Analysis of Elastoplastic Deformation of Soils," Geotechnique, 29(1) : 1-34.

12. Popov, E.P. and Ortiz, M. (1979), "Macroscopic and Microscopic Cyclic Metal Plasticity," Proceedings, 3rd ASCE/EMD Specialty Conference, Austin, U.S.A.

13. Prager, W. (1955), "The Theory of Plasticity: A Survey of Recent Achievement," Proceedings, Inst. Mech. Eng., London, Vol. 169, pp. 41-57.

14. Prevost, J.H. (1977), "Mathematical Modeling of Monotonic and Cyclic Undrained Clay Behaviour," Inst. J. Num. and Analyt. Mech. in Geomechanics, Vol. 1, No. 2, pp. 195-216.

15. Prevost, J.H. (1978), "Anisotropic Undrained Stress-strain Behaviour of Clays," Jour. of the Geotechnical Engineering Division, ASCE, Vol. 104, No. GT8, Proc. Paper 13942, August, pp. 1075-1090.

16. Prevost, J.H. (1979), "Mathematical Modeling of Soil Stress-strain Strength Behaviour," 3rd Int. Conf. on Numerical Methods in Geomechanics, Aachen, 2-6 April 1979, pp. 347-361.

17. Roscoe, K.H. and Burland, J.B. (1968), "On the Generalized Stress-Strain Behaviour of Wet Clays," in Engineering Plasticity, Cambridge Univ., pp. 535-604.

18. Taylor, Peter W. and Larkin, Thomas J. (1978), "Seismic Site Response of Nonlinear Media," Jour. of the Geotechnical Engineering Division, ASCE, Vol. 104, No. GT3, Proc. Paper 13597, March, pp. 369-383.

APPENDIX II - NOTATION

a = semi-major axis of Roscoe-Burland surface.

a_0 = semi-major axis of yield surface.

e_0 = void ratio.

e = initial void ratio

$f^{(m)}$ = m^{th} yield surface

g = plastic potential

G = shear modulus

H_m = slope of stress-strain curve in plastic range

H'_m = plastic modulus associated with m^{th} yield surface

K_m = yielding parameter for m^{th} yield surface

n_{ij} = components of normal to yield surface

OCR = over-consolidation ratio

p = mean-normal effective stress

q = deviator stress

S_{ij} = deviatoric stresses

$\alpha_{ij}^{(m)}$ = centre of mth yield surface in deviatoric plane

$\beta_{ij}^{(m)}$ = centre of mth yield surface on hydrostatic axis

γ = constant

γ_{ij} = engineering shear strain

δ = distance between yield and bounding surfaces

δ_0 = max. distance between yield and bounding surfaces

ε_{ij} = total deviatoric strains

ε_{ij}^{e} = elastic deviatoric strains

ε_{ij}^{p} = plastic deviatoric strains

σ_{ij} = stress components

σ_0 = initial yield stress

σ_r = isotropically work-hardened yield stress

Analysis of Soil Response With Different Plasticity Models

by E. Mizuno[1] and W. F. Chen[2], M. ASCE

1. INTRODUCTION

The mechanical behavior of soil or rock is greatly complicated by a variety of material characteristics, such as heterogeneity and anisotropy. While these characteristics play a critical role in the microstructural behavior of geotechnical materials requiring the consideration of parameters at the microstructure level, it is possible to treat the material as a continuum to characterize gross material behavior. In recent years, linear and nonlinear elastic models, viscoelastic models, and elastic-plastic models have been used to analyze geotechnical engineering problems. Although the models such as hyperelastic or hypoelastic can represent the phenomena associated with dilatancy and hardening or softening of soil behavior, the effect of plastic strain induced during loading cannot be predicted within the framework of an incremental Hooke's law with variable moduli which are functions of the stress and/or strain.

Current research in soil constitutive modelling is moving towards the development of three-dimensional stress-strain relations based on the principles of plasticity as well as linear or nonlinear elasticity. Although many constitutive laws have been proposed so far, there has been little evaluation of them with respect to general problems in geomechanics.

In the present paper, the classical plasticity soil models such as Coulomb, Drucker-Prager criterion, and an advanced model (cap model) are reviewed and then discussed with respect to their advantages and limitations for solving problems in geomechanics. These models are formulated within the context of incremental plasticity theory according to a non-associated flow rule as well as an associated flow rule.

Finally, to demonstrate the variation in analytical solutions using the different plasticity models, two problems are considered. These include (1) the case of a shallow stratum of clay with a smooth strip footing, and (2) the compression of sand under uniaxial strain condition. The finite element method was used in the analyses.

2. GEOMETRICAL REPRESENTATION OF STRESS STATE

The state of stress at a point inside a soil medium can be completely defined by the stress tensor σ_{ij} in a three dimensional space. In a matrix notation, the components of the stress tensor can be written in the Cartesian coordinate system (1,2,3) as

---

[1] Research Assistant of Civil Engineering, Purdue University, West Lafayette, IN.

[2] Professor and Head of Structural Engineering, Purdue University, West Lafayette, IN.

$$\sigma_{ij} = \begin{vmatrix} \sigma_{11} & \sigma_{12} & \sigma_{13} \\ \sigma_{21} & \sigma_{22} & \sigma_{23} \\ \sigma_{31} & \sigma_{32} & \sigma_{33} \end{vmatrix} \quad (1)$$

where each component of the stress tensor σ_{ij} acts on a surface normal to the i-axis, and in the direction of the j-axis. Stress components σ_{ii} (σ_{11}, σ_{22} and σ_{33}) are called normal stresses, and components σ_{ij} ($i \neq j$) are called shear stresses. In addition, under the classical assumptions of nonpolarity, $\sigma_{ij} = \sigma_{ji}$.

In general, the stress tensor can be decomposed into two parts: (1) the hydrostatic pressure part, where the off-diagonal terms are identically zero and the diagonal terms are equal to mean normal stress; and (2) the deviatoric part S_{ij}. Thus,

$$\sigma_{ij} = \frac{1}{3} I_1 \delta_{ij} + S_{ij} \quad (2)$$

where I_1, the first invariant of the stress tensor, is the sum of the diagonal stress components σ_{11}, σ_{22}, σ_{33}, and δ_{ij} is the Kronecker delta. The hydrostatic pressure and the deviatoric stress, respectively, cause volumetric change and shape change of the material element.

Point A in Fig. 1 represents the state of stress at a material point in the principal stress coordinate system (σ_1, σ_2, σ_3). Stress vector OA can be decomposed into OB in the ξ-axis which is called hydrostatic axis ($\sigma_1 = \sigma_2 = \sigma_3$) and BA in the deviatoric plane which is perpendicular to the ξ-axis. The component vector OB represents the mean normal stress $p = I_1/3$ and the component vector BA represents the deviatoric stress S_{ij}. The lengths of OB and BA are $\sqrt{3} p$ and $\rho = \sqrt{S_{11}^2 + S_{22}^2 + S_{33}^2}$, respectively. These values are related to the first invariant of stress tensor $I_1 = 3p$ or the octahedral normal stress $\sigma_{oct} = p$ and the second invariant of deviatoric stress tensor $J_2 = \rho^2/2$ or octahedral shear stress $\tau_{oct} = \rho/\sqrt{3}$.

If the stress vector $\vec{OA}$ is viewed from the hydrostatic ξ-axis, the actual length and direction of BA can be seen in the π-plane ($\sigma_1 + \sigma_2 + \sigma_3 = 0$) as shown in Fig. 2. The angle θ in the figure is defined by Lode as

$$\theta = \frac{1}{3} \cos^{-1}(\frac{3\sqrt{3}}{2} \frac{J_3}{J_2^{3/2}}) \quad (3)$$

where $J_3 = \frac{1}{3} S_{ij} S_{jk} S_{k\ell}$ is the third invariant of the deviatoric stress.

Any state of stress can be expressed in terms of I_1, J_2, and J_3 in the principal stress space.

3. PLASTICITY MODELS FOR GEOTECHNICAL MATERIAL

3.1 Coulomb Model

The Mohr-Coulomb criterion is certainly the best known failure criterion in soil mechanics. This criterion, which was proposed for geotechnical materials

SOIL RESPONSE ANALYSIS

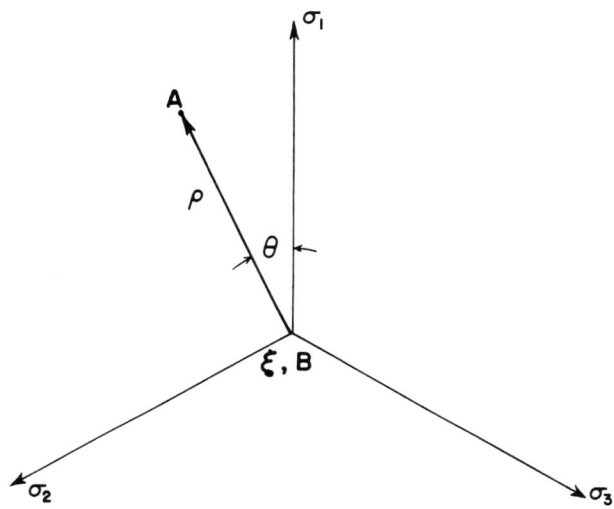

Fig. 2   Stress on π Plane.

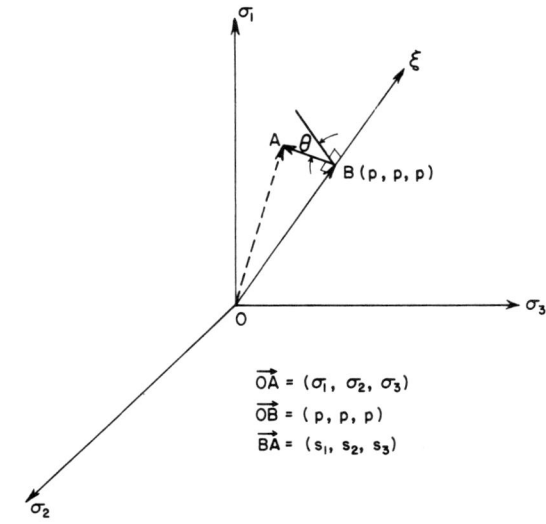

$\vec{OA} = (\sigma_1, \sigma_2, \sigma_3)$
$\vec{OB} = (p, p, p)$
$\vec{BA} = (s_1, s_2, s_3)$

Fig. 1   Stress State in Principal Stress Space.

much earlier than that of Tresca's and von Mises's yield criteria for metal, is the first type of failure criterion that takes into consideration the effect of hydrostatic pressure. This criterion states that failure occurs when the shear stress τ and the normal stress σ acting on any element in the material satisfy the linear equation

$$\tau + \sigma \tan \phi - c = 0 \qquad (4)$$

where c and ϕ denote the cohesion and the angle of internal friction, respectively.

As shown by Shield [12], Coulomb's failure is an irregular hexagonal pyramid in the principal stress space. A cross sectional shape of this pyramid in the π-plane is shown in Fig. 3. For some problems in geomechanics, Chen and Drucker [3] proposed a condition combining the Coulomb criterion with a small tensile strength cut-off:

$$\sigma = f_t \geq 0 \qquad (5)$$

This modified Coulomb criterion is defined by three material constants, namely: c, ϕ, and f_t.

Even though the Coulomb criterion as mentioned above is simple in graphical forms, the Coulomb surface exhibits corners or singularities in the three dimensional generalization. The resulting general yield or failure function with singularities gives rise to some difficulty in numerical analysis.

In addition, the Coulomb criterion neglects the influence of intermediate principal stress on shear strength. Nevertheless, for the most part, this criterion has in the past been used by necessity and simplicity to obtain reasonable solutions to important, practical problems in geotechnical engineering.

3.2 Drucker-Prager Model

General

For practical purposes, an approximation of the yield surface with singularities by a smooth surface is often made in the elastic-plastic finite element analysis under more general stress condition. The Drucker-Prager perfectly plastic model [8] can be considered as a first attempt to approximate the well-known yield or failure criterion of the Coulomb criterion by a simple smooth function. This criterion is expressed as a simple stress invariant function of the first invariant of stress tensor I_1 and the second invariant of the deviatoric stress tensor J_2 together with two material constants α and k. It has the simple form

$$\alpha I_1 + \sqrt{J_2} = k \qquad (6)$$

where the constants α and k may be related to Coulomb's material constants c and ϕ in several ways.

If α is zero, Eq. 6 reduces to the well-known von Mises yield condition for metal. When ϕ is zero, the Coulomb criterion reduces to the Tresca

SOIL RESPONSE ANALYSIS   119

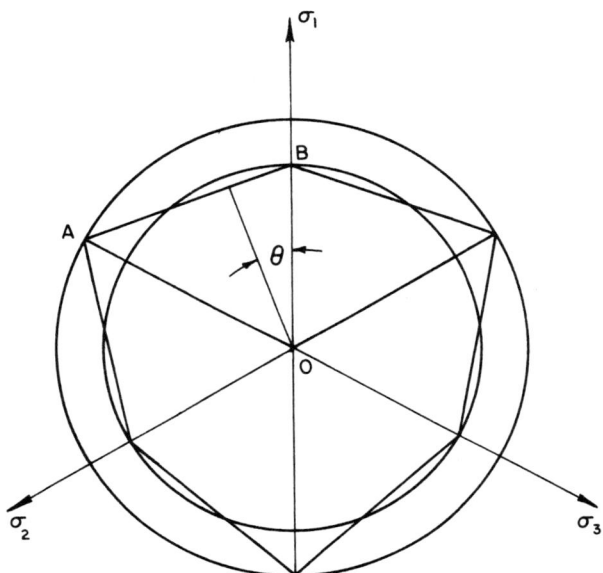

Fig. 3   Shape of Yield Criterion on π-Plane

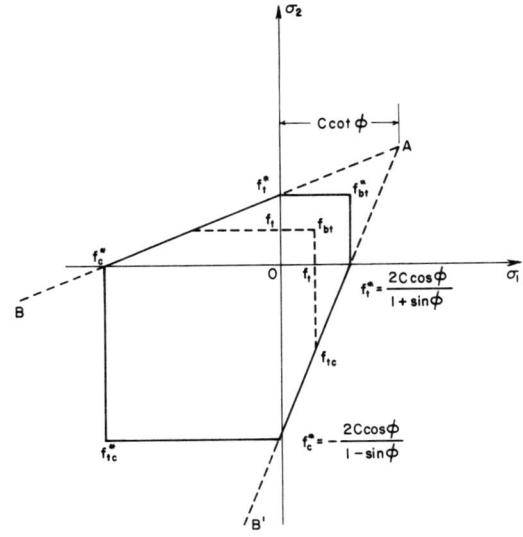

Fig. 4   Shape of Coulomb Criterion in Biaxial Stress Space.

criterion for metal. In a sense, the von Mises criterion may be considered as an approximate version of the Tresca yield criterion.

The Drucker-Prager model cannot predict plastic volumetric strain during hydrostatic loading. To improve this, an extended von Mises model with convex end cap was proposed by Drucker, Gibson, and Henkel [7]. However, the failure surface of this model results in a much greater dilatancy prediction than that observed in experiments. As a result, modified failure or yield surface which will be asymptotically parallel to I_1-axis under high value of I_1 was subsequently proposed by DiMaggio and Sandler [6].

Drucker-Prager Material Constants

The Drucker-Prager criterion is a simple extension of the well-known von Mises yield criterion which includes the linear term of the hydrostatic component I_1 of the stress tensor σ_{ij}. (Eq. 6). This extended von Mises yield function, as viewed in three dimensional principal stress space, is a cone with the space diagonal as its axis, while the Coulomb criterion is a pyramid with an irregular hexagonal base and the space diagonal as its axis.

In the three dimensional principal stress space, the Drucker-Prager criterion can be matched with the apex of the Coulomb criterion and either point A or B on its π-plane as shown in Fig. 3. In the former case, the cone circumscribes the hexagonal pyramid, and the material constants α and k are obtained as

$$\alpha = \frac{2 \sin \phi}{\sqrt{3}(3 - \sin\phi)}$$

$$k = \frac{6c \cos \phi}{\sqrt{3}(3 - \sin\phi)}$$

(7)

These material constants are the same as those given by Zienkiewicz [14]. The latter case results in an inner cone and the corresponding constants are

$$\alpha = \frac{2 \sin \phi}{\sqrt{3}(3 + \sin\phi)}$$

$$k = \frac{6c \cos \phi}{\sqrt{3}(3 + \sin\phi)}$$

(8)

In the biaxial stress space with $\sigma_3 = 0$, the shape of the Coulomb criterion yield surface is an unsymmetrical hexagon (solid line) as shown in Fig. 4. The Drucker-Prager or extended von Mises criterion is an off-center ellipse when parameter α is less than $\frac{1}{2\sqrt{3}}$ but becomes a parabola or a hyperbola when it is equal to or exceeds $\frac{1}{2\sqrt{3}}$, respectively.

To relate the Drucker-Prager constants α and k with those of the Coulomb criterion c and φ in this space, two conditions are needed. In the present case of plane stress, we can match, for example, the simple tensile strength f_t^* and simple compressive strength f_c^* of the two criteria. The material constants α and k of Drucker-Prager are determined as

$$\alpha = \frac{1}{\sqrt{3}} \sin \phi$$

$$k = \frac{1}{\sqrt{3}} c \cos \phi \tag{9}$$

There are several ways to match the two criteria and the corresponding material constants. As shown in Table 1, material constants matched with the biaxial tensile strength f^*_{bt} and the simple compressive strength f^*_c or with the biaxial compressive strength f^*_{bc} and the simple tensile strength f^*_t are identical to those obtained previously in three dimensional matching.

The simple tensile strength f^*_t of Coulomb criterion, however, is found generally smaller than the actual strength obtained from a simple tension test. Hence, the modified Coulomb criterion with a tension cut-off was proposed [3]. This modified Coulomb criterion is drawn in Fig. 4 where tension part (solid line $f^*_{bt} - f^*_t$) is cut off by the broken line ($f_{bt} - f_t$).

The material constants α and k matched with the modified Coulomb criterion are listed in Table 2. In Table 2, the parameter β is the ratio f^*_c/f_t, i.e., the ratio of the simple compressive strength to the simple tensile strength. On the other hand, if the Drucker-Prager and Coulomb criteria are expected to give identical limit loads (plastic collapse loads) for the plane strain case, then, the following two conditions (1) same limit load; and (2) plane strain, must be used. The broken lines AB and AB´ in Fig. 4 are the projection of Coulomb criterion in the biaxial stress space for the plane strain case. It should be noted that stress σ_3 is now not zero. In this case, material constants α and k are given by [5,8]:

$$\alpha = \frac{\tan \phi}{\sqrt{9 + 12 \tan^2 \phi}}$$

$$k = \frac{3c}{\sqrt{9 + 12 \tan^2 \phi}} \tag{10}$$

In the case of metal, the simple tension yield value is usually used to determine the material constant k for both Tresca and von Mises criteria. It follows that the maximum deviation between Tresca and von Mises predictions occurs in pure shear which cannot exceed 15%. In the case of soils, rock, or concrete, the replacement of Coulomb or other well-known criteria with singularities by the smooth Drucker-Prager function is more complicated. For example, if simple tension and simple compression stress states are used to relate Drucker-Prager constants α and k with that of Coulomb's c and ϕ, the matching may be reasonable in the tension-compression domains for plane stress condition, but they may result in a significantly different prediction on plastic collapse loads for their applications to the plane strain problems. The assessment of various possible matching is given in Ref. [4].

Table-1 Material Constants Matched with Coulomb Criterion (Plane Stress).

Matching Points	α	k
f_t^*, f_c^*	$\dfrac{1}{\sqrt{3}} \sin \phi$	$\dfrac{2}{\sqrt{3}} c \cos \phi$
f_{bt}^*, f_c^*	$\dfrac{2 \sin\phi}{\sqrt{3}\,(3-\sin \phi)}$	$\dfrac{6 c \cos\phi}{\sqrt{3}\,(3-\sin \phi)}$
f_t^*, f_{bc}^*	$\dfrac{2 \sin\phi}{\sqrt{3}\,(3+\sin \phi)}$	$\dfrac{6 c \cos\phi}{\sqrt{3}\,(3+\sin \phi)}$
f_{bt}^*, f_{bc}^*	$\dfrac{1}{2\sqrt{3}} \sin \phi$	$\dfrac{2}{\sqrt{3}} c \cos \phi$

Table-2 Material Constants Matched with the Modified Coulomb Criterion (Plane Stress).

Matching Points	α	k
f_{bt}, f_c^*	$\dfrac{\beta-1}{\sqrt{3}\,(\beta+2)}$	$\dfrac{\sqrt{3}\beta}{\beta+2} f_t$
f_{bt}, f_{bc}^*	$\dfrac{\beta-1}{2\sqrt{3}\,(\beta+1)}$	$\dfrac{2\beta}{\sqrt{3}\,(\beta+1)} f_t$
f_t, f_c^*	$\dfrac{\beta-1}{\sqrt{3}\,(\beta+1)}$	$\dfrac{2\beta}{\sqrt{3}\,(\beta+1)} f_t$
f_t, f_{bc}^*	$\dfrac{\beta+1}{\sqrt{3}\,(2\beta+1)}$	$\dfrac{\sqrt{3}\beta}{2\beta+1} f_t$

$\beta = f_c^*/f_t$

3.3 Advanced Cap Models

General

The introduction of a spherical end cap to the Drucker-Prager model was made by Drucker, Gibson, and Henkel [7] to control the plastic volumetric change of soils, or dilation. Since then, several strain hardening plasticity models based on the critical state concept were developed by Cambridge group [13], and a specific Cam-Clay model based on normally consolidated or lightly over-consolidated clay was suggested by Roscoe, Schofield, and Thurairajah [9]. In recent years, the cap models have been further modified and refined by DiMaggio [6], Sandler [11], and Baladi [1,10].

Below we describe a simple plane cap model [2] and an elliptic cap model.

Simple Plane Cap Model

The loading functions for this model consist of tension cutoff limit and hydrostatic hardening function in addition to the usual Drucker-Prager type of failure function (Fig. 5). The merits of this model are the improvement of the hardening function under hydrostatic loading, limited tensile strength, and limited dilatancy. The loading function of this model consists of the following three surfaces;

(i) Drucker-Prager type of yield surface for loading and failure,

$$F_L = \alpha I_1 + \sqrt{J_2} - k(\overline{e}^P) = 0 \qquad (11)$$

where $\overline{e}^P$ is the effective plastic strain which will be defined in Section 4.

(ii) Compression plane cap surface,

$$F_C = I_1 - x(\varepsilon_{kk}^P) = 0 \qquad (12)$$

where x is the hardening function which causes the plastic volumetric change $d\varepsilon_{kk}^P$. The location of the cap x is related to the plastic volumetric strain ε_{kk}^P according to the following relation:

$$\varepsilon_{kk}^P = W(e^{Dx} - 1) \qquad (13)$$

where D and W are material constants; and

(iii) Tension cutoff limit plane,

$$F_t = I_1 - T = 0 \qquad (14)$$

where T is tension cutoff limit.

The amount of plastic volumetric strain predicted during shearing cannot be reduced at high pressure range of I_1 because the failure surface is still of the Drucker-Prager type with a straight line in the $I_1 - \sqrt{J_2}$ space. Neither is this model accurate in predicting the behavior of the material under shear loading conditions.

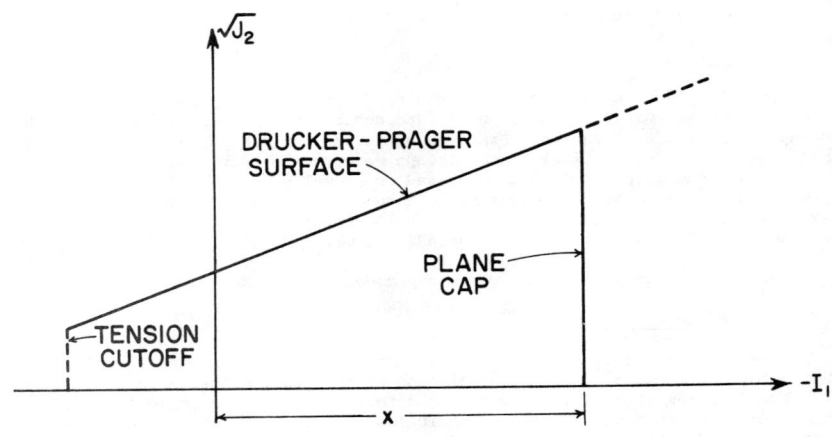

Fig. 5 Plane Cap Model.

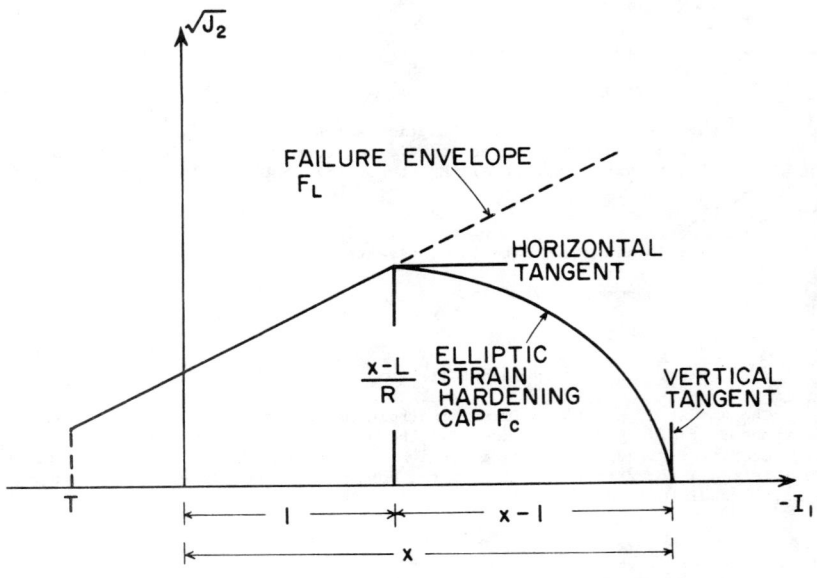

Fig. 6 Elliptic Cap Model.

Elliptic Cap Model

The schematic shape of this model is drawn in Fig. 6. Loading functions consist of the following three parts:

(i) A failure function may be assumed as the same as Drucker-Prager function (Eq. 11). Alternatively, the following function is assumed by Sandler [11] and Baladi [1]

$$F_L = \sqrt{J_2} - (A - Ce^{BI_1})$$ (15)

for the failure envelope in which A, B, and C are material constants;

(ii) The strain-hardening cap function has the form of a quarter of ellipse

$$F_c = (I_1 - \ell)^2 + R^2 J_2 - (x - \ell)^2 = 0$$ (16)

in which x is hardening function which depends on plastic volumetric strain, ℓ is the value of I_1 at the center of the elliptic cap, and R is the ratio of the major to minor axis of the elliptic cap which may be a function of ℓ; and

(iii) Tension cutoff limit plane is introduced in a similar manner as that of plane cap model.

$$F_t = I_1 - T = 0$$ (17)

This model can prevent much dilatancy on the failure surface under high hydrostatic pressure I_1. Although this model can predict not only strain-softening of soils, but also strain-hardening, the model cannot predict exactly the hysteresis loop under shear loading. This is because the hardening function in this model is assumed to be controlled only by plastic volumetric strain. In some cases, this assumption contradicts the experimental data.

4. THEORETICAL DEVELOPMENT OF CONSTITUTIVE EQUATION

The plasticity models proposed so far contain: (1) a "failure envelope" on which the material element will fail or harden due to shear stress; and (2) a "hardening function" by which the rate of plastic volumetric change will be controlled in terms of the magnitude of I_1.

In general, the function of the failure envelope or hardening function can be expressed by

$$F = F[\sigma_{ij}, x(\varepsilon_{ij}^p), k(\overline{e^p})]$$ (18)

in which σ_{ij}, $x(\varepsilon_{ij}^p)$, and $k(\overline{e^p})$ are respectively the stress tensor, the hardening parameter which is a function of plastic strain tensor ε_{ij}^p, and the material constant k which is a function of the effective plastic strain $\overline{e^p}$.

According to the incremental theory of plasticity, the total strain increment $d\varepsilon_{ij}^T$ is assumed to be the sum of the elastic strain increment $d\varepsilon_{ij}^e$ and the plastic strain increment $d\varepsilon_{ij}^p$, i.e.:

$$d\varepsilon_{ij}^T = d\varepsilon_{ij}^e + d\varepsilon_{ij}^P \tag{19}$$

The elastic stress-strain law is

$$d\sigma_{ij} = K\, d\varepsilon_{kk}^e \delta_{ij} + 2G\, d e_{ij}^e \tag{20}$$

where K and G are bulk modulus and shear modulus, respectively. Sometimes, K and G are treated as functions of stress tensor σ_{ij} and the plastic volumetric strain tensor ε_{ij}^P, and de_{ij}^e is the elastic deviatoric incremental strain tensor. Eq. 20 can be rewritten as

$$d\sigma_{ij} = C_{ijk\ell}^E\, d\varepsilon_{k\ell}^e \tag{21}$$

where $C_{ijk\ell}^E$ is elastic stiffness tensor expressed as

$$C_{ijk\ell}^E = \lambda(\delta_{ij}\delta_{k\ell}) + \mu(\delta_{ik}\delta_{j\ell} + \delta_{i\ell}\delta_{jk}) \tag{22}$$

where λ and μ are Lamé constants. These constants are related to K and G according to the following:

$$\lambda = K - \frac{2}{3}G \quad \text{and} \quad \mu = G \tag{23}$$

Using Eq. 19, we can express Eq. 21 as

$$d\sigma_{ij} = C_{ijk\ell}^E (d\varepsilon_{k\ell}^T - d\varepsilon_{k\ell}^P) \tag{24}$$

Plastic strain incremental tensor $d\varepsilon_{ij}^P$ as given by flow rule is

$$d\varepsilon_{ij}^P = d\lambda\, \frac{\partial \psi}{\partial \sigma_{ij}} \tag{25}$$

where ψ is plastic potential function and $d\lambda$ is positive scalar function

If the yielding or failure function is taken as a potential function ψ, the normality condition is satisfied. This is known as the material with an associated flow rule. In the case of the potential function ψ being different from the failure function F, the material is one with a non-associated flow rule. Although the normality condition is not always satisfied in soils, the associated flow rule has generally been used in the past for simplicity.

During plastic flow, the yield or hardening, function F (Eq. 18) should satisfy the consistency condition dF = 0. That is,

$$dF = \frac{\partial F}{\partial \sigma_{ij}} d\sigma_{ij} + \frac{\partial F}{\partial \varepsilon_{ij}^P} d\varepsilon_{ij}^P + \frac{\partial F}{\partial \bar{e}^P} d\bar{e}^P = 0 \tag{26}$$

Now we define effective strain $d\bar{e}^P$ as

$$d\bar{e}^P = h\sqrt{d\varepsilon_{ij}^P\, d\varepsilon_{ij}^P} \tag{27}$$

where h is a constant depending on the yield function. For a von Mises material, h is $\sqrt{\frac{2}{3}}$, and for a Drucker-Prager material h is $(\alpha + \frac{1}{\sqrt{3}}) \big/ \sqrt{3\alpha^2 + \frac{1}{2}}$. Substituting Eqs. 25 and 27 into Eq. 26, we have

$$dF = \frac{\partial F}{\partial \sigma_{ij}} d\sigma_{ij} + d\lambda \frac{\partial F}{\partial \varepsilon^P_{ij}} \frac{\partial \psi}{\partial \sigma_{ij}} + h \ d\lambda \frac{\partial F}{\partial e^P} \sqrt{\frac{\partial \psi}{\partial \sigma_{ij}} \frac{\partial \psi}{\partial \sigma_{ij}}} = 0 \quad (28)$$

Multiplying Eq. 24 by $\frac{\partial F}{\partial \sigma_{ij}}$, we obtain

$$\frac{\partial F}{\partial \sigma_{ij}} d\sigma_{ij} = \frac{\partial F}{\partial \sigma_{ij}} C^E_{ijk\ell}(d\varepsilon^T_{k\ell} - d\lambda \frac{\partial \psi}{\partial \sigma_{k\ell}}) \quad (29)$$

Substituting into Eq. 28, and solving for $d\lambda$, we have

$$d\lambda = \frac{\frac{\partial F}{\partial \sigma_{ij}} C^E_{ijk\ell} \, d\varepsilon^T_{k\ell}}{\frac{\partial F}{\partial \sigma_{ij}} C^E_{ijk\ell} \frac{\partial \psi}{\partial \sigma_{k\ell}} - \frac{\partial F}{\partial x} \frac{\partial x}{\partial \varepsilon^P_{ij}} \frac{\partial \psi}{\partial \sigma_{ij}} - h \frac{\partial F}{\partial k} \frac{\partial k}{\partial e^P} \sqrt{\frac{\partial \psi}{\partial \sigma_{ij}} \frac{\partial \psi}{\partial \sigma_{ij}}}} \quad (30)$$

Therefore, Eq. 24 becomes

$$d\sigma_{ab} = \left[C^E_{abcd} - \frac{C^E_{abk\ell} \frac{\partial \psi}{\partial \sigma_{k\ell}} \frac{\partial F}{\partial \sigma_{ij}} C^E_{ijcd}}{\frac{\partial F}{\partial \sigma_{ij}} C^E_{ijk\ell} \frac{\partial \psi}{\partial \sigma_{k\ell}} - \frac{\partial F}{\partial x} \frac{\partial x}{\partial \varepsilon^P_{ij}} \frac{\partial \psi}{\partial \sigma_{ij}} - h \frac{\partial F}{\partial k} \frac{\partial k}{\partial e^P} \sqrt{\frac{\partial \psi}{\partial \sigma_{ij}} \frac{\partial \psi}{\partial \sigma_{ij}}}} \right] d\varepsilon^T_{cd} \quad (31)$$

Eq. 31 is the elastic-plastic constitutive equation with the non-associated flow rule. In the case of the associated flow rule, the ψ function is assigned equal to the failure or hardening function F in Eq. 31.

5. STIFFNESS MATRICES OF SOIL MODELS

In general, $\frac{\partial F}{\partial \sigma_{ij}}$ or $\frac{\partial \psi}{\partial \sigma_{ij}}$ can be written as

$$\frac{\partial F}{\partial \sigma_{ij}} = \frac{\partial F}{\partial I_1} \delta_{ij} + \frac{\partial F}{\partial J_2} S_{ij} + \frac{\partial F}{\partial J_3} t_{ij}$$

or (32)

$$\frac{\partial \psi}{\partial \sigma_{ij}} = \frac{\partial \psi}{\partial I_1} \delta_{ij} + \frac{\partial \psi}{\partial J_2} S_{ij} + \frac{\partial \psi}{\partial J_3} t_{ij}$$

where $t_{ij} = \frac{\partial J_3}{\partial \sigma_{ij}} = S_{ik}S_{kj} - \frac{2}{3} J_2 \delta_{ij}$

Substituting Eqs. 13 and 32 into Eq. 31, we obtain after some simplifications the following final form:

$$d\sigma_{ij} = \left[C^E_{ijk\ell} - \frac{H^*_{ij} H_{k\ell}}{H} \right] d\varepsilon^T_{k\ell} \quad (33)$$

where

$$H = 3AL(3\lambda + 2\mu) + 2B\mu(2MJ_2 + 3NJ_3) + 2C\mu(3MJ_3 +$$

$$NS_{ik}S_{kj}S_{i\ell}S_{\ell j} - \frac{4}{3}NJ_2^2) - \frac{\partial F}{\partial x}\frac{3L}{D(\epsilon_v^P + W)} -$$

$$h\frac{\partial F}{\partial k}\frac{\partial k}{\partial \bar{e}^P}\sqrt{3L^2 + 2M^2J_2 + 6MNJ_3 + N^2(S_{ik}S_{kj}S_{i\ell}S_{\ell j} - \frac{4}{3}J_2^2)},$$

$$H_{ii} = A(3\lambda + 2\mu) + 2\mu B S_{ii} + 2\mu C t_{ii} \quad \text{(no summation)},$$

$$H^*_{ii} = L(3\lambda + 2\mu) + 2\mu M S_{ii} + 2\mu N t_{ii} \quad \text{(no summation)},$$

$$H_{ij}(i \neq j) = 2\mu B S_{ij} + 2\mu C t_{ij},$$

$$H^*_{ij}(i \neq j) = 2\mu M S_{ij} + 2\mu N t_{ij},$$

$$A = \frac{\partial F}{\partial I_1}, \quad B = \frac{\partial F}{\partial J_2}, \quad C = \frac{\partial F}{\partial J_3}, \quad L = \frac{\partial \psi}{\partial I_1}, \quad M = \frac{\partial \psi}{\partial J_2},$$

and $N = \frac{\partial \psi}{\partial J_3}$.

Therefore, elastic-plastic constitutive tensor $C_{ijk\ell}^{EP}$ is given by

$$C_{ijk\ell}^{EP} = C_{ijk\ell}^{E} - \frac{H_{ij}^* H_{k\ell}}{H} \tag{34}$$

For the special case of plane strain condition, the elastic-plastic matrix reduces to

$$\begin{Bmatrix} d\sigma_{xx} \\ d\sigma_{yy} \\ d\tau_{xy} \\ d\sigma_{zz} \end{Bmatrix} = \begin{bmatrix} \lambda + 2\mu - \frac{H^*_{xx}H_{xx}}{H}, & \lambda - \frac{H^*_{xx}H_{yy}}{H}, & -\frac{H^*_{xx}H_{xy}}{H} \\ \lambda - \frac{H^*_{yy}H_{xx}}{H}, & \lambda + 2\mu - \frac{H^*_{yy}H_{yy}}{H}, & -\frac{H^*_{yy}H_{xy}}{H} \\ -\frac{H^*_{xy}H_{xx}}{H}, & -\frac{H^*_{xy}H_{yy}}{H}, & \mu - \frac{H^*_{xy}H_{xy}}{H} \\ \lambda - \frac{H^*_{zz}H_{xx}}{H}, & \lambda - \frac{H^*_{zz}H_{yy}}{H}, & -\frac{H^*_{zz}H_{xy}}{H} \end{bmatrix} \begin{Bmatrix} d\epsilon_{xx} \\ d\epsilon_{yy} \\ d\gamma_{xy} \end{Bmatrix} \tag{35}$$

As can be seen from Eqs. 34 and 35, the stiffness matrix is not symmetric for the non-associated flow rule case but becomes a symmetric one for the associated flow rule case. In Appendix 1, the equations of A, B, C, L, M, N, etc. are presented for the models of the Drucker-Prager, Coulomb, the plane cap and the elliptic cap under the associated as well as the non-associated flow rule assumptions.

6. NUMERICAL EXAMPLES

The present study investigates the application of different soil plasticity models in the analysis of geotechnical problems. The Coulomb or Drucker-Prager criterion is probably the most popular since it has only two material constants. For the Drucker-Prager criterion, a careful selection of material constants α and k is required so that they match with the well-known Coulomb criterion.

For the more refined cap models, it contains many material constants which require elaborate experimental work. These models may predict more accurately soil behavior than that of the former model.

In the present work, the finite element method has been applied to 1) the analysis of footing problem using the Drucker-Prager type of perfectly plastic model with different material constants and 2) the analysis of a problem of a uniaxially compressed sand using the different plasticity models described previously.

6.1 Shallow Stratum of Clay Loaded Smooth Strip Footing

As a first example, a plane strain problem of a clay layer ($\phi=20°$ and c = 10 psi) with a smooth strip footing on the surface as shown in Fig. 7 is analyzed. The Drucker-Prager criterion with different material constants (Table 1) is utilized in the present analysis. Boundary conditions and dimensions used in the analysis are given in Fig. 7. In the finite element procedure used herein, each rectangular element is defined by four constant strain triangles with a common node at the rectangular center. Three different types of material constants as given in Eqs. 7, 8, and 10 are used in the analysis with an associated flow rule. These constants are obtained from matching the Drucker-Prager model with the Coulomb model along the compressive meridian, the tensile meridian, and under plain strain conditions respectively. The values of material constants α and k are 0.149 and 12.25 lb/in^2, 0.118 and 9.74 lb/in^2, and 0.112 and 9.22 lb/in^2 respectively.

The complete load displacement response of the strip footing is shown in Fig. 8 where the applied pressure is plotted versus the centerline displacement directly beneath the load for each case. The circles plotted in Fig. 8 correspond to actual computed points obtained from the small deformation analysis. As can be seen, the analysis with material constants matched with the compressive meridian of Coulomb criterion in a three dimensional space results in a collapse load (365 psi) which is almost twice that of the other loads (158, 190 psi). The collapse load predicted by Drucker-Prager criterion matched in the plane strain condition is, as expected, the same as that of Coulomb criterion. This load is close to the solutions reported by Terzaghi and Prandtl. The analysis with material constats of Eq. 7 does not agree with the well-known solution of Terzaghi and Prandtl.

Fig. 9 shows the load-displacement curves predicted by the Drucker-Prager criterion with an associated and a non-associated flow rule. For the case of an associated flow rule, the material constants α and k (Eq. 10) obtained from matching Coulomb in plane strain conditions are used in the failure function F and the potential function $\psi = F$. For the case of an non-associated flow rule, the failure function is the same as that of an associated flow rule case, but a von Mises type of function (no plastic volumetric strain) is taken as the potential function. In Fig. 9, the load displacement curves for both cases are seen to be almost the same up to the applied load of 60 psi because the state of stress in all element at this load level is still within elastic region. Then, as load is gradually increased, their behavior may become quite different. The collapse load for the soil with a non-associated flow rule is found to be less than that of an associated flow rule material.

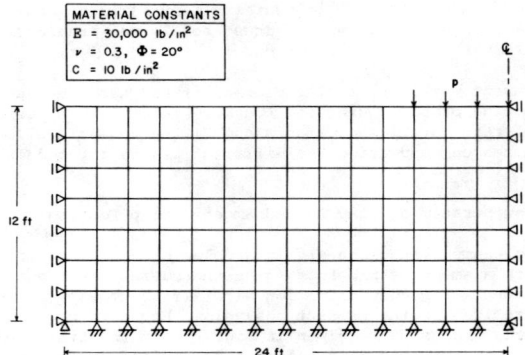

Fig. 7 Analytical Model of Shallow Stratum of Clay.

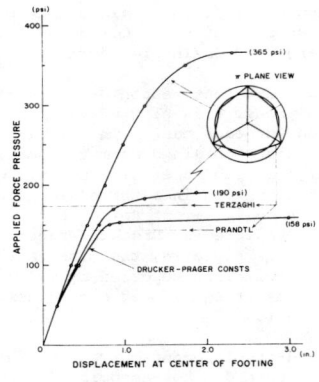

Fig. 8 Load-Displacement Curve (Plane Strain Condition).

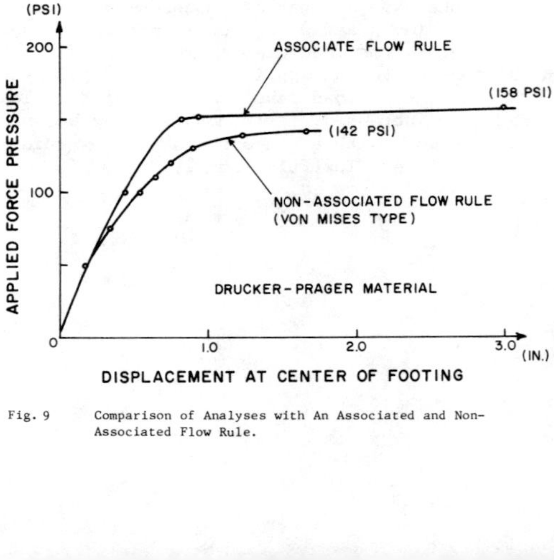

Fig. 9 Comparison of Analyses with An Associated and Non-Associated Flow Rule.

SOIL RESPONSE ANALYSIS 131

In Figs. 10 and 11, the velocity fields at collapse load are presented for both cases. The broken and solid lines on the figure are the outlines of the Terzaghi and the Prandtl velocity fields. The magnitude and direction of each velocity field is presented by arrows in which each displacement at the center of footing is taken as a normalized unit length. The numerically obtained velocity fields are seen in a fair agreement with that of the Terzaghi and the Prandtl fields. The magnitude of velocity in the radial shearing zone and near surface zone for a non-associated flow material is seen quite small compared with that beneath the footing. However, the magnitude of velocity for an associated flow material has an almost uniform value throughout the entire field.

6.2 Compression of Sand Under Uniaxial Strain Condition

The behavior of sand located on the surface under a uniaxial strain condition using different plasticity models is considered in this second problem. The models used in this analysis are the Drucker-Prager model, the Drucker-Prager model with a simple plane hardening cap, and the Drucker-Prager model with an elliptic hardening cap. An associated flow rule is used in this analysis.

The material constants of sand are as follows [1]: ϕ = the angle of internal friction = 49.093°, c = the cohesion = 0 psf, ν = the Poisson's ratio = 0.2736, E = Young's modulus = 841396 psf, R = the geometrical parameter in elliptic cap (in Eq. 16) = 4.33, W = the material constant (in Eq. 13) = 0.0075, and D = material constant = 6.781 x 10^{-5} ft^2/lb.

The load-displacement curves are shown in Figs. 12 and 13 for each model. The applied pressure is loaded up to 24000 psf and then unloaded to zero under a static condition. The solid lines in Figs. 12 and 13 present the same load-displacement curve predicted by the Drucker-Prager criterion without a cap. In this case, the load displacement relationship is linearly elastic under the loading as well as unloading path. As can be seen from the inset of Fig. 12, the stress path in $I_1 - \sqrt{J_2}$ space for this case is a linear line within a failure envelope.

The curve with small open circles in Fig. 12 shows the behavior predicted by Drucker-Prager type of material model with a simple plane cap under a loading condition. The load-displacement relationship is not linear from the beginning. This stems from the fact that a compaction of plastic volumetric strain caused by a simple plane cap influences the behavior of sand under a low level I_1. As the applied pressure becomes higher, however, the state of the compaction in sand approaches that of maximum compaction. Therefore, the slope of the load-displacement curve under a high level of the applied load becomes gradually identical to that of a behavior predicted by Drucker-Prager criterion with a cap.

The state of a stress in this case is situated at a corner which is the intersection of a failure envelope with a plane cap in $I_1 - \sqrt{J_2}$ space from the start of loading. As the load increases, the plane cap expands representing the corner behavior. In the case of unloading, the stress path is not within an elastic region but moves along the failure envelope to the origin. In the process of unloading, the stress path on the failure envelope produces volume expansion or dilatancy. As a consequence, the plane cap contracts gradually

132 SOIL STRESS STRAIN APPLICATIONS

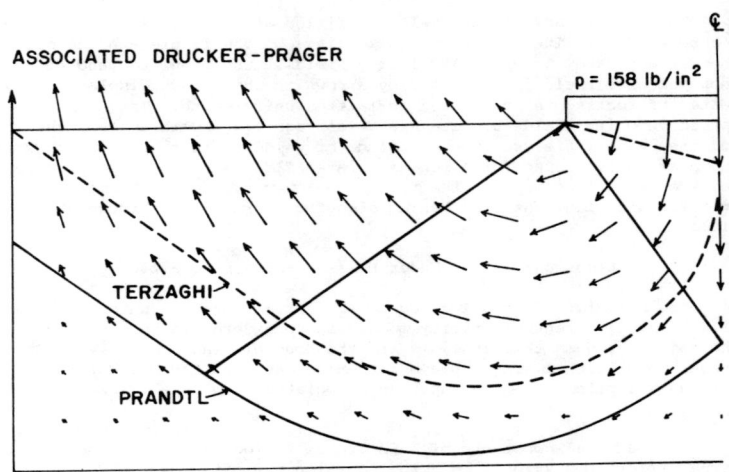

Fig. 10 Velocity Field at Numerical Collapse Load (158 lb/in^2)--
Associated Flow Rule and Small Deformation.

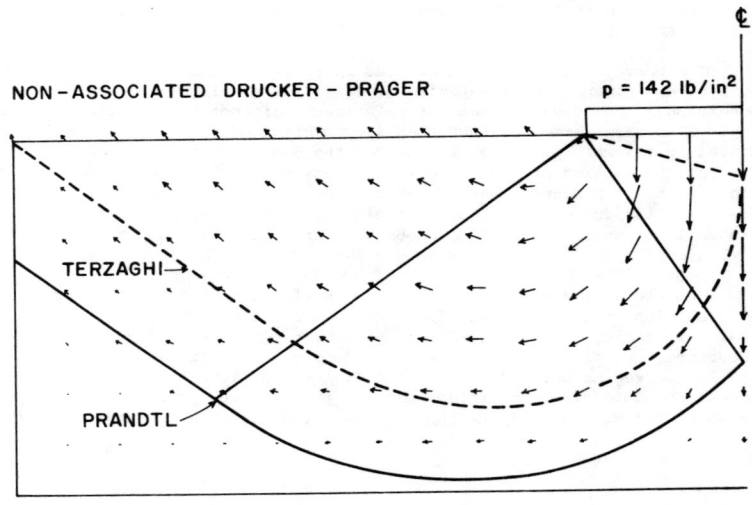

Fig. 11 Velocity Field at Numerical Collapse Load (142 lb/in^2)--
Non-Associated Flow Rule and Small Deformation.

SOIL RESPONSE ANALYSIS

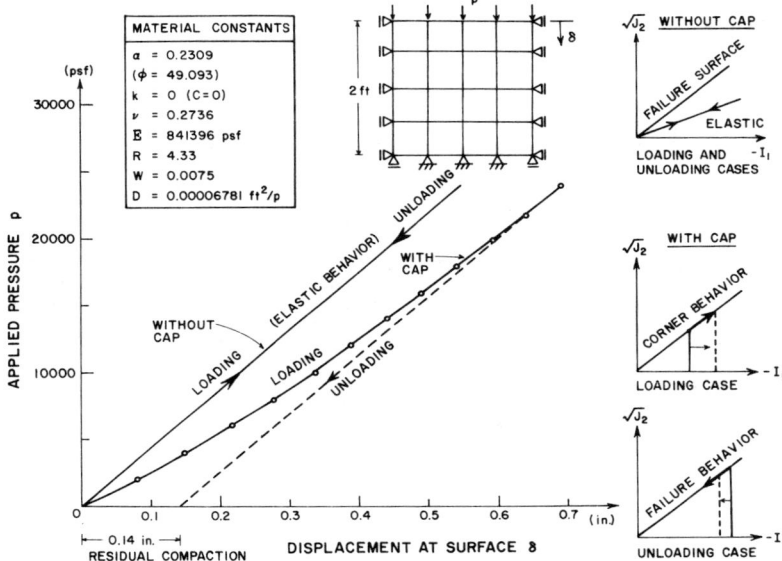

Fig. 12 Load-Displacement Curve by Drucker-Prager Criterion without or with A Plane Cap.

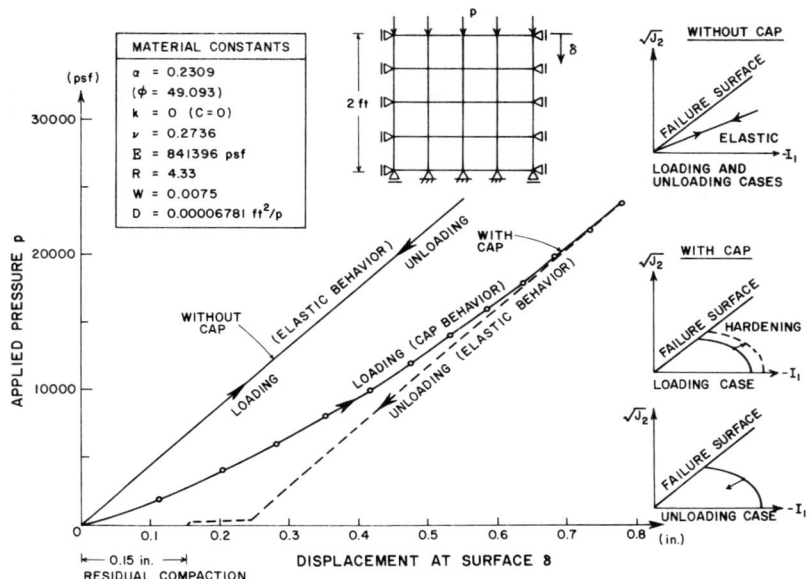

Fig. 13 Load-Displacement Curve by Drucker-Prager Criterion without or with An Elliptic Cap.

toward the direction of the origin. From the load-displacement curve shown in Fig. 12 with a broken line, it can be seen that the behavior for unloading is similar to an elastic behavior. Finally, a residual compaction of 0.14 in. at the surface results from this loading-unloading cycle.

In a similar manner to the case of a plane cap model, the load-displacement curve for loading and unloading conditions is drawn in Fig. 13 for an elliptic cap model. The displacement predicted by this model is larger than that by a simple plane cap model. With a high level of the applied load, the load-displacement curve has a similar character with a curved loading behavior and linearly elastic unloading behavior. The stress path in $I_1 - \sqrt{J_2}$ space during loading is seen moving within the failure envelope and situating on the elliptic cap. This cap expands due to the increase of hydrostatic pressure I_1. On the other hand, the stress path during unloading moves within the current elastic region while the elliptic cap remains fixed in the stress space. Consequently, the load-displacement curve presents a linearly elastic behavior, as shown in Fig. 13 with a broken line, until the stress path touches the failure envelope on the opposite side. Once the stress path reaches the failure envelope, a volume expansion or dilatancy is developed. As the result of this dilatancy, the cap contracts twoards the origin. After the completion of loading-unloading cycle, a residual compaction of 0.15 in. at the surface is predicted by elliptic-cap model which is very close to the value of 0.14 in. predicted previously by the plane-cap model.

7. SUMMARY AND CONCLUSIONS

7.1 Summary

The well-known Coulomb criterion of failure exhibits an irregular hexagonal pyramid in a three dimensional stress space. It is known that Coulomb criterion with its singularities is difficult to handle in a numerical analysis. The Drucker-Prager criterion is therefore used to approximate the well-known failure criterion such as the Coulomb or the modified Coulomb criterion by a simple smooth function. The material constants α and k are expressed in terms of the material constants c and ϕ for the case of matching with Coulomb criterion. On the other hand, these are expressed in terms of a simple tensile strength and a simple compressive strength for the case of matching with the modified Coulomb criterion. These earlier models, however, cannot predict the compaction before the failure of soils. To this end, caps are introduced to the criterion. Cap models can predict not only the behavior of soils during loading and compaction, but also can control the amount of dilatancy required by experimental data. The cap model gives a reasonable agreement with experimental observations [1].

These three models are formulated by the incremental plasticity theory according to a non-associated flow rule as well as an associated flow rule. The stiffness matrices of plasticity models are presented in a readily applicable form for finite element applications. The finite element method is then applied using these stiffness matrices derived in this paper for the analysis of the stratum of clay loaded by smooth footing and the analysis of a uniaxially compressed sand.

The Drucker-Prager type of perfectly plastic model is used in the analysis of footing problem. The analysis with three types of material constants is made using an associated flow rule. Three types of material constants are obtained from matching with the Coulomb criterion along the compressive meridian, along the tensile meridian, and under plane strain condition respectively. As expected, the collapse load predicted by the Drucker-Prager criterion matched in plane strain condition is the same as that of the Coulomb criterion. From this analysis, it is also found that the analysis based on the material constants matched with the Coulomb criterion along the compressive meridian does not agree with the bearing capacity solutions of Terzaghi and Prandtl.

Next, the analysis is carried out for an associated and a non-associated flow rule. In this case, the Drucker-Prager criterion with the material constants matched in plane strain condition is taken as a failure function, and a von Mises type of function is taken as potential function. It becomes clear that the collapse load by a non-associated flow rule is less than that by an associated flow rule. At these collapse load levels, velocity fields are found to be similar to the fields of Terzaghi and Prandtl for both flow rules. However, the load-displacement curves between both may be quite different.

The three different plasticity models are used in the analysis of a uniaxially compressed sand. The behavior predicted by the Drucker-Prager model without a hardening cap cannot explain the compaction of sand. Although the simple plane cap model and elliptic cap model predict different load-displacement curves and stress paths under loading and unloading, the residual compaction predicted by both models is almost the same.

7.2 Conclusions

Three plasticity-based constitutive models for geotechnical materials are examined from the viewpoints of formulation and numerical calculation, within the context of their use in the finite element analysis of soil-structure problems. The three constitutive models considered are the Drucker-Prager model, a plane-cap model, and an elliptic-cap model. The important point to be noted in using the plasticity models is the careful selection of the material constants. In order that the Drucker-Prager criterion represents a proper generalization of Coulomb or modified Coulomb criteria under general three dimensional stress state, its material constants α and k must be properly defined. These constants should not be treated as fixed values for all types of applications. Rather, their choice depends on the particular problems to be solved.

The constitutive relations of geotechnical materials play an important role in the analysis of soil-structure problems. The basic requirements for a soil model are: (a) the mathematical formulation results in a unique and stable stress-strain relationship, (b) the constitutive equation reflects the key characteristics of experimental data, (c) the mathematical relationship is defined by a very small number of parameters which can be determined from standard test data, and (d) the mathematical model should encompass the well-known Coulomb criterion as special case. The three models described in this paper satisfy all these requirements. Some of the aspects above have been examined numerically as well as analytically. These models have been incorporated in the finite element software system which has been developed at Purdue University for the analysis of earthquake-induced landslides. At the present state, it appears there is little need for further refinements of this type of model, since the present fluctuations of experimental measurements make it desirable that the constitutive models be kept relatively simple while their numerical implications on practical applications be examined more thoroughly. These analytical results should then be corrected with field data to assess the adequacy of these models in reproducing the important features of soil-structure interactions and to identify areas in which improved modelling of the geotechnical material properties is necessary and essential.

8. ACKNOWLEDGEMENTS

The research reported here was supported by the National Science Foundation under grant no. PFR-7809326 to Purdue University, West Lafayette, IN.

REFERENCES

[1] Baladi, G. Y. and Rohani, B., "An Elastic-Plastic Constitutive Model for Saturated Sand Subjected to Monotonic and/or Cyclic Loadings", Third International Conference on Numerical Method in Geomechanics, Aachen, 2-6 April 1979, pp. 389-404.

[2] Bathe, K. J., Snyder, M. D., and Cimento, A. P., "On Finite Element Analysis of Elasto-Plastic Response", Reprints of Papers at Conference "Engineering Applications of the Finite Element Method", Organized by A. S. Computers, P.O. Box 310, 1322 Hovik, Norway, May 9-11, 1979.

[3] Chen, W. F., "Limit Analysis and Soil Plasticity", Elsevier, Amsterdam, The Netherlands, 1975.

[4] Chen, W. F. and Mizuno, E., "On Material Constants for Soil and Concrete Models", Third ASCE/EMD Specialty Conference, 1979, pp. 539-542.

[5] Davidson, H. L. and Chen, W. F., "Non-Linear Analysis in Soil and Solid Mechanics", Numerical Methods in Geomechanics, ASCE, Edited by C. S. Desai, Vol. 1, 1976, pp. 205-216.

[6] DiMaggio, F. L. and Sandler, I. S., "Material Models for Granular Soils", Journal of the Engineering Mechanics Division, ASCE, Vol. 97, No. EM3, 1971, pp. 936-950.

[7] Drucker, D. C., Gibson, R. E., and Henkel, D. J., "Soil Mechanics and Work-Hardening Theories of Plasticity", Transactions, ASCE, Vol. 122, 1957, pp. 338-346.

[8] Drucker, D. C. and Prager, W., "Soil Mechanics and Plastic Analysis or Limit Design", Quarterly of Applied Mathematics, Vol. 10, No. 2, July 1952, pp. 157-175.

[9] Roscoe, K. H., Schofield, A. N., and Thurairajah, A., "Yielding of Clays in State Wetter than Critical", Géotechnique, Vol. 13, No. 3, 1963, pp. 211-240.

[10] Sandler, I. S., DiMaggio, F. L., and Baladi, G. Y., "Generalized Cap Model for Geological Materials", Geotechnical Division, ASCE, Vol. 102, No. GT.7, 1976, pp. 683-699.

[11] Sandler, I. S. and Melvin, L. B., "Material Models for Geological Materials in Ground Shock", Numerical Methods in Geomechanics. Edited by C. S. Desai, Vol. 1, 1976, pp. 219-231.

[12] Shield, R. T., "On Coulomb's Law of Failure in Soils", J. Mech. Phys. Solids, 4(1), 1955a, pp. 10-16.

[13] Schofield, A. N. and Wroth, P., "Critical State Soil Mechanics", McGraw-Hill, New York, 1968.

[14] Zienkiewicz, O. C., "The Finite Element Method", McGraw-Hill, 1978 (Chapter 18).

[15] Zienkiewicz, O. C., Humpheson, C., and Lewis, R. W., "Associated and Non-Associated Visco-Plasticity and Plasticity in Soil Mechanics", Géotechnique, 25, No. 4, 1975, pp. 671-689.

Appendix I--Material Stiffness of Different Models

The stiffness matrix of each model is described for the associated flow rule case and for the non-associated flow rule case.

1. Associated flow rule case

   1.1 Coulomb criterion

    Coulomb criterion in terms of stress invariants is given by Zienkiewicz [15].

$$F = \psi = I_1 \sin \phi + \frac{3(1 - \sin \phi) \sin \theta + \sqrt{3}(3 + \sin \phi) \cos \theta}{2} \sqrt{J_2} - 3c \cos \phi = 0 \tag{A-1}$$

where $\theta = \frac{1}{3} \cos^{-1}(\frac{3\sqrt{3}}{2} \frac{J_3}{J_2^{3/2}})$

Taking derivatives of (A-1) with respect to I_1, J_2, and J_3, we obtain

$$\frac{\partial F}{\partial I_1} = A = L = \sin \phi,$$

$$\frac{\partial F}{\partial J_2} = B = M = \frac{3(1 - \sin \phi) \sin \theta + \sqrt{3}(3 + \sin \phi) \cos \theta}{4} J_2^{-1/2} +$$

$$\frac{3\sqrt{3} J_3 [3(1 - \sin \phi) \cos \theta - \sqrt{3}(3 + \sin \phi) \sin \theta]}{8 J_2^2 \sin 3\theta} \tag{A-2}$$

$$\frac{\partial F}{\partial J_3} = C = N = -\frac{\sqrt{3}[3(1 - \sin \phi) \cos \theta - \sqrt{3}(3 + \sin \phi) \sin \theta]}{4 J_2 \sin 3\theta},$$

$$\frac{\partial F}{\partial x} = 0, \text{ and } \frac{\partial k}{\partial \bar{e}^p} = 0 \text{ } (k = c \text{ is assumed to be constant})$$

1.2 Drucker-Prager criterion

The equation is written again as

$$F = \psi = \alpha I_1 + \sqrt{J_2} - k(\bar{e}^P) = 0 \tag{A-3}$$

where $k(\bar{e}^P)$ is a hardening function of effective plastic strain $\bar{e}^P$.

Therefore, in a similar manner to the previous case,

$$A = L = \alpha,$$

$$B = M = \frac{1}{2\sqrt{J_2}} \tag{A-4}$$

$$C = N = 0, \quad \frac{\partial F}{\partial x} = 0, \text{ and } \frac{\partial F}{\partial k} = -1$$

It should be noted that $\frac{k}{\bar{e}^P}$ is obtained from experimental data.

1.3 Plane cap hardening function

From the function F_c in Eq. 12,

$$A = L = 1, \quad B = M = 0, \quad C = N = 0,$$

$$\frac{\partial F}{\partial x} = -1, \text{ and } \frac{\partial F}{\partial k} = 0 \tag{A-5}$$

1.4 Elliptical cap hardening function

From the function F_c in Eq. 16,

$$A = L = 2(I_1 - \ell), \quad B = M = R^2, \quad C = N = 0,$$

$$\frac{\partial F}{\partial x} = -2(x - \ell), \text{ and } \frac{\partial F}{\partial k} = 0 \tag{A-6}$$

2. Non-associated flow rule case

In the non-associated flow rule case, a failure or hardening function F is not identical to that of a potential function ψ. As an example, for the analysis with the Coulomb criterion, the same type of Coulomb function or Drucker-Prager function may be taken as a potential function ψ. Combining (A-2) and (A-4) through (A-6), the different stiffness matrix can be derived by users.

Shallow Penetration of Marine Foundations

by

James D. Murff*, M.ASCE and Terrell W. Miller*

Introduction

In offshore engineering applications it is frequently necessary to install objects on the seafloor. Shallow foundations (embeddment depth less than the minimum footing dimension) are often employed in such operations either as the permanent installation or as a temporary base pending pile driving. Mooring bases, subsea production systems, and temporary production platform supports (mudmats) are typical examples. When such emplacements must be made on very soft clay soils, which are commonly encountered offshore, penetration of the footing below the mudline may be significant. In such cases operational requirements may necessitate fairly accurate predictions of the footing penetration below the mudline.

This study focuses on the problem of predicting penetration in normally consolidated to underconsolidated clays in which the rate of increase with depth of the undrained strength is large with respect to the footing size. Indeed the method presented may technically be applied to any soil profile however it is believed to be most appropriate for conditions where the near surface soil is relatively soft resulting in a squeezing type deformation.

A number of different techniques might be employed to develop this prediction. An intuitive method is to use the familiar bearing capacity equation [7] to determine the failure load at various footing embedments. These results could then be associated with a load-penetration relationship. This procedure would however require the use of weighted average strength. Since the shear deformations implicit in the bearing capacity equation occur to depths below one-half the footing diameter and extend to various distances to the side of the footing, it is likely to be difficult to determine a generally applicable weighting function for various combinations of strength profiles and footing configurations. Another approach would be to attempt to solve the complete (boundary value) problem numerically; however, this would likely be very time consuming and expensive. Considering the difficulty in accurately characterizing soil behavior we suspect the "complete" approach would not necessarily be more accurate than the approximate method described below.

In this study we favor a third approach, the use of limit analysis techniques, particularly the upper bound method [9]. The soil is idealized as a rigid, perfectly plastic, incompressible solid. The failure mechanisms are then selected intuitively as deformation patterns consistent with the squeeze type failure. To assess the load penetration behavior, the failure load for

* Research Associate, Exxon Production Research Company, Houston Texas

various assumed footing depths is determined. The change in strength profile due to the deformation of the soil can be approximately accounted for by using the deformation patterns of the selected mechanism.

Before proceeding directly to a discussion of the upper bound method and our approach, it is pertinent to point out that the bearing capacity equations are derivable using limit analysis techniques. We are using the same methodology to derive limit loads for more generally occuring soil conditions and, in a way, are exchanging intuition in deriving "proper" deformation patterns for "good" guesses in deriving methods of averaging soil strengths for use in the bearing capacity equation. However the method we present should have more general applicability since we know from the bounding theorem that the lowest value of the limiting load for a given penetration will be the best estimate for that case.

The Upper Bound Method

In order to apply the upper bound method we are idealizing the soil as a rigid plastic material. To model undrained behavior, i.e. purely cohesive strength, we will use the von Mises yield criterion [6],

$$J_2^{1/2} - k = 0 \qquad (1)$$

where J_2 is the second invariant of the stress deviation tensor and k is a material constant associated with strength. For plastic analysis of metals, k is usually interpreted as the ultimate strength in pure shear, and for the von Mises yield criterion k is equal to $\sigma_0/\sqrt{3}$, where σ_0 is the uniaxial compressive strength. This relationship of k to σ_0 is in contrast to that in the Tresca (or maximum shear stress) criterion typically used in soil plasticity where $k = \sigma_0/2$. One needs to be mindful of these differences when interpreting results of analyses using these two criteria. Historically, most undrained analyses have used the Tresca criteria because of the simplicity of its use in 2-D problems. However, for 3-D problems the von Mises criterion becomes much simpler. Because of this and the approximate nature of the undrained analysis the von Mises criterion seems to be a reasonable choice.

For the von Mises criterion the rate at which energy per unit volume is dissipated, D, in a deforming region in terms of strain rates, $\dot{\varepsilon}_{ij}$, is [8]

$$D = k \sqrt{2\dot{\varepsilon}_{ij}\dot{\varepsilon}_{ij}} \qquad (2)$$

where repeated subscripts indicate summation. For a slip surface the dissipation per unit area is [2]

$$D = kV_R \qquad (3)$$

where V_R is the relative velocity difference across the slip surface.

The upper bound theorem of plasticity states, [4]: "Collapse must occur if for any compatible flow pattern considered as plastic only, the rate at which external forces do work on the bodies equals or exceeds the rate of internal dissipation." To apply the theorem we (1) construct a kinematically admissible collapse mechanism (2) equate the rate of energy dissipation in the soil to the rate at which the unknown external loads are doing work and (3) solve for the unknown loads. This calculation is carried out using the concept of virtual work. To be consistent with the literature we adopt the terminology of virtual velocities instead of displacements. This results in a straightforward solution to kinematic (i.e., upper bound) problems as will be shown later. In reality time is used only as a scale factor and velocity is equivalent to virtual displacement since material behavior is not time dependent. Returning to virtual work concepts, one develops an upper bound solution by assuming virtual velocity fields consistent with the velocity boundary conditions. These fields may be continuous and/or discontinuous resulting in volumetric and/or surficial energy dissipation. The resulting dissipation is equated to the virtual work being done by the external forces, and by this formulation the magnitude of these forces are determined.

According to the theorem the collapse load determined by this method is an upper bound to the actual collapse load (for the conditions of the idealized problem). However, it has been our experience [9] that upper bounds often give reasonable predictions for bearing capacity problems as compared to the empirically derived correction factors to the bearing capacity equation. They are particularly useful in assessing the relative effects of a varying soil strength profile. A lower bound method which involves constructing admissible stress fields (i.e. stresses that nowhere exceed yield and are in equilibrium with the applied loads) is available in plasticity theory. It is generally more difficult to apply however and predictions are often so divergent from upper bounds that little new information is gained from the exercise. In the problem described herein over prediction of penetration may often be as troublesome as underprediction, thus poor lower bound results would certainly not provide an advantage.

Collapse Mechanisms

The idealized model selected consists of a rigid circular disc supported by a non-homogeneous isotropic soil. The disc is loaded vertically downward. As mentioned previously the soil is idealized as a rigid, perfectly plastic material obeying the von Mises yield criterion. Further, axisymmetric conditions are imposed.

The two collapse mechanisms considered both consist of three deforming regions. Fig. 1 shows a typical cross section of the regions with the material outside these three regions taken as rigid. The geometry of the regions is the same for both mechanisms however the respective velocity (or deformation) fields within each region are, of course, different. The rectangle ABB'A', region 1, is a cross section of the cylindrical deforming zone directly below the footing (footing shown shaded). Region 2, ABC and A'B'C', is the cross section of a triangular deforming toroid zone; region 3, ACDE and A'C'D'E', the cross section of a rectangular deforming toroid zone. Two parameters, h and r_o, dictate the dimensions of the deforming region for a specified footing radius, r_i, and embedment depth, d. The parameters h and r_o

are then varied to find the combination which yields the minimum collapse load F. This value is the best solution i.e. the upper bound closest to the exact solution. The appendix provides the details of the construction of the velocity fields for the two mechanisms, however a brief description seems warranted here.

There are three orthogonal velocity components that must be determined for the complete specification of the velocity field at each point. Because of axial symmetry the tangential velocity component, v_θ, is necessarily zero and the other two components, v_r and v_z, are not functions of θ. We can then consider an r-z cross section as shown in Fig. 1 as representative. The functional form of v_r and v_z were assumed (intuitively to attain the best bound). Incompressibility, (this condition is a direct consequence of the assumed yield function), the velocity boundary conditions and normal velocity compatibility requirements between regions provide enough information to determine the coefficients to completely specify the velocities.

Mechanism 1 is constructed to provide a solution for soil profiles that consist of a uniform soft layer overlying a stronger layer. Region 1 is assumed to deform maintaining its cylindrical shape. Regions 2 and 3 are selected to be compatible with region 1 as described in the Appendix. Fig. 2(a) shows a typical velocity field resulting from the assumed mechanism.

Mechanism 2 is constructed to provide a solution for soil profiles that contain a gradual increase in strength with depth. In this case the upper part of the deforming region flares out, concentrating deformation in the weaker soil, whereas the lower part displaces very little. Fig. 2(b) depicts this velocity field.

Having constructed the velocity fields, the strain rate fields within deforming regions and the relative slip velocities at boundaries for given values of h and r_o can be determined in a straightforward manner from the well known strain rate-velocity relationships (equivalent to strain-displacement relationships). The energy dissipation rates can then be calculated for a specific point within the field using equation 2 or 3 as appropriate. In this solution both types of deformation are considered. The total dissipation is then determined by integration of the local rates over the appropriate volume or surface. For irregular soil profiles, which can vary radially as well as vertically, numerical integration is usually necessary. The total dissipation rate can then be written as

$$\dot{I} = I v_o \tag{4}$$

where v_o is the virtual velocity of the footing and I is a function of geometry of the problem and the strength profile but not of v_o. The upper bound theorem is then applied, i.e. in this case,

$$F v_o = I v_o \tag{5}$$

MARINE FOUNDATIONS

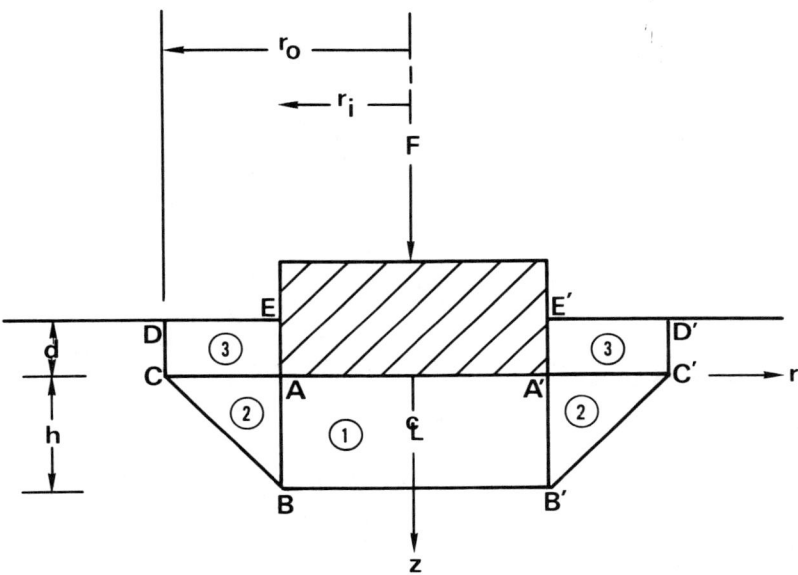

Figure 1 - Configuration of Velocity Fields for Collapse Mechanisms

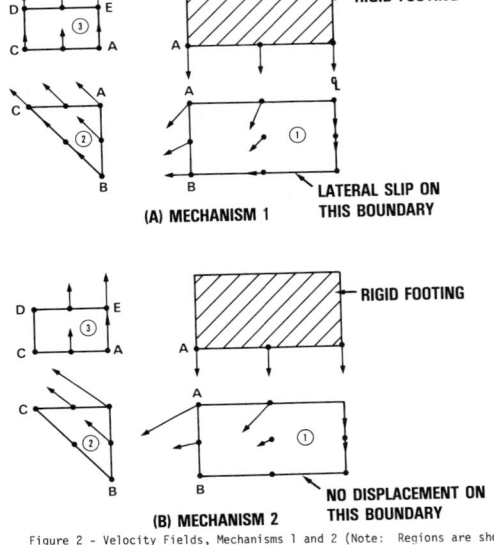

Figure 2 - Velocity Fields, Mechanisms 1 and 2 (Note: Regions are shown separated at common boundaries for clarity; Arrow lengths indicate velocity magnitude relative to arbitrary footing velocity)

The virtual velocity terms cancel and we can solve for F, i.e.

$$F = I \tag{6}$$

These calculations can be carried out for a large number of combinations of h and r_o to find the minimum value of F or an optimization program can be used. The computer requirements to optimize the solution typically amount to only a few seconds even when numerical integration is required.

Parameter Studies

Parameter studies were carried out for the case of linearly increasing soil strength with depth using both mechanisms 1 and 2 for smooth (no adhesion or frictionless) and rough (perfect adhesion, energy dissipated at kV_r on the footing surface) footings. The results (optimized solutions) are summarized in Fig. 3 which is a plot of dimensionless collapse pressure vs dimensionless rate of strength increase. Fig. 3 shows that for the smooth case mechanism 2 is the better solution and that for the rough case mechanism 1 is the better solution. This is consistent with the assumptions used in constructing the mechanisms.

To calibrate these solutions it would be desirable to have some exact solutions for comparison. The only solutions known to us for the circular footing are for the uniform case ($\rho=0$). The exact solutions for the collapse pressure of a smooth and rough footing are $2.85\sigma_0$ and $3.03\sigma_0$ respectively (see references 5 and 10); the best upper bound solutions for this case, as shown in Fig. 3 are $3.24\sigma_0$ (5.62k) and $4.01\sigma_0$ (6.95k) giving errors of approximately 14 and 32 percent respectively.

Exact solutions are available for the linearly increasing strength case for a strip footing [3]. These solutions are plotted in Fig. 4 along with the best solutions for the circular footing from Fig. 3. Note that the abcissa must be non-dimensionalized in a different way for the strip and circular footings, i.e. using a different characteristic dimension. For the strip footing solution shown, B is the footing width. When plotted as shown the collapse pressure for the circular footing is approximately 1.2 times that for the strip footing. Note that the selection of the abcissa shown is somewhat arbitrary and the factor 1.2, typically used as the shape factor in the bearing capacity equation, is simply fortuitous since it is a function of the particular plot.

The results presented here seem intuitively reasonable. Furthermore for the few known solutions they seem to be in good agreement. The available evidence therefore is encouraging but additional checking, particularly at large values of strength increase, is needed. It is important to emphasize that the method presented has the advantage of generality. It can easily be applied to irregular strength profiles that would be very difficult to analyze exactly. With additional confirmation it is likely that these mechanisms will be able to the used with greater confidence.

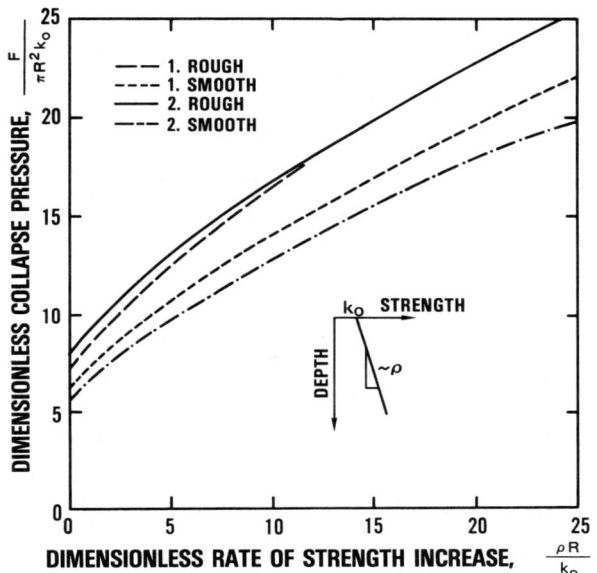

Figure 3 - Collapse Pressures for Linearly Increasing Strength with Depth

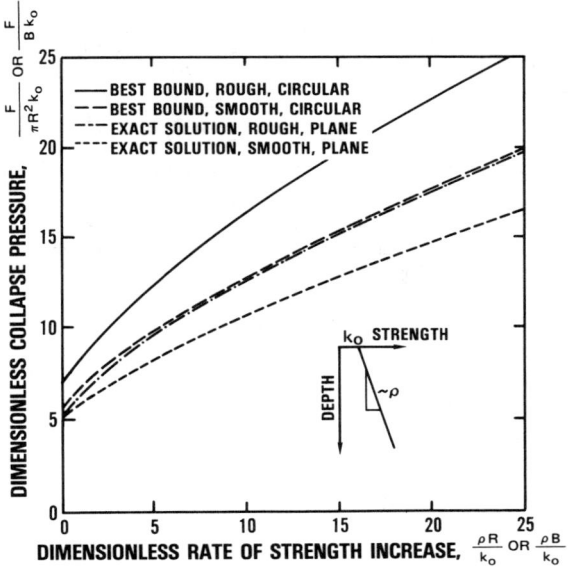

Figure 4 - Collapse Pressures for Circular and Strip Footings

Case History

The case study presented here concerns the prediction of the penetration of mudmats for the Hondo offshore oil production platform [1]. The structure was installed in a location where the near surface soil is a soft silty clay.

Fig. 5 shows a plan view of the platform mudmats. The irregular footings shown were idealized for analysis as circular footings of equivalent area. We believe the error introduced thru this approximation is small. Of greater importance here is the footing sizes compared to the rate of strength increase. The inset in Fig. 6 shows the interpreted strength profile which is based on miniature vane shear tests, unconfined compression and unconsolidated undrained triaxial compression tests, and other indirect indicators such as liquid limits and water contents.

Fig. 6 shows the predicted penetration vs depth performance of the platform base. The structure was ballasted to a net weight on bottom of 3500 kips and achieved a penetration of 1.5 feet as shown in the figure. Assuming a rough footing (complete adhesion at every depth) the predicted penetration is .8 feet, considerably short of the actual value. It seems likely however that the very soft soil near the surface was carried down with the footings to some depth, thus making the smooth approximation more reasonable. The agreement is much better for the smooth assumption.

A further correction was made by using the velocity field to approximate the change in the strength profile due to the deformation of the soil. This correction was made only for the soil below the plane of the footing. No attempt was made to correct for the radial non-homogenieties due to deformation since this affect appears to be very small. This idealization further improves the prediction. To make the strength correction the velocity field at a particular footing depth was assumed to hold during a footing penetration of .25 feet. The optimum solution (with a new velocity field) was then determined for the next increment and so on. Note that the velocity field is derived based on small strain assumptions and hence finite steps introduce cumulative errors. We believe that the errors introduced in this prediction are fairly small however the process could obviously not be continued indefinitely. Great care should therefore be taken when applying this correction procedure.

Summary and Conclusions

A general method for determining the collapse load for a circular footing on a cohesive soil has been developed using the upper bound method of plasticity. The method is much easier and less costly to implement than more exact procedures. Further it compares well with the available known solutions and with the case history presented. Additional studies are needed to provide verification for a wider range of conditions, thus care should be exercised until the general validity of the solution can be established.

Acknowledgements

We would like to express our appreciation to G. A. Nystrom for his assistance in various aspects of this work.

MARINE FOUNDATIONS

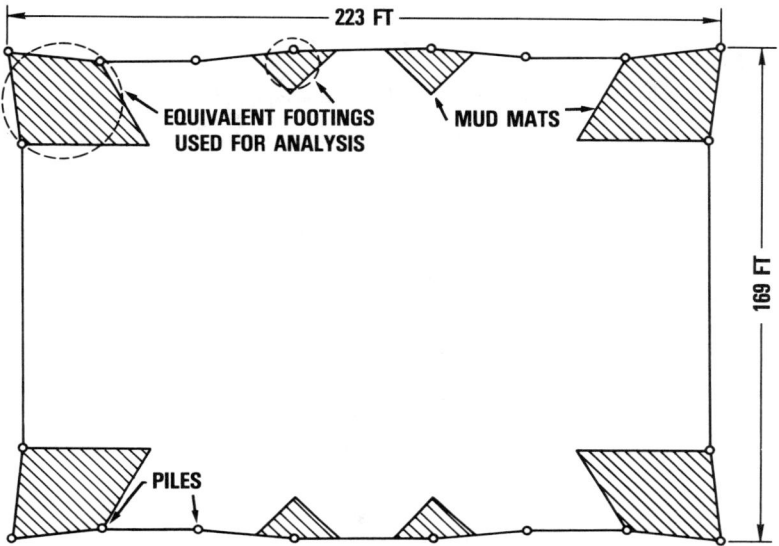

Figure 5 - Plan View of Platform Base Showing Mud Mats

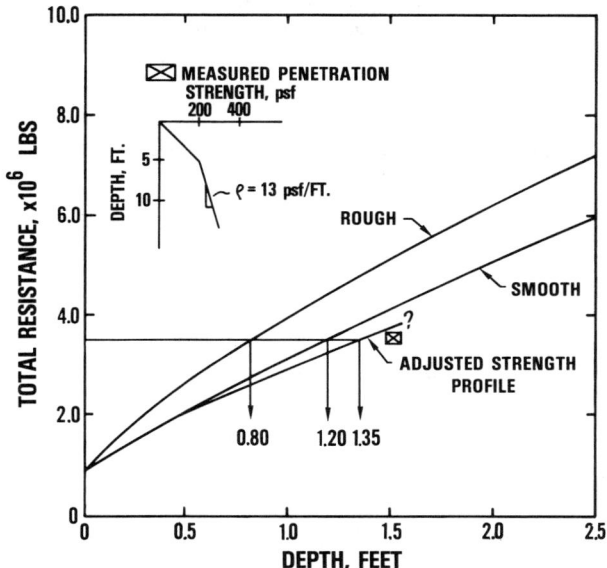

Figure 6 - Results of Case Study

Appendix - Derivation of Velocity Fields

1. Mechanism 1

In each region (Fig. 1) there are two unknown velocity components: v_r and v_z. ($v_\theta = 0$ due to axisymmetry). The condition of incompressibility provides one equation. The second equation is obtained by assuming the form of one or both of the components.

Region 1

Assume that the vertical velocity component varies linearly from the footing-soil interface where it is v_0, the velocity of the footing, to the boundary B-B', where it is zero. This assumption provides the following equation,

$$v_z = v_0 (1 - \frac{z}{h}) \qquad (7)$$

Now incompressibility gives us the other needed condition

$$\dot{\varepsilon}_r + \dot{\varepsilon}_\theta + \dot{\varepsilon}_z = 0$$

or

$$\frac{\partial v_r}{\partial r} + \frac{v_r}{r} + \frac{\partial v_z}{\partial z} = 0$$

which is

$$\frac{\partial v_r}{\partial r} + \frac{v_r}{r} = \frac{v_0}{h}$$

Now in general $v_r = v_r(r,z)$ but we elect to simplify this further by assuming that $v_r = v_r(r)$. This assumption is allowable as long as the boundary conditions can be satisfied. The governing equation is then an ordinary differential equation

$$\frac{dv_r}{dr} + \frac{v_r}{r} = \frac{v_0}{h}$$

the solution of which is well known. Applying boundary conditions, $v_r = 0$ at $r = 0$ results in the following velocity

$$v_r = \frac{v_0}{2h} r \qquad (8)$$

The velocity field is thus determined by equations 7 and 8.

Region 2

A similar approach is used here. We first assume that both v_r and v_z are functions of r alone. Note this is consistent with the required normal velocity at the boundary AB between regions 1 and 2. Incompressibility then gives

$$\frac{dv_r}{dr} + \frac{v_r}{r} = 0$$

Applying the boundary condition imposed by the velocity field in region 1, i.e.

$$v_r = \frac{v_o}{2h} r_i \text{ at } r = r_i$$

v_r can then be determined uniquely, as

$$v_r = \frac{v_o r_i^2}{2hr} \tag{9}$$

Now along boundary BC the velocity must be tangent to the boundary because the material outside region 2 is rigid. This gives the requirement

$$\left.\frac{v_r}{v_z}\right|_{BC} = -\frac{r_o - r_i}{h}$$

Now since v_z and v_r are not functions of z they are constant for a given value of r thus

$$v_z = -\frac{v_r h}{r_o - r_i} = \frac{-r_i^2 v_o}{2r(r_o - r_i)} \tag{10}$$

Equations 9 and 10 thus constitute the velocity field.

Region 3

In region 3 we assume that $v_r = 0$. Therefore incompressibility provides the equation,

$$\frac{\partial v_z}{\partial z} = 0$$

or v_z is a function of r only. It is then only necessary to make v_z compatible with the vertical velocity in region 2. It then follows that

$$v_z = \frac{-r_i^2 v_o}{2r(r_o - r_i)} \tag{11}$$

the same as region 2.

2. Mechanism 2

Mechanism 2 employs the same regions as mechanism 1. Different assumptions are made in deriving the velocity fields of course.

Region 1

Assume that the radial velocity decreases linearly from the soil-footing interface to the bottom of region 1 where it is zero. The radial velocity is then

$$v_r = (1 - \frac{z}{h})f(r)$$

where $f(r)$ is some, as yet unknown, function of r. Now assume that v_z is a function of z only which is consistent with the vertical velocity imposed by the footing and the lower rigid boundary BB'. The incompressibility condition then gives

$$(1 - \frac{z}{h})\frac{\partial f}{\partial r} + (1 - \frac{z}{h})\frac{f}{r} + \frac{\partial v_z}{\partial z} = 0$$

This can be rewritten as

$$\frac{\partial f}{\partial r} + \frac{f}{r} = \frac{h}{h - z}\frac{\partial v_z}{\partial z}$$

The left side of this equation is a function of r only, the right side is a function of z only. Thus each side must be equal to the same constant, say c_1. This results in two ordinary differential equations,

$$\frac{df}{dr} + \frac{f}{r} = c_1$$

$$\frac{h}{h - z}\frac{dv_z}{dz} = c_1$$

Solving these equations and applying the boundary conditions

$$v_z = v_o \quad \text{at } z = 0$$
$$v_z = 0 \quad \text{at } z = h$$
$$v_r = 0 \quad \text{at } r = 0$$

gives the following velocity field

$$v_r = (1 - \frac{z}{h})\frac{v_o r}{h} \tag{12}$$

$$v_z = (1 - \frac{2z}{h} + \frac{z^2}{h^2})v_o \tag{13}$$

Region 2

Assume that the radial velocity varies linearly from $z = 0$ to the lower boundary (r constant) and that it varies linearly from AB to BC (z constant). v_r must also satisfy the boundary condition along AB to be compatible with region 1 and along BC where all velocity components are assumed to be zero. We can then write

$$v_r = \frac{v_0 r_i}{h}(1 - \frac{z}{h})(\frac{r_b - r}{r_b - r_i})$$

where r_b is the radius to the boundary. This can be further simplified to

$$v_r = \frac{v_0 r_i}{h}[\frac{r_0 - r}{r_0 - r_i} - \frac{z}{h}] \quad (14)$$

Incompressibility provides the needed equation

$$\frac{\partial v_z}{\partial z} = \frac{v_0 r_i}{h}[(\frac{1}{r_0 - r_i}) - (\frac{r_0/r - 1}{r_0 - r_i} - \frac{z}{rh})]$$

Note that v_z is a function of r and z and thus integrating with respect to z results in an arbitrary function of r that can be evaluated based on the condition $v_z = 0$ along the BC boundary. The unknown velocity v_z is then

$$v_z = \frac{v_0 r_i}{h}\{[z - h(\frac{r_0 - r}{r_0 - r_i})]\cdot[\frac{2 - r_0/r}{r_0 - r_i}] + \frac{1}{2rh}[z^2 - h^2(\frac{r_0 - r}{r_0 - r_i})^2]\} \quad (15)$$

Region 3

As before we assume that $v_r = 0$ and $v_z = v_z(r)$. The vertical velocity is then that required for compatibility with region 2, i.e. the same as region 2 at $z = 0$,

$$v_z = v_0 r_i [-(\frac{r_0 - r}{r_0 - r_i})\cdot(\frac{2 - r_0/r}{r_0 - r_i}) - \frac{1}{2r}(\frac{r_0 - r}{r_0 - r_i})^2] \quad (16)$$

References

1. Bardgette, J. J. and Irick, J. T., "Construction of the Hondo Platform in 850 Ft of Water in the Santa Barbara Channel", *Preprints, Offshore Technology Conference*, OTC 2959, May 1977.

2. Chen, W. F., *Limit Analysis and Soil Plasticity*, Elsevier Publishing Co., Inc., New York, N.Y., 1975.

3. Davis, E. H. and Booker, J. R., "The Effect of Increasing Strength with Depth on the Bearing Capacity of Clays", *Geotechnique* 23, No. 4, 1973, pp. 551-563.

4. Drucker, D. C. "Coulomb Friction, Plasticity, and Limit Loads," *Journal of Applied Mechanics*, ASME, Vol. 21, 1954, pp. 71-74.

5. Eason, G. and Shield, R. T., "The Plastic Indentation of a Semi-Infinite Solid by a Perfectly Rough Circular Punch", *ZAMP*, Vol. 11 1960, pp. 33-43.

6. Fung, Y. C., *Foundations of Solid Mechanics*, Prentice Hall, Inc., Englewood Cliffs, N.J., 1965.

7. Hansen, J. B., "A Revised and Extended Formula for Bearing Capacity," *Bulletin No. 28*, The Danish Geotechnical Institute, Copenhagen, Denmark, 1970.

8. Murff, J. D., "Upper Bound Analysis of Incompressible, Anisotropic Media," *Proceedings, Fifteenth Annual Meeting of the Society of Engineering Science Inc.*, Gainesville, Florida, 1978, pp. 521-526.

9. Murff, J. D. and Miller, T. W., "Foundation Stability on Non Homogeneous Clays", *Journal of the Geotechnical Engineering Division*, ASCE, Vol. 103, No. GT10, Proc. Pager 13302, Oct., 1977, pp. 1083-1095.

10. Shield, R. T., "On the Plastic Flow of Metals Under Conditions of Axial Symmetry," *Proceedings of the Royal Society, Series A*, Vol. 233, 1955, pp. 267-287.

Expansion of Cavities in Sand

by

H.B. Poorooshasb* and B. Lelievre**

INTRODUCTION

Expansion of cavity solutions have contributed to the understanding and formulation of many geotechnical problems including the bearing capacity of deep foundations, (Vesic (1975)), determination of soil properties, (Gibson and Anderson (1961), Wroth and Hughes (1973) cratering by explosives (Vesic (1965) and the breakout resistance of anchors (Vesic (1971)).

In studies by Vesic (1972) and Baligh (1976) it is assumed that the soils surrounding the cavity are brought to a state of failure within a sphere of radius Rp and that beyond this distance the medium behaves as a linearly elastic material. Both authors take into account the compressibility of the soil using an iteration technique within the failure zone.

In the present paper the medium is assumed to be rigid plastic with a plastic potential ψ and a yield function F where $\psi \neq F$. The initial hydrostatic pressure acting on the medium is assumed to be p and the cavity pressure p_o.

From this study it appears that the system is initially unstable. That is, due to structural collapse of the soil skelton even in the case of a dense sand, an increase in p_o causes indefinite deformations in an infinite soil medium; a phenomenon which appears to have not been appreciated before. The instability proceeds until the state of the stress within the mass readjusts itself to the condition commonly referred to as the at rest earth pressure (k_p) condition. Beyond this point the system is stable although the increase in p_o may bring the state of the elements surrounding the cavity to the "failure" (corresponding to the ϕ_{max}) or even the "critical" (corresponding to the ϕ_{cr}) conditions.

Key Words: Sands, Yield Surface, Plastic Potential, Cavity Expansion, Initial Instability, Soil Expansion and Contraction, Mobilized Angle of Friction

---

* Professor, Concordia University, Dept. of Civil Engg., 1455 de Maisonneuve Blvd. West, Montreal, Quebec, Canada H3G 1M8.

** Associate Professor, University of Waterloo, Dept. of Civil Engg., Waterloo, Ontario, Canada.

The Constitutive Model

Using spherical coordinates, the state of stress, assumed effective, is defined by the two non-zero components σ_r and σ_θ; the radial and tangential components respectively.

According to Poorooshasb et al (1967) and with this choice of stress parameters, the mode of deformation of a cohesionless granular medium can be expressed by the set of equations

$$d\varepsilon_r = h(\eta) \frac{\partial \psi}{\partial \sigma_r} <dF>$$

$$d\varepsilon_\theta = h(\eta) \frac{\partial \psi}{\partial \sigma_\theta} <dF> \quad (1)$$

where $d\varepsilon_r$ and $d\varepsilon_\theta$ are strain increment components, h is the hardening function, ψ is the plastic potential and F the yield function. (Since the material is assumed to be rigid plastic the elastic components of strains are zero and hence $d\varepsilon_{ij} = d\varepsilon_{ij}^p$).

These authors who postulated the existence of potential function ψ and in the form

$$\psi = \sigma_\theta \, \bar{\psi}(\eta) \quad (2)$$

further suggested that

$$F = \eta = \sigma_r/\sigma_\theta \quad (3)$$

as a convenient formulation for the yield function. It was pointed out that such a formulation did, infact, suffer from certain inconsistencies.

For example, it was indicated that in a constant stress ratio test, say k_0 test,

$$dF = d\eta = 0$$

and hence the theory predicted no plastic flow. Yet small, but measurable, plastic strains were always encountered.

Poorooshasb (1971) proposed a revised formulation which fitted the experimental data more closely. According to this, equations 1 & 2 remained, essentially, unchanged whilst equation 3 was rewritten in the form

$$F = \eta + m \ln \sigma\theta \quad (4)$$

where m is a material constant.

In Fig. 1,a are shown the family of F = Constant and ψ = Constant curves for a typical granular medium, satisfying conditions 2 & 3. In Fig. 1,b the same family of curves are represented assuming a value of m = 0.5 in equation 4. Before closing this section it is worth mentioning that all the above formulations are rational if the medium possesses a unique critical voids ratio. Otherwise they are violating a fundamental principle of similarily first proposed by Poorooshasb and Roscoe (1963).

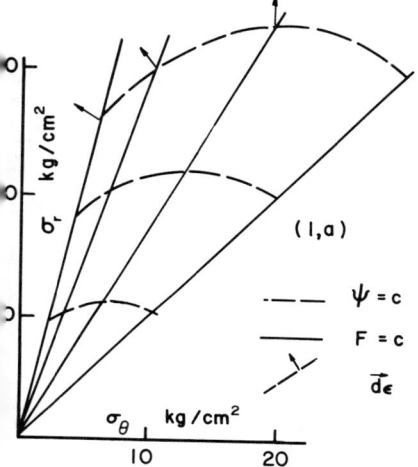
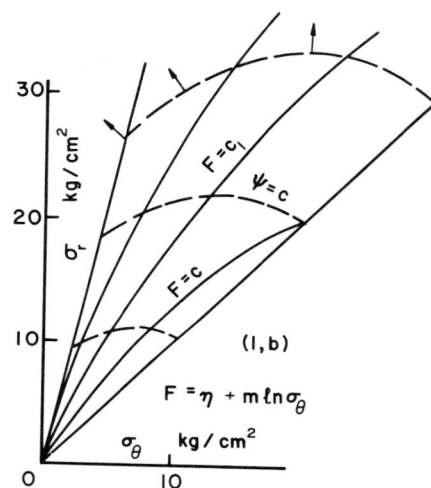
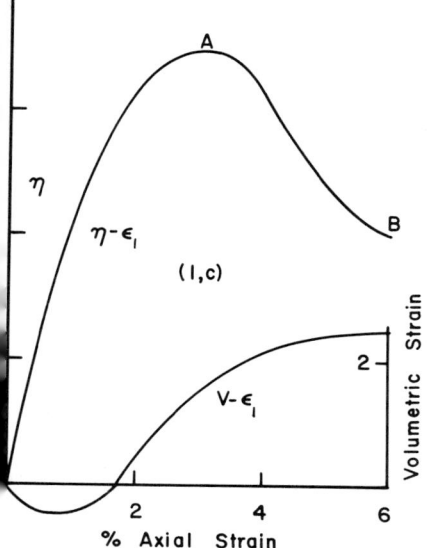

Fig.1,a - Yield and plastic potential formulation by Poorooshasb et al (1967).

Fig.1,b - Formulation by Poorooshasb (1971).

Fig.1,c - "Stress - Strain" curve used in this paper.

Fig. 1 - Soil Properties

Formulation of the Problem

In the absence of body forces σ_r and σ_θ must satisfy the equation

$$\frac{d\sigma_r}{dr} = \frac{2}{r}(\sigma_\theta - \sigma_r) \qquad (5)$$

and strain components ε_r and ε_θ (assumed small) the equation

$$\frac{d\varepsilon_\theta}{dr} = \frac{1}{r}(\varepsilon_r - \varepsilon_\theta) \qquad (6)$$

Since $\sigma_r = \eta\sigma_\theta$, then

$$\frac{d\sigma_r}{dr} = \sigma\theta\frac{d\eta}{dr} + \eta\frac{d\sigma_\theta}{dr}.$$

which upon substitution and rearrangement of terms in equation 5 yields

$$\frac{1}{\sigma_\theta} \cdot \frac{d\sigma_\theta}{dr} = \frac{1}{\eta}[\frac{2}{r}(1-\eta) - \frac{d\eta}{dr}] \qquad (5,a)$$

But, from a combination of equations 1, 2 and 4

$$\frac{d\varepsilon_\theta}{dr} = h(\eta)[\overline{\psi}(\eta) - \eta\overline{\psi}'(\eta)]<\frac{d\eta}{dr} + m\frac{1}{\sigma_\theta}\frac{d\sigma_\theta}{dr}>$$

which upon substitution for $\frac{1}{\sigma_\theta} \cdot \frac{d\sigma_\theta}{dr}$ from (5,a) reduces to

$$\frac{d\varepsilon_\theta}{dr} = F_1(\eta)\,[(\eta-m\frac{d\eta}{dr} + \frac{2m}{r}(1-\eta)] \qquad (7)$$

where

$$F_1(\eta) = \frac{1}{\eta}h(\eta)\quad[\overline{\psi}(\eta) - \eta\overline{\psi}'(\eta)]$$

Similarly it may be shown that

$$\frac{d\varepsilon_r}{dr} = F_2(\eta)\,[(\eta-m)\frac{d\eta}{dr} + \frac{2m}{r}(1-\eta)] \qquad (8)$$

where

$$F_2(\eta) = \frac{1}{\eta}h(\eta)\quad\overline{\psi}'(\eta)$$

Differentiating both sides of equation 6 with respect to r and substituting, in the resulting equation, for $\frac{d\varepsilon_\theta}{dr}$ and $\frac{d\varepsilon_r}{dr}$ from equations 7 and 8 the governing equation is obtained in the form

$$r^2\frac{d}{dr}[(\eta-m)F_1(\eta)\frac{d\eta}{dr}] = \left\{2m[(\eta-1F_1'(\eta)+F_1(\eta)]+(\eta-m)[F_2(\eta)-2F_1(\eta)]\right\}\cdot$$

$$\cdot r\frac{d\eta}{dr} + 2m(\eta-1)\,[F_1(\eta)-F_2(\eta)] \tag{9}$$

For m = 0, corresponding to the yield locii of Fig. 1,a, equation 9 assumes a somewhat simpler form:

$$r\frac{d}{dr}[\eta F_1(\eta)\frac{d\eta}{dr}] = \eta[F_2(\eta)-2F_1(\eta)]\frac{d\eta}{dr} \tag{10}$$

Equation 9 (or 10 for m = 0) may be used to evaluate η as a function of r. Equations 5,a and 3 are then used to find the stress components σ_r and σ_θ. Noting that

$$\varepsilon_r = \frac{du_r}{dr}$$

where u_r is the radial displacement vector, equation 8 may be used to compute U_{r_0}; the expansion of the cavity.

Discussion

For convenience equation 9 is rewritten in the form

$$r^2\frac{d}{dr}[G_1(\eta)\frac{d\eta}{dr}] = G_2(\eta)r\frac{d\eta}{dr} + G_3(\eta) \tag{9,a}$$

Where $G_1(\eta) = (\eta-m)F_1(\eta)$

$G_2(\eta) = 2m[(\eta-1)F_1'(\eta) + F_1(\eta)] + (\eta-m)[F_2(\eta)-2F_1(\eta)]$

and $G_3(\eta) = 2m(\eta-1)[F_1(\eta) - F_2(\eta)]$

Let

$$r\frac{d\eta}{dr} = \xi(\eta) \tag{11}$$

This is permissible in view of the self similarity of the problem. Then equation (9,a) may be rewritten as

$$r^2\frac{d}{dr}[\frac{1}{r}G_1(\eta)\xi(\eta)] = G_2(\eta)\xi(\eta) + G_3(\eta) \tag{12}$$

Denoting $G_1(\eta)\xi(\eta)$ by the auxiliary variable $X(\eta)$ equation (12) reduces to

$$r\frac{dX}{d\eta} \cdot \frac{d\eta}{dr} - X = \frac{G_2 X}{G_1} + G_3$$

which upon rearrangements of terms and also remembering that $r\frac{d\eta}{dr}G_1 = \xi(\eta)G_1(\eta) = X(\eta)$ reduces to

$$X\frac{dX}{d\eta} - (G_1 + G_2)X = G_1 G_3 \tag{13}$$

and

$$\ln(r/r_0) = \int_{\eta_0}^{\eta} \frac{G_1(\eta)}{X(\eta)} d\eta \qquad (14)$$

where η_0 represents the value of σ_r/σ_θ at points in the immediate vicinity of the cavity boundary ($r = r_0$). From equations 13 and 14 the variation of η with r may be obtained through the auxiliary variable $X(\eta)$. If the nature of the soil is such that functions $G_1(\eta)$, $G_2(\eta)$ and $G_3(\eta)$ may be represented by simple analytical expressions then equation (13) and (14) are integrated directly; otherwise they are evaluated numerically.

Reverting to equation (14) it can be shown that for those values of η which are less than k_0, where k_0 is the coefficient of at rest earth pressure for the sand, then the right hand side of equation (14) is negative. This means that no real value of r can be found: the system is unstable and the magnitude of U_{r_0}, the expansion of the cavity is indefinite in an infinite medium.

To gain insight a simple explanation of this phenomenon is given. Consider a normal triaxial test specimen of dry sand subjected to a chamber pressure of σ_c ($\sigma_1 = \sigma_3 = \sigma_c$). Let at this stage the vertical boundary of the specimen be constrained form movement (freeze the chamber fluid around the specimen say). Now smallest increase in the vertical stress component σ_1 (even a small perturbation) would cause collapse of the sample until the value of σ_3 assumes a value equal to $k_0\sigma_1$. Since in the spherical expansion problem the particles are constrained to move in a radial direction only then a situation similar to the one just described arises at the initial stages of loading.

Because of this phenomen it is necessary to **as**sume that the volume of sand being considered is finite which leads to the concept of radius of influence and sphere of influence. The radius of influence r_i is defined as the radius beyond which the effect of the cavity expansion is not felt. The sphere of influence is a sphere with a radius r_i. (This radius is analogous to the radius of influence used in the analysis of the yield of the wells). In the subsequent analysis a value of r_i equal to $10r_0$ is assumed.

Variation of ϕ_m with r

Once the system has obtained stability i.e., $\sigma_\theta/\sigma_r = k_0$ everywhere within the sphere of influence, further increase of η_0 would produce definite expansion of the cavity. In Fig. 2 are shown the curves representing the variation of $\eta(r)$ for several values of η_0 and based on the data provided in Fig. 1,b with a value of parameter m = 0.5. The hardening function, h of equation (1), has been so chosen as to be compatible with the stress strain curve shown in Fig. 1,c and the plastic potential curves of Fig. 1,b (or 1,a).

Let ϕ_m represent the mobilized angle of friction within r_i,

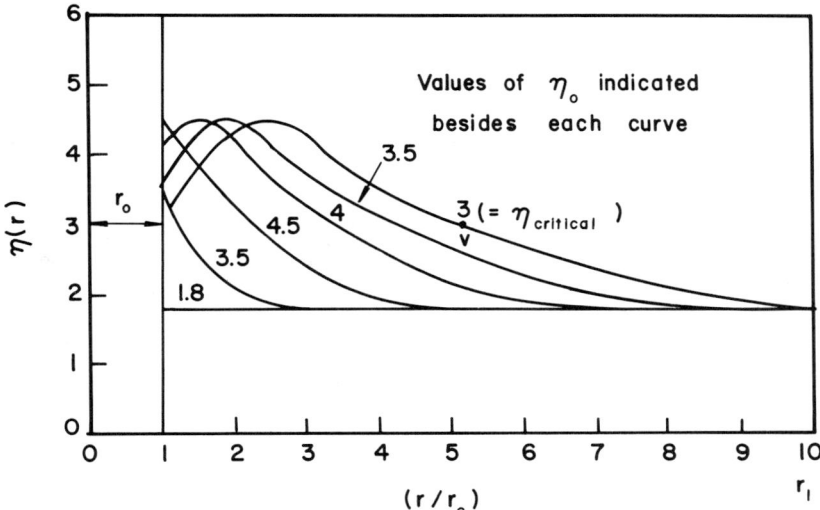

Fig. 2 - Variation of $\eta(r)$ for several values of η_o

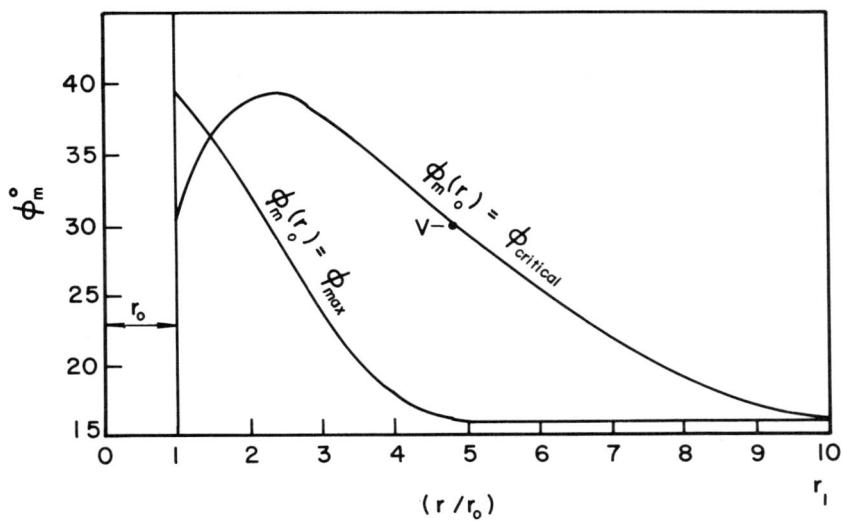

Fig. 3 - Variation of mobilized angle of friction with (r).

then

$$\eta(r) = \frac{\sigma_r}{\sigma_\theta} = \frac{1 + \sin\phi_m}{1 - \sin\phi_m}$$

or

$$\phi_m(r) = \sin^{-1}\frac{\eta(r) - 1}{\eta(r) + 1}$$

In Fig. 3 variation of the mobilized angle of friction with r is shown for the two cases of $\phi_m(r_0) = \phi_{critical}$ and $\phi_m(r_0) = \phi_{max}$. An interesting feature of this diagram is that the maximum angle of friction is not necessarily mobilized at the elements in the immediate vicinity of the cavity boundary. Indeed the cavity may be subjected to increasing pressures even after the state where these elements have "failed". That is, while the elements adjacent to cavity may have become unstable yet the system as a whole is stable. The results shown in Fig. 3 are also in variance with the basic assumptions made by Vesic (1972) who assumes a constant value of ϕ and by Baligh (1976) who allows some variation in the magnitude of ϕ. The variation allowed by Baligh, however, is due to curvature of the Mohr-Coulomb failure surface and differs, in principle, from the results presented in this paper.

Variation of Stress Components

In Fig. 4 is shown the variation of stress component $\sigma_r(r)$. To obtain $\sigma_\theta(r)$ the ordinates of this figure may be divided by the corresponding η values of Fig. 2.

An interesting feature of Fig. 4 is that the cavity pressure $\sigma_r(r_0)$ sustained at $\phi(r = r_0) = \phi_{critical}$ is nearly twice of that sustained at $\phi(r_0) = \phi_{max}$.

Stress Path and Volumetric Strains

The stress path followed by an element in the immediate vicinity of the cavity boundary and the corresponding volumetric strains experienced by the same element are shown in Fig. 5. Note that in this figure the stress component σ_θ is represented by the ordinate, σ_r by the abscissa (in variance with coordinates used in Fig. 1,a and Fig. 1,b) so that a direct comparison with Fig. 5 of Vesic (1972) paper may be made. While some similarities in the shape of the stress paths followed exist between the two figures (Fig. 5 of this paper and Fig. 5 of Vesic's paper) the "volumetric strain-stress paths" are in direct conflict. A detailed discussion of these discrepancies is beyond the scope of the present paper. It is merely stated that Vesic's theory appears to be applicable, at best, to a precompressed soil, i.e. to a soil which has been first subjected to a very high hydrostatic pressure and then unloaded. In this way the initial instability discussed above is removed, the soil is within a yield surface (say $F = C_1$ of Fig. 1,b) and hence application of the elastic theory in the initial loading stages, as assumed by Vesic and Baligh, appears

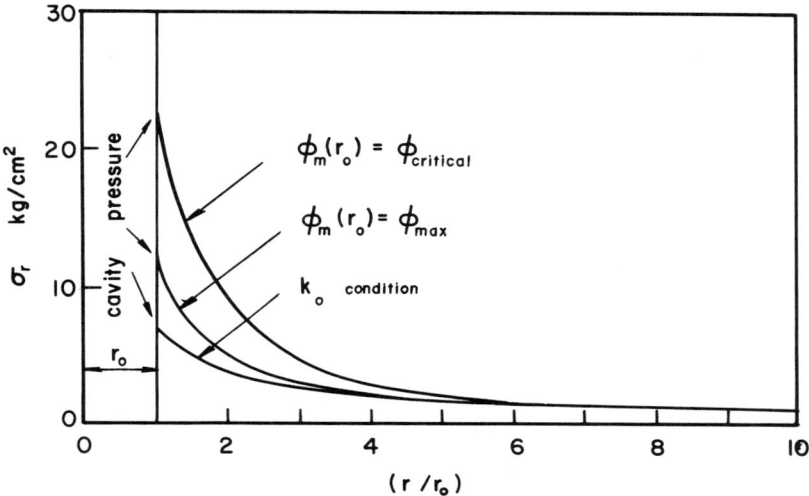

Fig. 4 - Variation σ_r (r) for various cavity pressures.

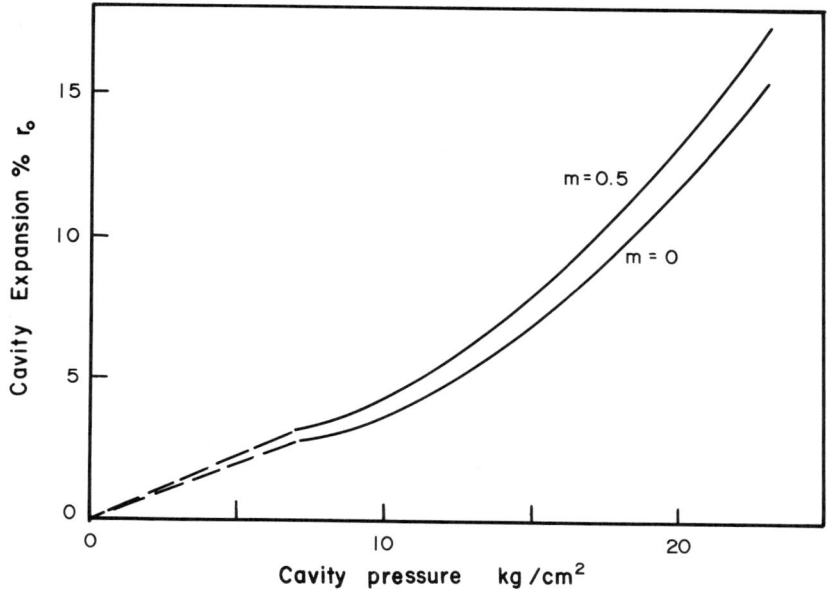

Fig. 6 - Expansion of cavity boundary as a function of cavity pressure.

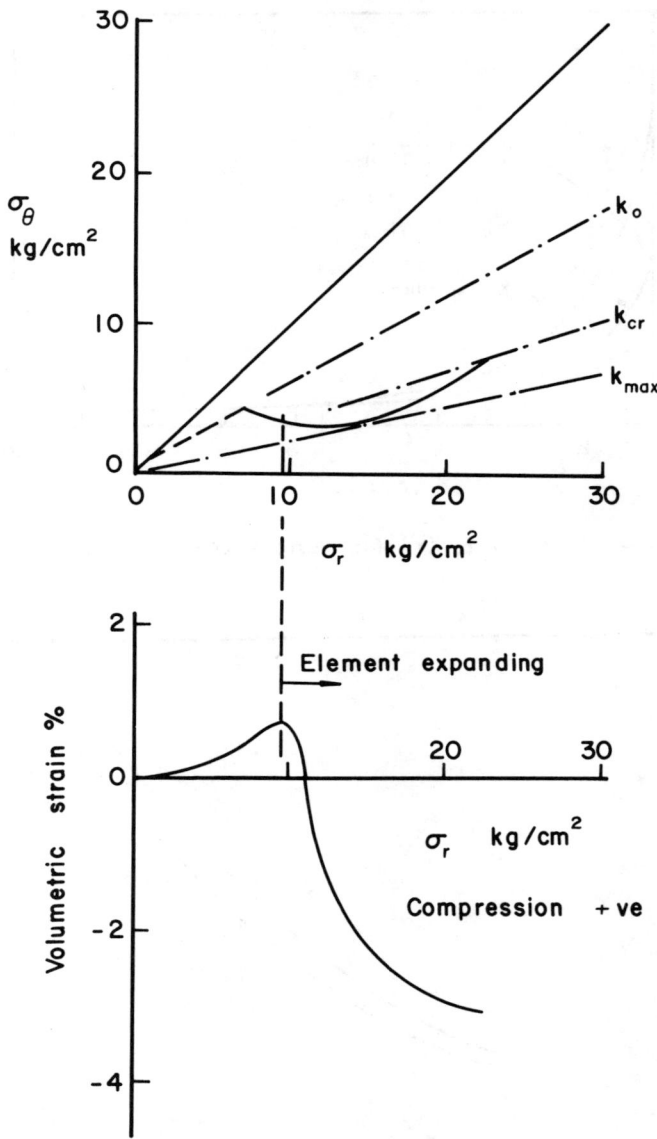

Fig. 5 — Stress path for and volumetric strain of an element adjacent to the cavity boundary.

to be in order.

Finally it is noted that the volumetric expansion of elements near to the cavity boundary is accommodated by contractions of elements further away for which the value of ϕ_m is less then ϕ_{cr} (elements situated to the right of point (V) of Fig. 3 when $\phi(r_0) = \phi_{critical}$).

Expansion of Cavity

Since
$$\frac{du_r}{dr} = \varepsilon_r$$

then from equation (8)

$$u_r = \int_{r_0}^{r_i} dr \left\{ \int_{\eta_0}^{\eta_i} F_2(\eta) [\eta-m] d\eta + 2m \int_{r_0}^{r_i} \frac{1-\eta(r)}{r} F_2[\eta(r)] dr \right\}$$

where $F_2[\eta(r)]$ is now a known function of r since $\eta(r)$ is known. When m = 0 this expression reduces to simple form

$$u_r = \int_{r_0}^{r_i} \int_{\eta_0}^{\eta_i} F_2(\eta) \, \eta \, d\eta \, dr$$

In Fig. 6 the expansion of cavity radius expressed as percentage of the initial radius is plotted for the two cases of m = 0 and m = 0.5. As may be seen the influence of m which is an indication of the curvature of the yield surface is not very strong. For example at a cavity pressure of 12.6 kg/cm^2 (at which pressure $\phi(r_0) = \phi_{max}$) the simple model (m = 0) predicts a volumetric expansion of 16.3% while the modified model with an m value of 0.5 predicts 18.6%. In other words the simple model of the stress strain relationship underestimates the expansion by about 12% as compared to the value obtained from the modified model. Thus, at least in a certain class of problems, the convenient formulation of Poorooshasb et al (1966, 67) is sufficiently accurate. This fact has been utilized by Meyerhof (1980) and by Poorooshasb, Meyerhof and Hanna (1980).

Conclusions

1. Expansion of a spherical cavity in a mass of sand subjected to an initial hydrostatic pressure is examined using a rigid-plastic constitutive model.
2. At the initial loading stages, the system goes through an instability phase until the pressure inside the mass attains the so called k_0 condition.
3. This instability phase would not be present in a precompressed (strain hardened) soil mass.
4. As the loading is further increased the sand elements surrounding the cavity may experience a range of values of mobilized angle of friction ϕ_m. For sufficiently high cavity pressures the value of ϕ_m in the immediate vicinity of the

cavity will be equal to $\phi_{critical}$ with ϕ_{max} mobilized at a certain distance away from the cavity boundary. Although the individual elements may be in an unstable (for even critical) state the mass as a whole is stable.
5. The curvature of the yield locii does not appear to have a strong influence on the magnitude of the cavity expansion.

Acknowledgment

The financial support from the National Science and Engineering Research Council of Canada is gratefully acknowledged.

References

(1) Baligh, Mohsen, M., "Cavity Expansions in Sands with Curved Envelopes" Journal of the Soil Mechanics and Foundation Division, ASCE, Vol. 102, No. GT11, Proc. Paper 12536, November 1976, pp. 1131-1146.
(2) Gibson, R.E., and Anderson, W.F., "In Situ Measurements of Soil Properties with the Pressuremeter", Civil Engineering and Public Works Review, Vol. 56, 1961, pp. 615-618.
(3) Meyerhof, G.G., "Limit Equilibrium Plasticity Application in Geotechnical Engineering", State of the Art Report, ASCE Annual Convention, Hollywood by the Sea, Florida, 1980.
(4) Poorooshasb, H.B., Meyerhof, G.G., and Hanna, A.M., "Settlement of a Strip Footing on a Two Layer Soil System", Paper to be presented at the 33rd Canadian Geotechnical Conference, 1980.
(5) Poorooshasb, H.B., "Deformation of Sand in Triaxial Compression", Proc. Fourth Asian Regional Conference on Soil Mechanics and Found. Eng., Bangkok, Thailand, Volume 1, 1971, pp. 63-66.
(6) Poorooshasb, H.B., Holubec, I., Sherbourne, A.N., "Yielding and Flow of Sand in Triaxial Compression", Parts II and III, Canadian Geotechnical Journal, Vol. IV, No. 4, 1967, pp. 376-397.
(7) Poorooshasb, H.B. and Roscoe, K.H., "A Fundamental Law of Similarity in Soil Mechanics", Proc. Second Asian Regional Conference on Soil Mechanics and Foundation Engineering, Tokyo, Japan, Vol. 1, 1963.
(8) Vesic, A.S., "Cratering by Explosives as an Earth Pressure Problem", Sixth International Conference on Soil Mechanics and Foundation Engineering, Montreal, Canada, Vol. 2, 1965, pp. 427-431.
(9) Vesic, A.S., "Breakout Resistance of Objects Embedded in Ocean Bottom", Journal of the Soil Mechanics and Foundation Division, ASCE, Vol. 97, No. SM9, Proc. Paper 8372, Sept. 1971, pp. 1183-1206.
(10) Vesic, A.S., "Expansion of Cavities in Infinite Soil Mass", Journal of the Soil Mechanics and Foundation Division, ASCE, Vol. 98, No. SM3, Paper 8790, March 1972, pp. 265-290.
(11) Vesic, A.S., "Principles of Pile Design", Lecture Series on Deep Foundations Sponsored by Geotechnical Group, Boston Society of Civil Engineers/American Society of Civil Engineers in cooperation with Massachusetts Institute of Technology, Cambridge, Mass., 1975.

(12) Wroth, C.P., and Hughes, J.M.O., "An Instrument for the In Situ Measurement of the Properties of Soft Clays", Proceeding, 8th International Conference in Soil Mechanics and Foundation Engineering, Moscow, U.S.S.R., Vol. 1, pp. 487-494.

PLASTIC-LIMIT EQUILIBRIUM STATES IN SOIL MEDIA

Jean H. Prevost[1], M. ASCE

ABSTRACT

It is demonstrated that elastic-plastic failure states in soil media may be captured in finite element models by employing (1) the elastic-plastic material stiffness to form the global stiffness, (2) reduced/selective integration techniques to alleviate mesh "locking" due to incompressibility, and (3) in the case of symmetrical configurations, an imperfection in the form of a weak element. The technique is illustrated by applying it to analyze the formation of shear bands in tensile specimen, bearing capacity and slope stability problems.

INTRODUCTION

Plasticity theory provides the rational design basis whereby problems of ultimate strength can be solved exactly. Use of the theory for solving practical engineering problems requires (1) knowledge of the proper field and constitutive equations, and (2) availability of a general solution procedure applicable to the specific problem at hand. Early works in plasticity only dealt with the simplest class of plastic materials, viz, isotropic elastic-perfectly plastic materials. In that case, the behavior of the real material is idealized by assuming that it behaves like a linear isotropic elastic solid until the state of stress reaches a critical condition defined by the failure criterion, after which it flows plastically. However, although this theory is the simplest, no general analytical method could then be developed for solving general boundary value problems involving an isotropic elastic-perfectly plastic solid body. Until recently, exact solution had only been obtained for trivial problems which are one-dimensional or ones which involve proportional loading conditions. Various simplified methods of analysis have thus been proposed. The most widely used in geotechnical engineering practice are the method of characteristics and limit equilibrium techniques.

KEY WORDS: Analysis, bearing capacity, failure states, finite elements, Geotechnical engineering, numerical methods, Plasticity, shear bands, slip lines, slope stability, stress-strain behavior.

---

[1] Assistant Professor of Civil Engineering, Princeton University, Princeton, NJ 08544

In the method of characteristics (see e.g. Ref. [6]), the deformations undergone by the solid body prior to failure are neglected and a failure stress criterion is postulated. The stress conditions at failure in the soil mass, for statically determinate problems, can then be obtained simply by solving a set of nonlinear hyperbolic differential equations [21]. A velocity field can be associated to the set of stress characteristics by using the plastic flow rule, and the discontinuities in the resulting slip-line solution can be used to define the rupture surface. Exact analytical solutions for the set of stress characteristics are difficult to obtain except for simple cases, but a numerical procedure can be used to resolve that dilemma [9,21]. Field observations that failure of soil masses often occur along well-defined rupture surfaces have led to simplified practical limit equilibrium analysis procedures [1,2] which make use of simple static considerations.

Although these methods are widely used in geotechnical engineering practice, they neglect altogether the influence of the deformations undergone by the solid body onto the failure conditions, and cannot provide often necessary design informations about deformations prior to failure. The recent development of numerical techniques such as the finite element method has now rendered possible, in principle, the solution of any properly posed boundary value problems in continuum mechanics. Numerous numerical solutions for elastic-plastic problems have thus been proposed in the recent literature. However, it appears that many of these solutions are deficient

(1) in not converging towards a limit load (when such a limit load exists) but rather rising steadily and attaining values far in excess of the true limit load, and

(2) in not exhibiting localization of deformation phenomena when such localizations should occur.

The fact that the numerical solution does not exhibit a limit or too high a limit load indicates that the computed stiffness is larger than it should be and/or does not become singular when it should. A possible cause for this was pointed out in [15] and is related to the fact that special care must be taken in the numerical formulation in order to be able to handle the incompressible plastic flow which takes place at failure. New techniques have now been devised to deal successfully with incompressibility constraint requirements [10]. However, even when using these techniques, numerical solutions have not seemed capable of capturing localization phenomena, thus prompting several investigators [7,20], to devise special purpose finite element procedures.

It is the purpose of this paper to:

(1) investigate the ingredients necessary to capture failure states accurately by numerical methods, and

(2) to demonstrate that limit loads and localization phenomena can be captured successfully by finite element models.

The most salient features of the following derivations have already been reported in [18] but some are repeated here for completeness and further expanded.

For simplicity in this presentation attention is first restricted to small strains/deformations. As for notation, boldface letters denote vectors, second-order and fourth-order tensors in three dimensions, and a symbolic notation is used throughout (see e.g. Ref. [11]).

PRELIMINARIES

The form of the plasticity equations first proposed by Melan [14] in 1938 is given as follows:

$$\underline{d}^p = <L> \underline{P} \tag{1}$$

in which $\underline{d}$ = symmetric part of the velocity gradient; $\underline{P}$ = dimensionless symmetric second-order tensor normalized in such a way that $\underline{P}:\underline{P}=1$ and such that $\underline{P}$ gives the direction of plastic deformations, and a superscript p is used to denote plastic rate of deformations. In Eq. 1, L is the loading functions

$$L = \frac{1}{H'}, \underline{Q} : \underline{\dot{t}} \tag{2}$$

in which H' = plastic modulus; $\underline{\dot{t}}$ = material derivative of Cauchy stress; $\underline{Q}$ = dimensionless symmetric second-order tensor normalized in such a way that $\underline{Q}:\underline{Q}=1$ and such that $\underline{Q}$ is the outer normal to the yield surface in stress space; and the symbol <> denotes the MacCauley's bracket, viz., <L> = L if L $\geq$ 0, otherwise <L> = 0. For a plastic hardening case, H' > 0, whereas H' < 0 for a softening case. When H'=0, a perfectly-plastic case is obtained.

Before proceeding any further, it is of importance to note that, as a consequence of the normality rule, the plasticity equations are singular, i.e., cannot be inverted, as shown hereafter. From Eqs. 1 and 2,

$$H'd^p_{ab} = P_{ab}Q_{cd}\dot{t}_{cd} = K_{abcd}\dot{t}_{cd} \tag{3}$$

The resulting fourth-order tensor $\underline{K}$ is equal to the outer product of the two second-order tensors $\underline{P}$ and $\underline{Q}$, and therefore has 81 components. However, due to the symmetry of both P and Q tensors, there are at most 36 distinct coefficients. They may be displayed as the components of a 6×6 matrix as

$$K_{AB} = P_A Q_B \quad (A,B = 1,\ldots,6) \tag{4}$$

in which $P_A(Q_B)$ is the vector whose components are the six distinct components of the tensor $P(Q)$, i.e., $P_1=P_{11},\ldots,P_4=P_{12}=P_{21},\ldots,P_6=P_{31}=P_{13}$. The following theorem may then be proven:

THEOREM: The K matrix defined by $K_{AB}=P_A Q_B$ is singular and only has one distinct eigenvalue non-equal to zero.

PROOF: Let X_A denote the eigenvector associated with the eigenvalue λ such that

$$K_{AB} X_B = \lambda X_A \tag{5a}$$

Then (from Eq. 4)

$$P_A Q_B X_B = \lambda X_A \tag{5b}$$

Let

$$p = Q_B X_B \tag{5c}$$

Eq. (5b) then writes

$$p P_A = \lambda X_A \tag{5d}$$

Two cases are to be examined:

(1) $p = \lambda$. Eq. 5d then has the solution

$$X_A = P_A \tag{5e}$$

and $p = Q_B P_B$ is the scalar product of vectors Q and P with components Q_B and P_B. $\lambda = Q_B P_B$ is thus an eigenvalue of the matrix K_{AB} associated with the eigenvector $X_A=P_A$. This vector is unique, and the corresponding eigenvalue is of the order one.

(2) $p = 0$. Then $\lambda = 0$ and the X_A are related by the equation

$$Q_B X_B = 0 \tag{5f}$$

$\lambda = 0$ is thus a multiple eigenvalue of order 6-1=5, the corresponding eigenvectors being all normal to Q_B.

From the above, the resulting characteristic equation writes

$$D = \det(K_{AB} - \lambda \delta_{AB}) = k\lambda^{n-1}[Q_A P_A - \lambda] \quad (5g)$$

in which n=6 and k is a scalar function of n. When D is expanded, the coefficient of λ^n is $(-1)^n$ and thus $k=(-1)^{n-1}$ and finally

$$D = (-1)^5 \lambda^5 [Q_A P_A - \lambda] \quad (5h)$$

The value of the determinant associated with the $\underset{\sim}{K}$ matrix is given by the sum

$$\det(\underset{\sim}{K}) = \sum \pm K_{A1} K_{B2} \cdots K_{F6} \quad (5i)$$

where (A,B,...,F) is a permutation of (1,2,...,6) and either the plus (+) or minus(-) sign is used according to whether the permutation is even or odd. Now from Eq. 4,

$$\det(\underset{\sim}{K}) = \det(P_A Q_B) = Q_1 Q_2 \cdots Q_6 \left[\sum \pm P_A P_B \cdots P_F \right] = 0 \quad (5j)$$

which completes the proof.

The plasticity equations alone are therefore singular, and cannot be inverted to yield a purely plastic material stiffness. In order to be able to derive such a stiffness, the material's elasticity must also be taken into account. It is a fundamental assumption of the theory of plasticity that the total rate of deformation tensor in the plastic range may be decomposed into the sum of elastic and plastic rates of deformations as

$$\underset{\sim}{d} = \underset{\sim}{d}^e + \underset{\sim}{d}^p \quad (6)$$

in which the superscript e and p are used to denote the elastic and plastic rage of deformation components. Note that this decomposition is valid for both small and large deformation problems. The elastic rates of deformation are given by the following equation

$$\underset{\sim}{d}^e = \underset{\sim}{E}^{-1} : \underset{\sim}{\dot{t}} \qquad d^e_{ab} = E^{-1}_{abcd} \dot{t}_{cd} \quad (7)$$

in which $\underset{\sim}{E}^{-1}$ denotes the fourth-order elasticity tensor which, in general, must possess the following symmetries

$$E^{-1}_{abcd} = E^{-1}_{bacd} = E^{-1}_{abdc} = E^{-1}_{cdab} \quad (8)$$

e.g., for isotropic elasticity

$$E^{-1}_{abcd} = -\frac{\Lambda}{2G(3\Lambda + 2G)} \delta_{ab}\delta_{cd} + \frac{1}{4G}(\delta_{ab}\delta_{bd} + \delta_{ad}\delta_{bc}) \quad (9)$$

in which Λ, G = Lamé's constants.

In order to be able to separate the contributions of the elastic and plastic properties in the total deformations, it is commonly assumed that the elasticity of the material is isotropic and linear. Anisotropic and nonlinear effects are assumed to be due to the material's plasticity. The elastic-plastic flexibility tensor defined by Eqs. 1,2,6-9 is non-singular and can be inverted to yield the material's stiffness as

$$\dot{t} = \underset{\sim}{C}:\underset{\sim}{d} = \underset{\sim}{E}:\underset{\sim}{d} - \frac{\underset{\sim}{E}:\underset{\sim}{P}}{H' + \underset{\sim}{Q}:\underset{\sim}{E}:\underset{\sim}{P}} \underset{\sim}{Q}:\underset{\sim}{E}:\underset{\sim}{d} \qquad (10)$$

in which (from Eq. 9)

$$E_{abcd} = \Lambda \delta_{ab}\delta_{cd} + G(\delta_{ac}\delta_{bd} + \delta_{ad}\delta_{bc}) \qquad (11)$$

Note that when $\underset{\sim}{P} \neq \underset{\sim}{Q}$ (i.e., a non-associative plastic flow rule is used), the $\underset{\sim}{C}$ tensor does not exhibit the major symmetry and therefore leads to a non-symmetric stiffness matrix. On the other hand, when P=Q (i.e., an associative plastic flow rule is used), the C˜tensor possesses the major symmetry and leads to a symmetric̃ stiffness matrix. For the simple case of an isotropic elastic-plastic material which yields according to the Von Mises criterion [6], viz.

$$\frac{3}{2} \underset{\sim}{s}:\underset{\sim}{s} - k^2 = 0 \qquad (12)$$

where $\underset{\sim}{s}=\underset{\sim}{t}-\frac{1}{3}$ trace $(\underset{\sim}{t})\underset{\sim}{1}$ = deviatoric stress tensor, Eq. 10 simplifies to:

$$\dot{\underset{\sim}{t}} = 2G\underset{\sim}{d} + \Lambda \text{ (trace } \underset{\sim}{d})\underset{\sim}{1} - 2G\frac{2G/H'}{1+2G/H'} \frac{3}{2k^2} \underset{\sim}{s}(\underset{\sim}{s}:\underset{\sim}{d}) \qquad (13)$$

for an associative plastic flow rule.

For cases in which finite deformations take place, the elastic-plastic constitutive equations (Eq. 10) are written in the following form

$$\overset{\triangledown}{\underset{\sim}{t}} = \underset{\sim}{C}:\underset{\sim}{d} \qquad (14)$$

where $\overset{\triangledown}{\underset{\sim}{t}}$ denotes the Truesdell derivative viz.

$$\overset{\triangledown}{\underset{\sim}{t}} = \dot{\underset{\sim}{t}} + \underset{\sim}{t}\cdot\underset{\sim}{L} - \underset{\sim}{L}\cdot\underset{\sim}{t} + \underset{\sim}{t} \text{ div } \underset{\sim}{v} \qquad (15)$$

in which $\underset{\sim}{L}$ = grad $\underset{\sim}{v}$ = (spatial) velocity gradient tensor; $\underset{\sim}{v}$ = spatial velocity. The finite deformation form of the plasticity equation above was first proposed by Hill [4] and has been advocated by McMeeking and Rice [13].

Elastic-plastic equations of the above type (Eqs. 10 and 14) lead to the following equations for continuing equilibrium

$$\text{div}(D:L) + \dot{\rho}\, b = 0 \quad , \tag{16}$$

where ρ = density; b = body force vector; and in component form

$$D_{abcd} = \begin{cases} C_{abcd} & \text{for small deformations} \\ C_{abcd} + \frac{1}{2}(t_{bd}\delta_{ac} - t_{ad}\delta_{bc} - t_{cb}\delta_{ad} - t_{ac}\delta_{bd}) & \text{for finite deformations} \end{cases} \tag{17}$$

where δ_{ab} = Kronecker delta. It may be seen that D possesses the major symmetry which leads to a symmetric tangent stiffness matrix, if and only if C possesses the major symmetry (i.e., when an associative plastic flow rule is assumed).

LIMIT LOAD

A state of failure is reached when deformations start to occur under constant surface tractions and/or body forces. For the elastic-plastic solid body to eventually reach such a state, the material stiffness must be singular, i.e.

$$C:d = 0 \tag{18}$$

must possess a non-trivial solution, so that the global stiffness of the solid body may be singular.

THEOREM: If $Q:E:P \neq 0$, the material elastic-plastic stiffness is singular when $H'=0$.

PROOF: The material stiffness is singular when the C tensor is non-definite, i.e., when

$$X:C:X = 0 \quad \text{with } X \neq 0 \tag{19}$$

where X is a symmetric second-order tensor. Select $X=Q$, Eq. 10 then yields

$$Q:C:Q = H' \frac{Q:E:Q}{H' + Q:E:P} \tag{20}$$

and therefore when $H'=0$, $Q:C:Q = 0$ which completes the proof.

It is therefore apparent that if the correct limit state is to be detected by the numerical solution, it is helpful to use the elastic-plastic material stiffness rather than any other algorithmically convenient stiffness to form the global stiffness (as in initial-stress type methods). However, to advance the solution into the post-bifurcation regime requires algorithmic strategies which are not considered herein.

LOCALIZATION PHENOMENA

The basic theoretical principles for understanding the localization phenomenon are contained in Refs. [3-5,12,19,22] where it is shown that its existence in elastic-plastic solids is contingent upon the loss of ellipticity of the velocity equations of equilibrium, i.e., localization is to occur when [5]

$$\det(\underset{\sim}{n} \cdot \underset{\sim}{D} \cdot \underset{\sim}{n}) = 0 \qquad (21)$$

in which det = determinant, and $\underset{\sim}{n}$ = unit vector, whose orientation defines a *characteristic* curve for the equations of continuing equilibrium $(D_{abcd} V_{c,d})_{,a} = 0$, where $\underset{\sim}{v}$ = spatial velocity. For an elastic-plastic material, Eq. 19 imposes that in the small deformation regime [19]

$$\frac{H'}{2G} = 2 \underset{\sim}{n} \cdot \underset{\sim}{P} \cdot \underset{\sim}{Q} \cdot \underset{\sim}{n} - (\underset{\sim}{n} \cdot \underset{\sim}{P} \cdot \underset{\sim}{n})(\underset{\sim}{n} \cdot \underset{\sim}{Q} \cdot \underset{\sim}{n}) - \underset{\sim}{P} : \underset{\sim}{Q} \qquad (22)$$

$$- \frac{\Lambda}{\Lambda + 2G} [(\underset{\sim}{n} \cdot \underset{\sim}{P} \cdot \underset{\sim}{n}) - \text{trace } \underset{\sim}{P}][(\underset{\sim}{n} \cdot \underset{\sim}{Q} \cdot \underset{\sim}{n}) - \text{trace } \underset{\sim}{Q}]$$

It is of importance to note that the plasticity equations of equilibrium are the ones which lose ellipticity. This again suggests that if localization phenomena are to be captured by the numerical solution, it is helpful to use the elastic-plastic material stiffness rather than any other algorithmically convenient stiffness.

NUMERICAL EXAMPLES

In the following sections, a number of examples are presented which demonstrate that both limit loads and localization phenomena in elastic-plastic solid bodies may be accurately captured by a finite element solution of the velocity equations of equilibrium. For that purpose, the finite element code DIRT[6] is used. Elastic-plastic equations lead directly to the definition of tangent stiffness matrix (see e.g. [13]), and an incremental predictor-corrector type algorithm is adopted [8]. The element and material model libraries are modularized and may be easily expanded without alteration of the main code. The present element library contains a two- and a three-dimensional element and full finite deformation effects may be accounted for. A contact element is also available for two- and three-dimensional analysis. The present material library contains a linear elastic model and various elasto-plastic and soil models. Some features which are available in the program are

- o Both symmetric and non-symmetric matrix equation solvers
- o Reduced/selective integration procedures, for effective treatment of incompressibility constraints [10].

In the following calculations, the four node bilinear isoparametric element is used with the standard selective integration scheme [8,10].

1. Localization of Deformations into Shear Bands [18]

Numerical results which illustrate the phenomenon of localization of deformation into shear bands for a rectangular block constrained to plane deformations and subjected to tension in one direction are presented hereafter.

In order to make the present study quite specific, the material is modeled here as an incompressible isotropic elastic-plastic Prandtl-Reuss material (Eq. 13). Fig. 1a shows the two-dimensional finite element representation of the tensile specimen. The grid consists of 171 bilinear isoparametric rectangular elements. The specimen length to width ratio is equal to two, and 9 elements are placed across the width. Uniform longitudinal end displacements are prescribed, and no shearing tractions are applied. The lower left corner of the specimen is fixed, and the loading is accomplished by imposing increments of displacement at the upper end of the specimen.

For the particular case of the Prandtl-Reuss material, loss of ellipticity of the velocity equations of equilibrium in the small deformation regime is achieved simply by selecting a plastic modulus less or equal to zero, and in the following $H'/2G=-0.048$. The corresponding angle for the plane of localization is then 38.7°. The assumed stress-strain curve is shown in Fig. 1b.

In a first attempt to obtain localization, both the material properties and the end displacements were taken as uniform. These conditions resulted in smooth and continuously varying deformation patterns well into the softening range, but no localization occurred. This result may be interpreted by recalling that the loss of ellipticity is a <u>necessary</u> but not a sufficient condition for localization. Some type of non-uniformity (perturbation) is necessary in order to trigger the phenomenon. In the following, localization is achieved by introducing a weak element which plays the role of a local "imperfection" in the material properties. This element is located either at the center (series C) or at the side (series D) of the specimen as shown in Fig. 1, and its plastic modulus is such that $H'/2G=-1/3$. Typical results are shown in Fig. 2. The spreading of the plastic zone is indicated by a shaded area. In Fig. 2 the imperfection is located at the center of the specimen and the imperfection has the same strength as its surrounding. In Fig. 2a, the axial strain = .100%, and all the elements have yielded. Upon further loading, localization occurs along one shear band as shown in Fig. 2b, where

PLASTIC-LIMIT EQUILIBRIUM STATES 175

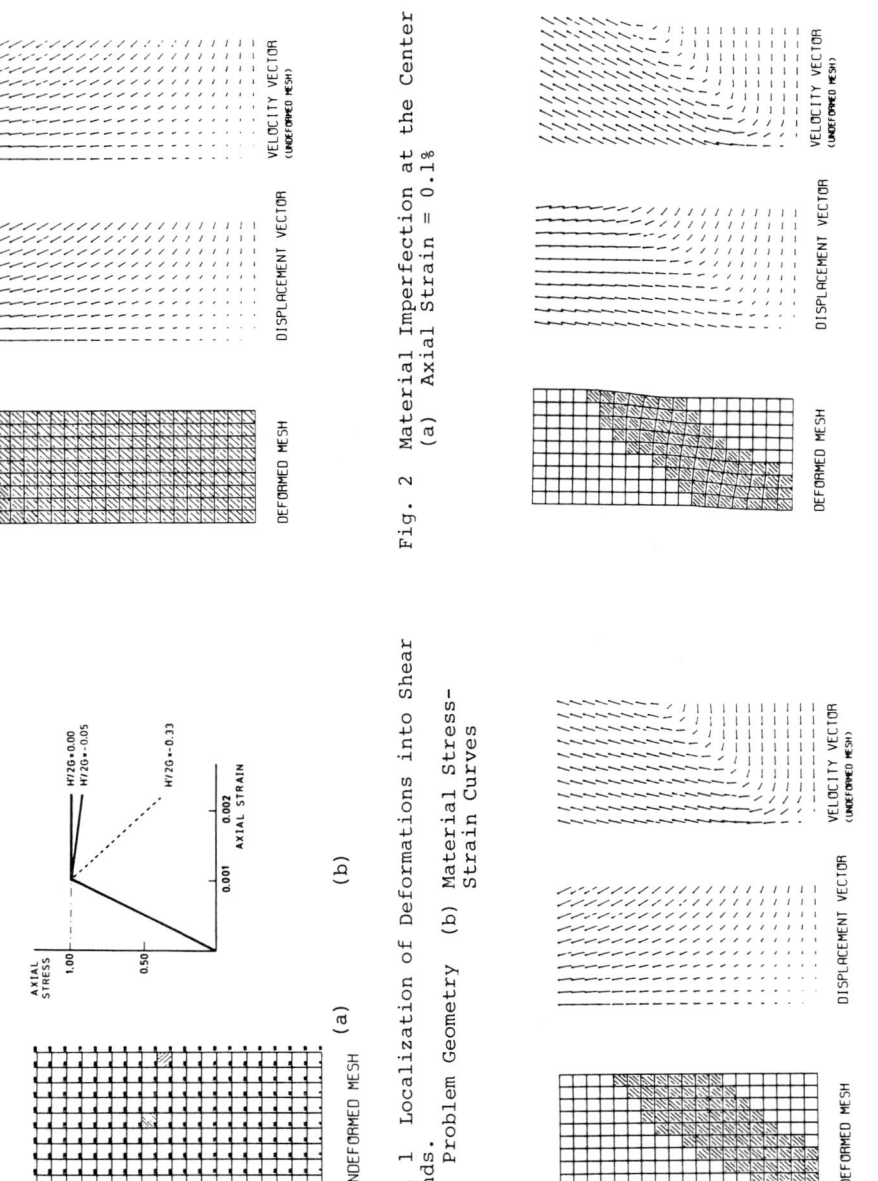

Fig. 1 Localization of Deformations into Shear Bands.
(a) Problem Geometry (b) Material Stress-Strain Curves

Fig. 2 Material Imperfection at the Center
(a) Axial Strain = 0.1%
(b) Axial Strain = 0.103%
(c) Axial Strain = 0.120%

the axial strain = .103%. This pattern is stable, and the subsequent failure of the specimen is illustrated by Fig. 2c (axial strain=.120%). Note that the angle of the shock line is very close to the predicted value (38.7°). Other examples of localization phenomena may be found in Ref. [18].

2. Bearing Capacity

In order to demonstrate that finite element solutions can capture failure states accurately, i.e., both limit loads and localization of deformation phenomena, numerical results for the classical punch problem [6,15,16] in both small and large deformation regimes are presented hereafter.

The material is the classical incompressible isotropic elastic-perfectly plastic Prandtl-Reuss material (Eq. 13 in which H'=0), and Fig. 3 shows the two-dimensional finite element representation of the problem geometry and the notation. The punch is represented by a strip of elements ten-thousand times stiffer than the supporting medium. Loading is achieved by the centered vertical force F. The computed load-displacement curves are shown in Fig. 4 where $c=k/\sqrt{3}$ = simple shear strength. Note that in the case of small deformations (G/c=1000) the theoretical failure load is very accurately captured by the numerical solution. Fig. 5 shows the computed velocity field at failure when the medium is initially perfectly homogeneous. Note that again localization could not occur because of the symmetry of the loading, geometry and homogeneity of the material properties. Fig. 6 shows the computed velocity field at failure [18] when a small inhomogeneity has been introduced by placing two weak elements (H'/2G=-1/3) in the line of the foundation as shown in Fig. 3. The load-displacement curve in that case remains identical to the one obtained for the homogeneous deposit. However, note that at failure in that case, localization of the deformations take place and that the computed velocity field very accurately follows the classical slip line solution first proposed by Prandtl [17]. It is of interest to note that this slip-line solution was obtained for both smooth and frictional interface conditions between the punch and supporting foundations.

Fig. 4 also shows the computed load-displacement curve when large deformations take place in the foundation (G/c=50). Localization of the deformations in that case occurs at a load about 15% higher than in the small deformation case. Fig. 7a shows the computed deformed configuration at failure, and Fig. 7b shows the corresponding velocity field plotted in the deformed geometry.

It should be noted that the use of weak elements to trigger the localization of the deformations at failure seems to be necessary for symmetric problems only. This is

PLASTIC-LIMIT EQUILIBRIUM STATES

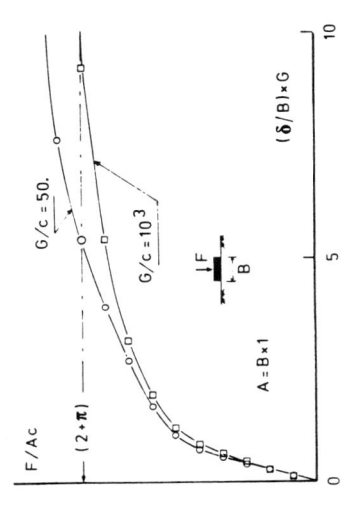

Fig. 4 Computed Load-Displacement Curves

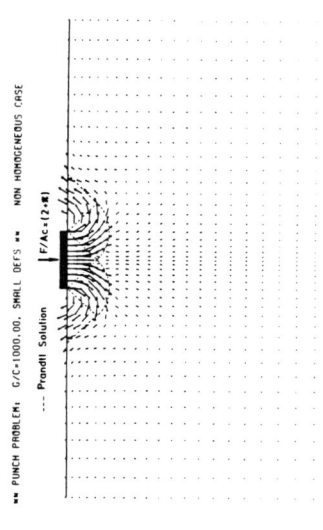

Fig. 6 Velocity Field at Failure -- Non-Homogeneous Case

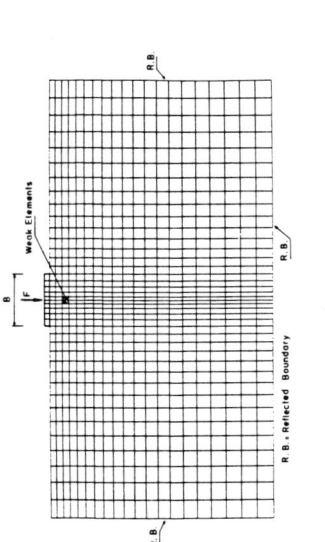

Fig. 3 Rigid Punch - Problem Geometry

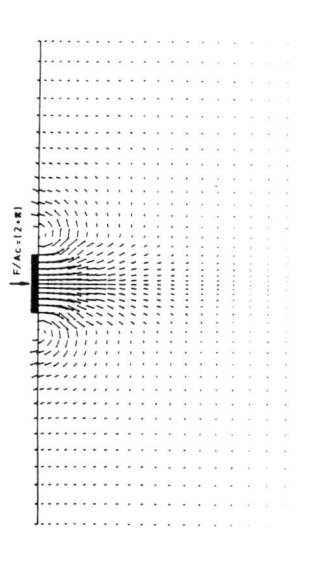

Fig. 5 Velocity Field at Failure -- Homogeneous Case

178 SOIL STRESS STRAIN APPLICATIONS

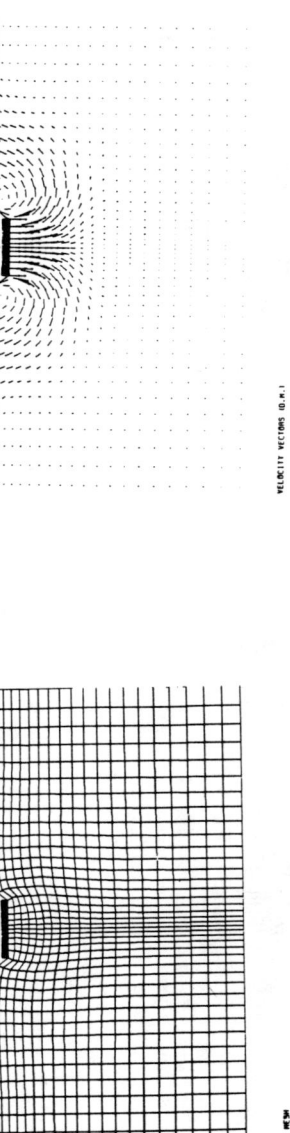

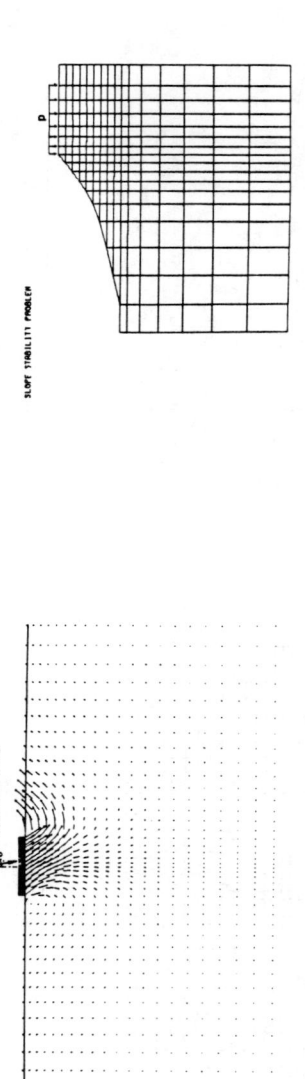

Fig. 7 Rigid Punch Problem - Large Deformations Case

(a) Computed deformed configuration at localization

(b) Computed velocity field at localization plotted in deformed configuration

Fig. 8 Rigid Punch Problem Eccentric and Inclined Load

Fig. 9 Slope Stability -- Problem Geometry

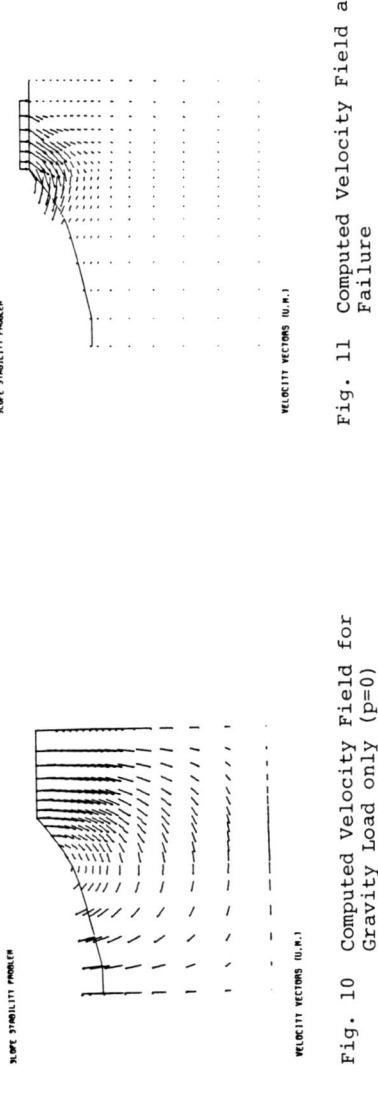

Fig. 10 Computed Velocity Field for Gravity Load only (p=0)

Fig. 11 Computed Velocity Field at Failure

illustrated by Fig. 8 which shows the computed localized velocity field at failure for a rigid punch subjected to an inclined and eccentric load without using weak elements in the foundation.

3. Slope Stability

The material is again the classical incompressible isotropic elastic-perfectly plastic Prandtl-Reuss material, and Fig. 9 shows the two-dimensional finite element representation of the problem geometry and the notation. Loading is achieved by (1) applying the gravity load, and (2) increasing the pressure p. Fig. 10 shows the computed velocity field under gravity load only. Failure occurs at $p=3.57$ as predicted by the slip line solution obtained by Sokolovski [21], and Fig. 11 shows the corresponding velocity field which accurately follows the slip line solution presented in [21].

SUMMARY AND CONCLUSIONS

It is shown in this report that finite element solutions of elastic-plastic boundary value problems can accurately represent failure states.

ACKNOWLEDGEMENTS

The author is most grateful to B.E. Hjorth who helped to perform some of the computations. Computer time was provided by Princeton University Computer Center.

REFERENCES

[1] Chen, W.F., Limit Analysis and Soil Plasticity, Scientific Publishing Co., Elsevier, Amsterdam, The Netherlands, 1975, p. 638
[2] Fellenius, W., Erdstatische Berechnungen, (Revised Edition) W. Ernst and Sohn, Berlin, 1939.
[3] Hadamard, J., "Lecon our la Propagation des Ondes et les Equations de l'Hydrodynamique," Paris, Chapter 6, 1903.
[4] Hill, R., "A General Theory of Uniqueness and Stability in Elastic-Plastic Solids," J. Mech. Phys. Solids, Vol. 6, 1958, pp. 236-249.
[5] Hill, R., "Acceleration Waves in Solids," J. Mech. Phys. Solids, Vol. 10, 1962, pp. 1-16.
[6] Hill, R., The Mathematical Theory of Plasticity, Oxford University Press, London, 1950, p. 355.

[7] Hodge, P.G., and Van Rij, H.M., "A Finite Element Model for Plane Strain Plasticity," *Journal of Applied Mechanics*, ASME, Vol. 46, No. 3, 1979, pp. 536-542.

[8] Hughes, T.J.R. and J.H. Prevost, "DIRT II-A Nonlinear Quasi-static Finite Element Analysis Program," California Institute of Technology, Pasadena, CA, August 1979.

[9] Ko, H.Y., and Scott, R.F., "Bearing Capacities by Plasticity Theory," *Journal of the Soil Mechanics and Foundation Engineering*, ASCE, Vol. 99, No. SM1 (Proc. Paper 9497), January 1973, pp. 25-43.

[10] Malkus, D.S. and T.J.R. Hughes, "Mixed Finite Element Methods -- Reduced and Selective Integration Techniques: A Unification of Concepts," *Computer Methods Appl. Mech. Eng.*, Vol. 15, No. 1, 1978, pp. 63-81.

[11] Malvern, L.E., *Introduction to the Mechanics of a Continuous Medium*, Prentice-Hall, Inc., Englewood Cliffs, NJ 1969.

[12] Mandel, J., "Conditions de Stabilite' et Postulate de Drucker," in *Rheology and Soil Mechanics*, Eds. J. Karavtchenko and P.M. Sireys, Springer-Verlag, 1966, pp. 58-68.

[13] McMeeking, R.M. and J.R. Rice, "Finite Element Formulation for Problems of Large Elastic-Plastic Deformation," *Int. J. Solids Structures*, Vol. 11, 1975, pp. 601-616.

[14] Melan, E., "Zur Plastizitat des zaumlichen kontinuums," *Ing.-Azch.*, Vol. 9, 1939, pp. 116-126.

[15] Nagtegall, J.C., Parks, D.M., and Rice, J.R., "On Numerically Accurate Finite Element Solutions in the Fully Plastic Range," *Computer Methods in Applied Mechanics and Engineering*, Vol. 4, 1974, pp. 153-177.

[16] Prager, W., and Hodge, P.G., *Theory of Perfectly Plastic Solids*, Dover Publications, NY, 1968, p. 264.

[17] Prandtl, L., "Uber die Haerte plasticher Koerper," *Goettinger Nachr. Math-Pshys.*, Kl 1920, pp. 74-85.

[18] Prevost, J.H., and Hughes, T.J.R., "Finite Element Solution of Elastic-Plastic BVP," *J. Appl. Mech.*, ASME, 1980 (to appear).

[19] Rice, J.R., "The Localization of Plastic Deformations," in Theoretical and Applied Mechanics, *Proceedings*, 14th IUTAM Congress, Delft, The Netherlands, North Holland Publishing Co., 1976, pp. 207-220.

[20] Van Rij, H., and Hodge, P.G., "A Slip Model for Finite-Element Plasticity," *Journal of Applied Mechanics*, ASME, Vol. 45, No. 3, 1979, pp. 527-532.

[21] Sokolovskii, V.V., *Static of Granular Media*, (Translated from Russian by J.K. Lusher), Pergamon Press, London, England, 1965.

[22] Thomas, T.Y., *Plastic Flow and Fracture in Solids*, Academic Press, Inc., 1961.

THE APPLICATION OF
GENERALIZED STRESS-STRAIN RELATIONS

by

John T. Christian, F. ASCE
Senior Consulting Engineer
Stone & Webster Engineering Corporation
Boston, Massachusetts

ABSTRACT

Following symposia at Montreal and Chicago, at which the theoretical and experimental status of non-linear constitutive relations was investigated, a final pair of sessions deals with their application in practice. The practical application of a constitutive model involves not only its use for calculations but also its implementation in an appropriate computer program. This raises several programming problems. The history of the development of constitutive relations has focussed attention on the plastic behavior of soil, with emphasis on kinematic, anisotropic hardening in the most recent models.

The present state of the art of practical application of generalized stress-strain relations favors simple models such as hyperbolas and several forms of eslaticity. It is important that the initial state of stress and the stress history be described correctly and that their effects on the values of the parameters in the constitutive relations be understood. The success of several of the recent efforts in predicting or recovering observed behavior encourages further use in practice.

The advanced constitutive relations, which were the major focus of the Montreal symposium, have had very little practical application to field problems. However, they have passed well beyond the stage of speculative or qualitative description and promise to deal effectively with loading conditions different from those available in the laboratory.

Introduction

The present collection of papers were prepared for the last in a series of meetings that have probed the current state of knowledge of constituent relations for soils. Under the sponsorship of the U. S. National Science Foundation and the Canadian National Science and Engineering Research Council, a three day symposium at McGill University in May, 1980, investigated what constitutive relations are available, how they are formulated, and how well they can be used to predict experimental behavior. A one day symposium at the Chicago convention of the American Society for Testing and Materials in June, 1980, discussed laboratory testing procedures as they are used to establish constitutive relations and to determine the values of their parameters. The present proceedings are concerned with the practical application of the theoretical and experimental results, in limit plasticity calculations and as generalized stress-strain relations. As the three meetings dealt with one broad topic, their proceedings must be taken as a whole. Thus, for example, the detailed exposition of the use of a particular stress-strain relation is best found in the material from the Montreal symposium and is not repeated here. In addition, the interested reader should consult the proceedings of the Third International Conference on Numerical Methods in Geomechanics, held in Aachen, Germany, in 1979.

The application of a constitutive theory involves at least two distinct procedures. One is the use of the theory to predict the behavior of a constructed facility in the field, to analyze its observed behavior, or to design a new facility. This implies that the constitutive theory has been implemented in some formal method of analysis, which usually involves a finite element computer program. The other procedure is the conversion of a theoretical constitutive model into a form that can be used in a finite element program and the writing of that program. A few comments should be made about this essential first step in the application of constitutive relations to real problems.

Use of Constitutive Models in Finite Element Programs

All static finite element programs based on the displacement method have as their central procedure the solution of a system of simultaneous, linear, algebraic equations of equilibrium:

$$[K]\{U\} = \{P\} \tag{1}$$

where $\{U\}$ represents the displacements or increments of displacements, $\{P\}$ represents the loads, and $[K]$ is the stiffness. The stiffness is computed from the well known equation:

$$[K] = \int_V [B]^T [C] [B] \, dV \tag{2}$$

in which $[B]$ is the matrix relating strains to nodal displacements and $[C]$ is the matrix relating stress to strain.

Non-linearity of stress-strain relations or deviations from initial linearity are handled in two ways. The $[C]$ matrix can be changed at each step of the loading to reflect the current state of stress and strain. This will lead to a new stiffness for each element, and the global stiffness $[K]$ must be reassembled for the entire problem at each step. Alternatively, $[C]$ can be kept constant and the deviations incorporated by additional fictitious and self-equilibrating forces in $\{P\}$. The initial stress method is an example of the latter approach. In either case it is usually necessary to iterate several times for each step of loading to ensure that the non-linear relations are satisfied.

The choice between these two approaches depends in part on properties of the constitutive relation. Some relations lend themselves to rapid calculation of [C] and [K]; others are more easily expressed as a deviation from linear elasticity, that is, a changed {P}. Even when there is no obvious reason for preference, each method has its partisans. The author believes that, in the absence of other compelling motivation, updating [C] and [K] is preferable for four reasons. First, although each iteration is less time consuming and costly in a method that changes {P} and does not change [K], more iterations are usually necessary to achieve the same degree of convergence. Therefore, there is likely to be little or no saving in total computational effort as a result of not changing [K]. Second, because convergence is faster in methods that change [K], there tends to be significantly less average error even when the same convergence criterion is used (75). Third, the organization and logic of the computer program are usually somewhat simpler when [K] is updated, and the programs are consequently easier to check and to modify. Fourth, Prevost and Hughes (62) have shown that reassembly of the correct tangent stiffness is necessary if failure states are to be recovered in finite element analyses.

In either procedure, each system of simultaneous equations for each step is an expression of a problem in linear elasticity. The stiffness may be unsymmetrical, and the terms in [C] may be computed by a very non-linear set of relations, but the actual problem being solved in the computer consists of a set of linear equations. This has two important consequences.

First, the linearization of a non-linear process will introduce some errors into the solution. If iterations are not performed to obtain a good secant fit to the non-linear process, the solution will continue to diverge from the true one. The ease with which these iterations can be formulated and the rate they converge can vary substantially among the various constitutive relations.

Second, loading and unloading are governed by the same linear equations. If there is a difference between behavior in loading and unloading, it cannot be built directly into the linear equations. The computer program must assume a mode of behavior for each element, solve the equations under this assumption, check whether the assumption for each element is valid, and then repeat the solution process if all assumptions are not correct. Performing such checks for all elements can be expensive, and cases can occur in which the computer program searches for a long time among a set of conflicting assumptions without finding a satisfactory pattern of loading and unloading.

There are many other programming considerations. Constitutitve relations that require the storage of a great deal of information on the history of stress and strain for each element can require exorbitant amounts of computer memory. Because of the large memories and virtual memories of modern computers, this is less of a problem than it used to be, but it is still not inconsequential.

Brief Review of Some Theory of Constitutive Relations

The diversity and complexity of soil behavior does not always conform to the procrustean bed of continuum mechanics. The distinctions among the different types of behavior are necessarily somewhat arbitrary and have evolved in part to explain the behavior of other engineering materials. Nevertheless, the classical definitions of continuum mechanics provide useful distinctions between different phenomena, and without such a conceptual framework, discussion of stress-strain properties becomes extremely difficult. Most models now in practical use assume infinitessimal strains, as is done in the following simplified summary.

Fundamental distinctions exist between elasticity, viscosity, and plasticity. Elastic materials are those whose relations between stress and strain are reversible and have no time delay. Linear elasticity applies when the relation is linear. Viscous materials are those whose stress-strain relations also involve time. For example, the strain response to an applied stress might increase over a period of time. Linear visco-elasticity implies that the relation between stress and strain is linear for equal values of time, but the dependence of either stress or strain on time is usually non-linear. Plastic materials are those that exhibit some irreversible behavior. The usual example involves application and removal of a prescribed stress and observation of some unrecovered strain after the cycle is complete. Obviously, combinations of behavior can also be listed, such as visco-plasticity, and real materials show all three types of behavior.

A distinction is also made between deformation and incremental theories. A typical deformation theory describes the relation between the strains and the state of stress; an example is the hyperbolic relation proposed by Kondner (42) and elaborated by Duncan and Chang (29). An incremental theory describes the relation between the increments of strain and the states of stress and strain in the material. The total strains must be found by integrating the increments. Most formulations derived from plasticity theory are of the incremental type.

The concept of a yield surface, yield criterion, or yield function, f, is central to plasticity theory. It is a function of the stress, but other paramters such as strain may be included. If the stresses are on the yield surface or satisfy the yield function (f = 0), the material is undergoing plastic strains. If the stresses are within the yield surface or are less than those needed to satisfy the yield function (f < 0), only elastic strains occur. The stresses cannot exceed the yield criterion. Elastic strains can occur along with the plastic ones; the yield criterion can move about as the result of plastic strains; and one can construct plastic formulations so that the stresses always satisfy the yield criterion and so that there are always plastic strains. Strain hardening exists when the yield criterion depends on the past plastic strain history as well as the state of stress. The special case of strain softening occurs when the yield surface contracts as a result of plastic strain. When f depends on stress only, there is perfect plasticity.

Normality and Stability:

The directions of the plastic strain rates or increments are classically governed by the plastic potential, g, so that increments of plastic strain, $\dot{\epsilon}_{ij}^p$, are proportional to the gradient of the plastic potential:

$$\dot{\epsilon}_{ij}^p = \Lambda \frac{\partial g}{\partial \sigma_{ij}} \qquad (3)$$

This is called the normality rule or the associated flow rule, and it is part of the classical theory of plasticity described by Hill (34) as well as others. In the simplest case the plastic potential and the yield criterion are the same. For strain hardening and softening materials the factor Λ in equation (3) can be computed from the hardening or softening rule. For perfectly plastic materials it may be indeterminate, or it may be found from requirements of compatibility with elastic strains.

Drucker (23) defined a stable material as essentially one from which work could not be extracted irreversibly in a closed cycle of loading and unloading. He than showed that for such a material the yield criterion is the plastic potential with an associated flow rule. In other words, normality applies to the yield surface. A further consequence is that both the upper bound and lower bound theorems of limit plasticity are valid. Since they are the basis for the plastic design of structures as well as bearing capacity and slope stability calculations, this is a very powerful result.

Friction:

The first failure criterion of any practical importance was the Mohr-Coulomb rule for soils (18):

$$\tau = c + \bar{\sigma} \tan \phi \qquad (4)$$

Coulomb applied it in what was essentially a limit analysis that has continued to be one of the practical foundations of geotechnical engineering. It is interesting that Coulomb's failure criterion and its application predated the simpler Tresca and von Mises criteria. Drucker and Prager (24) proposed that the Mohr-Coulomb relation could be generalized into three dimensions by relating the square root of the second invariant of the deviator stress to the first invariant of the stress:

$$J_2^{1/2} = \alpha + kI_1 \qquad (5)$$

They then showed that the assumption of normality with respect to this function implies a substantial volumetric plastic strain rate as long as any plastic strains are occurring. This continuous, irreversible expansion is not observed experimentally, although dilation does take place to a lesser degree and for a more limited range of strain in dense soils. Therefore, either normality does not apply, or there is something wrong with the use of this sort of relation as a yield criterion, or both. In a later paper Drucker (25) demonstrated that assemblages of frictional particles are not necessarily stable in his sense, and therefore one would not expect normality to apply to the frictional failure relation.

The Mohr-Coulomb law has been programmed as a yield criterion with and without the volumetric strain rates induced by normality (13). Non-symmetric stiffness matrices occur if normality is relaxed. Figure 1 shows the pattern of deformations calculated under a strip footing when the plastic volumetric strains are suppressed. Figure 2 shows the results when normality is enforced and the plastic volumetric strains occur. The impact of the continuous dilation is quite clear.

It is worth noting that the Drucker-Prager relation was proposed for its mathematical simplicity so as to examine some of the consequences of classical plasticity theory. Bishop (6) pointed out that it does not describe the failure states observed in laboratory tests, and should not be used to describe real soils.

Isotropic Capped Models:

The search for a way out of this dilemma led Drucker, et al. (26) to propose that the true yield surface of a soil is not the Mohr-Coulomb failure envelope but is instead a "cap" over the open end of the Mohr-Coulomb (or Drucker-Prager) wedge. The cap expands as the result of virgin compression, which must involve plastic strains because part of the strain occurring during virgin isotropic compression is irrecoverable. Figure 3 is a conceptual presentation of the essential idea. As irreversible volumetric strains occur, the yield cap expands. Any reversal of the stress path within the cap causes the soil to behave elastically. A stress path moving out from a point on the surface of the cap causes plastic volumetric and deviatoric strains normal to the cap, and the cap moves in such a way that the final state of stress is on the cap in its new position. Elaboration of the basic idea led to the basic theoretical and experimental work of Roscoe and his colleagues (65, 68) and the further development of the various Cambridge models (58), including the Cam clay model. Although there were earlier applications of versions of the capped model (12), the most successful uses have been those of Sandler, Baladi, and their fellow workers (3, 66).

Kinematic Hardening Models:

One problem with the capped models in their original form is that they involve isotropic hardening. That is, the cap expands uniformly as the result of plastic strain. In real materials it is observed that plastic shearing strains tend to make the material harden in the direction of the shearing strain while it appears to become softer for reversed stresses. This effect cannot be described by isotropic hardening, but it can be represented by kinematic hardening, in which the yield criterion translates instead of expanding. For example, in Figure 4a, isotropic hardening causes the yield function to expand as shearing increment causes the stress to move from point 1 to point 2. Upon reversal of the load, the yield stress in the reversed direction is found to be at point 3, which is located

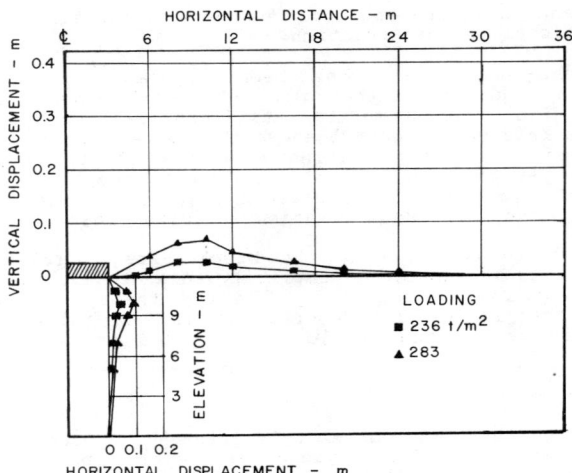

Displacements — Non-Dilatant Mohr-Coulomb Material (ref. 13)

FIGURE 1

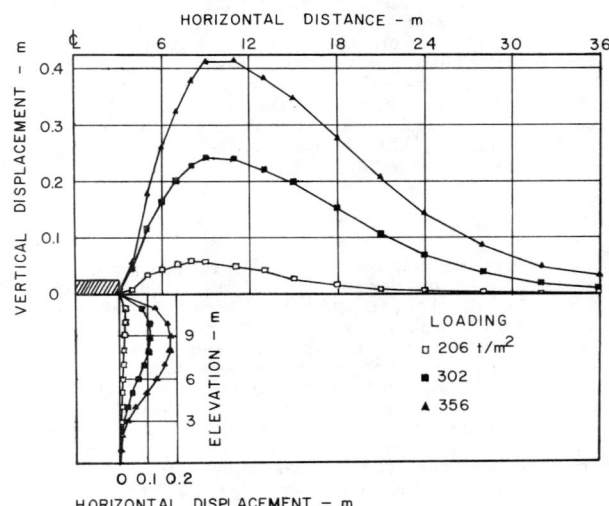

Displacements — Dilatant Mohr-Coulomb Material (ref. 13)

FIGURE 2

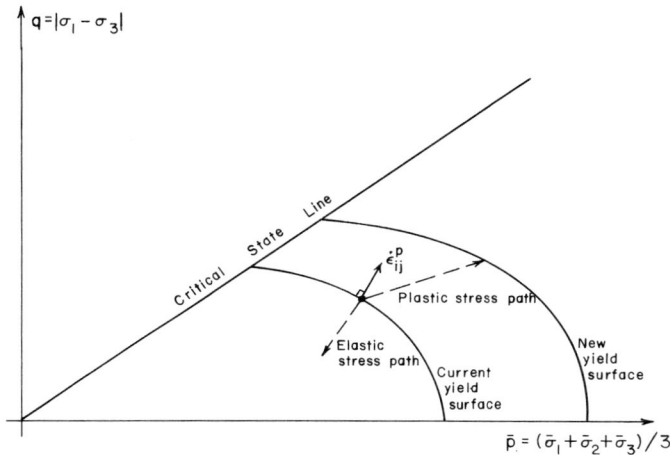

Capped Plastic Model

FIGURE 3

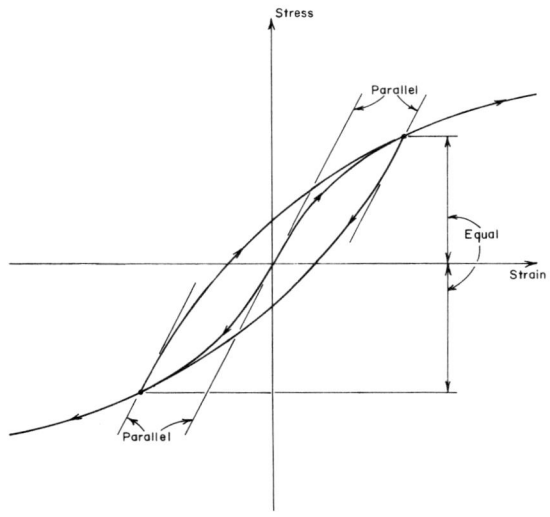

Masing Relation

FIGURE 5

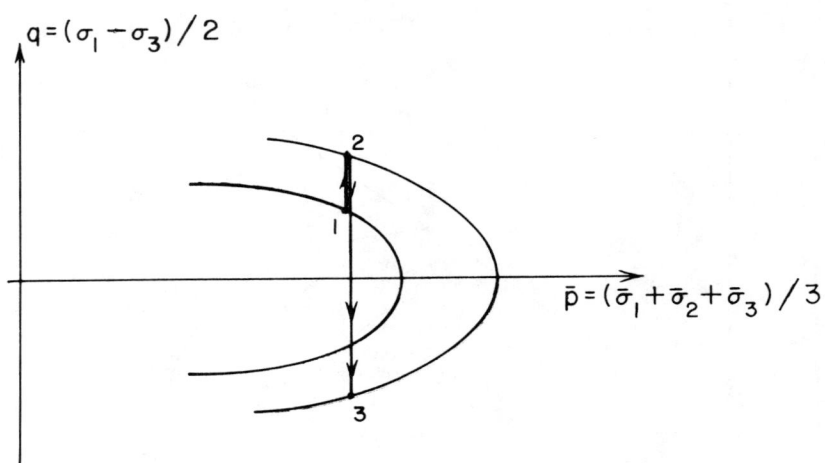

(a) Isotropic Hardening

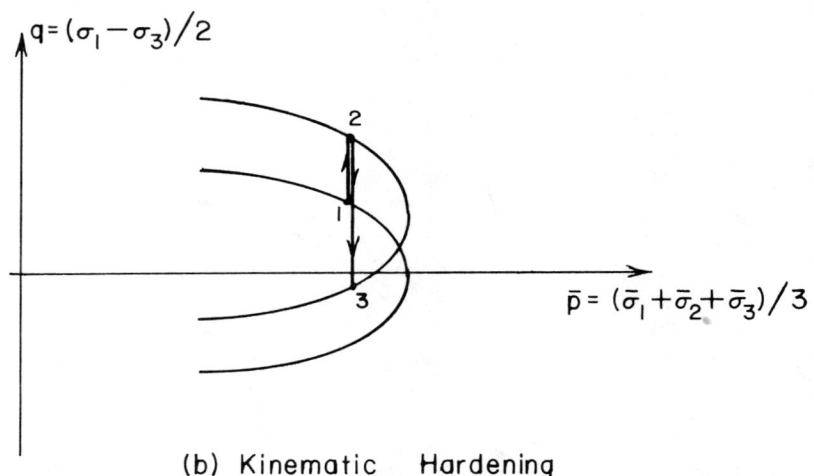

(b) Kinematic Hardening

Hardening Rules

FIGURE 4

symmetrically to point 2. The material has become stronger in both directions. A more realistic behavior is shown in Figure 4b, where the yield surface translates instead of expanding. Now, the magnitude of the shear stress at yielding for the reversed load (point 3) is less that at point 2.

Prevost (59, 60) and Mroz, et al. (53, 54, 55) have proposed different versions of kinematic strain hardening models. These involve families of nested yield criteria or functional relations describing how the criterion translates during shearing strain. The latest versions of the capped model used by Sandler, Baladi, and their colleagues include kinematic hardening. Dafalias and Herrmann (19) have developed what seems to be a particularly attractive model that uses concepts very similar to those of plasticity theory, although they emphasize that their relation differs in some important respects from conventional plasticity theory. As of this writing it appears that several researchers are describing essentially the same phenomena by different but closely related mathematical formulations.

Other Models:

Simpler plasticity formulations have been used to study a number of problems. Strip footings on yielding soil have been analyzed for undrained conditions with constant shear strength and varying intial stresses (36), with anisotropic yield stress (4), and with strain softening properties (35). The expansion of cylindrical cavities similar to those caused by the pressiometer have been studied for strain softening soil (61) and for both constant shear strength (Tresca) and capped models (7). Simpson et al. (70) proposed a three staged model that moves from linear elasticity to a yielded state described by the Cam clay theory and applied it to several idealized problems. Lade (47, 48) has created a different plasticity model with both associated and non-associated flow rules and a failure envelope that resembles what is observed experimentally for sands.

Although plasticity theory and its close relations have been the most popular vehicles for developing non-linear constitutive relations for soils, others have been used. Deformation theories based on empirically observed relations between stress and strain have been very widely used. The best known of these is the hyperbolic relation applied in finite element analyses by Duncan (29,42). Drnevich (21,22) has expanded the theory to account for anisotropy and sample disturbance. Kavazanjian and Mitchell (39) have combined an earlier visco-elastic theory with the hyperbolic relation and other empirical relations to develop a constitutive relation that shows considerable promise. The endochronic model has also been proposed for soil (5), but its application seems to require considerable empirical effort to establish which parameters should be used and how to find their values.

A detailed description, or even a listing, of all the versions of non-linear formulations that have been proposed in recent years is beyond the scope of this summary. However, the position papers for the Montreal symposium in May, 1980, present quite a comprehensive view of the current state of that theoretical art. Table 1 summarizes the features of the major models presented; Ko and Sture (42) have compared the

TABLE 1

CONSTITUTIVE MODELS DISCUSSED AT MONTREAL

Predictor	Model Type
Wroth	Isotropic strain hardening capped Cam clay model
Lade	Plasticity relations with and without normality for yield criterion based on third invariant of stress; applicable to sands
Baladi/Sandler	Capped model with anisotropic kinematic hardening
Dafalias	Bounding surface model similar to anisotropic kinematic hardening
Prevost	Anisotropic kinematic hardening with isotropic hardening cap
Duncan	Hyperbolic deformation relation
Chen	A - Hyperelastic relation based on fourth order relation expression for strain energy B - Drucker-Prager or Mohr-Coulomb strength relation with linear elasticity
Kavazanjian/Mitchell	Viscous model with empirical relations for non-linearity
Adachi	Visco-plastic theory with yield similar to anisotropic kinematic hardening theory
Gudehus	Homogeneous states of stress and strain with curve fitting
Bazant	Endochronic relation with empirical parameters

formulations of several relations in some detail. It can be seen that about half these models are based on some form of plasticity theory and that most of these are elaborations of the concepts of anisotropic kinematic hardening with an isotropically hardening cap. Today the engineer proposing to use a non-linear constitutive relation for soil, whether based on strain hardening (or softening) plasticity theory or on some other approach, has available a bewildering variety of options. The proceedings of the Montreal symposium and the state of the art report by Ko and Sture are excellent places to learn what exists and how to use it.

It should be borne in mind that different models are intended for different purposes. The conceptual models, such as the various strain hardening models and particularly the Cam clay model, are most valuable in providing insight into the behavior of soil but may be difficult to use in practice, while the empirical ones, such as the hyperbolic model, are very effective in practical application but provide little or no information on the fundamentals of soil behavior. Either type of model can be used to reproduce the loading curve from a particular test. The versatility of a model is shown when it is applied to stress conditions other than those from which it was derived, and it is a major purpose of the sophisticated conceptual models to achieve as much versatility as possible.

The tests used in conjunction with the Montreal symposium showed that, in trained hands, the empirical models of Kavazanjian and Mitchell and of Duncan as well as the anisotropic kinematic hardening models could predict the behavior of soils in monotonically loaded tests. When the models were applied to a test in which the principal stresses rotated continuously through more than 360 degrees, none performed well. Arthur et al. (1) have emphasized that the rotation of stresses presents experimental and theoretical problems that remain to be solved.

Comments on the Current Use of Non-Linear Relations

Hyperbolic Models:

While there has been much activity in developing and publishing new constitutive relations, there have been relatively few published papers describing the application of these to predicting the behavior of actual facilities or describing the use of the constitutive models in design. Most practical applications have involved either linearly elastic models with some consideration of the effects of the initial state of stress or variations on the hyperbolic stress-strain model. Starting with the original work of Duncan and Chang (29) and Clough and Duncan (15), the model has been successfully used to study reinforced earth (11), excavations and retaining structures (14, 16, 56, 57), culverts (28), tunnels (40, 72), and embankments (44). By adding a detailed examination of the stress history and its effects on the soil properties, Ladd and his colleagues (33, 45, 69) have been able to describe the behavior of several embankments on very difficult soils such as soft organic clays and varved clays. Even though many of the cases described are actually investigations after the fact, these users have demonstrated that the model can be used with some confidence.

There seem to be several reasons for this state of affairs. The hyperbolic models have been available for about ten years and are incorporated into a number of computer programs. The models and the programs are understood by many people, including those who do not work primarily with numerical methods or constitutive relations. The parameters are reasonably comprehensible in the context of soil mechanics. They can also be measured without extraordinary testing equipment in most cases.

A more subtle and fundamental reason for the success of the hyperbolic model arises from its range of practical use: at stress levels less than about 75% of failure. This is the range in which a great many geotechnical problems occur. The hyperbolic model provides a ready means of establishing the equivalent elastic moduli, such as Young's modulus or the shear modulus, and compensating for the effect of in situ stresses. In the standard formulations the modulus is related to the value of the minor effective confining stress and to the ratio between the actual shear stress and the shear stress at failure. These can be accounted for in the direct formulation of the hyperbolic relations, or more sophisticated methods can be used to refine the input (21, 22, 46, 69). In many practical cases the changes in modulus as computed by the hyperbolic relation, or most other models, are not great during the course of the practical loading program. It could be argued that, once the apparent moduli are corrected for the effects of confining stress and initial shear stress, the subsequent effects are so small that a program that made no further changes would be adequate for many practical problems. For this class of problems the model need not describe very accurately what happens near failure or under large changes in stress; it is sufficient to get the effects of the in situ conditions properly accounted for.

There have, of course, been many successful applications of the hyperbolic models in cases where some non-linearity of the response of the soil was important. Simon et al. (69) and Kulhawy et al. (44) described several such instances. However, it remains true that the hyperbolic model works best when the deviations from linear response under the working loads are small and when the stress paths are reasonably close to those that can be reproduced in the laboratory. When significant non-linear response is to be anticipated, it is difficult to use the hyperbolic model effectively.

Other Elastic or Deformation Models:

When the effects of the in situ initial stress and the stress history are properly accounted for, simple elastic or deformation models can be used. The modulus can be related to a fractional power of the effective octahedral stress or calculated from a study of the experimental relation between the state of stress and the deformability. Such linearized models have been very effective in the study of track ballasting systems (9), culverts (10), tunnels (31), buried concrete pipe (43), and dams (52). Schmidt and Grantz (67) extended such models to include the effects of virgin compression and rebound. Leathers and Ladd (49) used a detailed study of the stress history and its effects to normalize the soil properties, obtaining equivalent linearly elastic properties for the analysis of an embankment on varved clay.

As suggested above under "Hyperbolic Models," the reasons for the successful use of of these simpler elastic deformation models are the great significance of the in situ stress and stress history and the limited range of stress and strain observed in many practical problems. When these conditions obtain, a simple model may be adequate. However, it is essential that great care be taken in establishing the elastic parameters.

When the time dependent effects of consolidation are to be considered, most analyses rely on linearly elastic models (30, 51, 57). There are a few examples of the use of non-linear effective stress-strain relations in consoliddation analyses (8, 20), and these usually employ simple functional relations between effective stress and strain.

In this context it is worth noting that Lee and Idriss (50) concluded from comparative analytical studies that the effects of non-linearity are not very important in the analysis of the behavior of earth dams.

Capped Models:

The capped models developed by Sandler et al. (66) were first used to analyze problems of non-linear response under large overpressures caused by explosive shocks. In these cases the non-linearity may be quite significant. The original capped models, with isotropic hardening, describe the strain hardening that occurs under the large first pulses of loading, and their use has been refined by a series of comparisons between theoretical predictions and field observations. In the process the capped models were refined and improved.

The successful application of the capped models was helped by the fact that there is usually only one large pulse in problems involving explosive loads. This tends to expand the cap, and the rest of the response, which is usually of less interest anyway, occurs under substantially elastic conditions. Therefore, the inability of the earlier capped models to describe the effects of cyclic yielding with kinematic hardening was not a detriment. As the use of the capped models has spread to other problems, they have been modified to include kinematic effects to the extent that developers of capped models with kinematic hardening and of anisotropic, strain hardening models of the Prevost or Mroz type have at times agreed informally that their models are becoming very similar to each other.

Other Plastic Models:

Early applications of fully plastic models, except for the capped ones mentioned above, used very simple yield criteria that did not consider the frictional component of the strength of the soil. These could only be applied to undrained behavior of saturated clays, but some useful results were be obtained. For example, the significance of the initial state of stress in computing plastic deformations under strip footings (36) was demonstrated. Also, the relation between plastic yield and the development of excess pore pressure was predicted (36) and related to field observations (37) using such simple models.

Cyclic Loading:

As models incorporating frictional behavior, volumetric yielding, and kinematic hardening have developed, they have also become complicated. This has limited their use to problems where such non-linearities are truly significant. The cyclic loading of soils has been one such area. The effects of cyclic loading on the deformation of clays and on the development of excess pore pressure in sands are of concern for earthquake conditions and for structures loaded by ocean waves. Most of the earthquake analyses for cyclic loading in which a non-linear stress-strain relation was used throughout the analysis have dealt with the one-dimensional vertical passage of shear waves through a column of soil. The preferred model has been a Ramberg-Osgood or Masing relation (17, 32, 38, 64, 71). In these models the loading, unloading, and reloading are related to an original backbone curve of stress versus strain so that the initial slope of any loading or reloading path is the same and the hysteresis loops fit the backbone curve (Figure 5).

Because of the complexities and the uncertainties in defining Masing relations for two- or three-dimensional problems, they have been used primarily for one-dimensional conditions. The case of offshore structures loaded by waves cannot be simplified to one-dimensional propagation, so more sophisticated models are appropriate. Prevost's model has been applied successfully to analyze observed displacements of models of offshore foundations under cyclic loads (63). There is a great deal of work to be done in refining the use of anisotropic plasticity models, but the early results look promising.

There is inevitably some error in any model. Part of the error is in the calculation of the exact strains recovered after a cycle of load. Since the overall effect of a pattern of cyclic loading is the sum of the effects of the individual cycles, it is to be expected that relatively minor errors in predicting the strain for one cycle will add up to create large errors for the entire sequence of loading. For this reason many engineers have been skeptical of the predictive use of non-linear models that required the individual cycles to be analysed. Such detailed following of the loading sequence can also be expensive in computer time.

As an alternative, it has been proposed that the effects of a number of cycles of stress or strain be described by a realtion similar to the S-N curves used in fatigue analysis. Then the consequences of a given sequence of waves would be computed by adding the effects of the groups of waves at each level. Andersen (2) described one of the first such techniques. It uses a computational scheme a little more complicated that that just described, but it is simple enough to be used in engineering design. Further investigation reveals that the number of cycles plays a role similar to that of time in visco-elastic theory. Dumas and Lee (27) and Marr and Christian (51) have used this device to predict the development of strains under many cycles of load. Marr and Christian were concerned in particular with the strains that remained after the cyclic loading had ceased and with cases in which there was not initial symmetry of the horizontal loads. These techniques permit the cyclic loading to be described by spectra of waves at different intensities, which can be compatible with the design waves developed by hydrologic investigators.

Viscous Models:

Although secondary compression is one of the earliest forms of non-linear soil behavior to be recognized, there has been less attention paid to viscous models for soil than to models based on plasticity theory. Part of this is because a realistic description of viscous effects requires non-linear visco-elastic descriptions of the behavior. Such models are difficult to employ in numerical calculations without large computer storage and long runs. Kavazanjian and Mitchell (39) have proposed one rather general visco-plastic model that has been used in some practical applications. It was quite successful in prediction the laboratory results at the Montreal symposium. The endochronic model (5) uses an internal fictitious time and can be applied to both time dependent and time independent problems by appropriate definition of the parameters.

Conclusions

There has been a great deal of progress in the development and understanding of the constitutive relations that underlie the behavior of soils. Most of this has been concentrated in the experimental and theoretical areas, and the Montreal symposium gives some hope that the investigators in these two areas are relating their efforts to each other. The application of the newer models has lagged somewhat, as one would expect. There is substantial overlap between the models, particularly where the anisotropic, strain hardening, and non-associated flow rule plasticity theories are concerned. Part of the benefit of the current series of meetings has been the exchange of information about what is involved in the various theories.

There is no one best model. Each model works best in an application for which it was developed, and it may not work at all in another one. This is true of other materials as well as soil; the intelligent analyst does not try to model everything but only those aspects of the behavior that are relevant.

The present state of the art of practical application of generalized stress-strain relations favors simple models such as hyperbolas and several forms of elasticity. It is important that the initial state of stress and the stress history be described correctly and that their effects on the values of the parameters in the constitutive relations be understood. Although there is by no means universal agreement that the present methods of describing these effects are correct (73, 74), the success of several of the recent efforts in predicting or recovering observed behavior encourages further use in practice.

The advanced constitutive relations, which were the major focus of the Montreal symposium, have had very little practical application to field problems. However, they have passed well beyond the stage of speculative or qualitative description and promise to deal effectively with loading conditions different from those available in the laboratory. The effort involved in applying them to the prediction or understanding field behavior will be substantial, requiring a major attempt at communication on the part of all concerned. It is to be hoped that the results of such cooperation will soon be seen.

References

1. Arthur, J. R. F., Chua, K. S., Dunstan, T., and Rodriguez del C., J. I., "Principal Stress Rotation: A Missing Parameter," *Journal of the Geotechnical Engineering Division*, ASCE, Vol. 106, No. GT4, April, 1980, pp. 419-433.

2. Andersen, K., "Behaviour of Clay Subjected to Undrained Cyclic Loading," *BOSS'76, Behaviour of Offshore Structures*, Proceedings, Trondheim, Norway, 1976, Vol. I, pp. 392-403.

3. Baladi, G. Y., and Rohani, B., "Elastic-Plastic Model for Saturated Sand," *Journal of the Geotechnical Engineering Division*, ASCE, Vol. 104, No. GT4, April, 1979, pp. 465-480

4. Ballester, F., and Sagaseta, C., "Anisotropic Elastoplastic Undrained Analysis of Soft Clays," *Geotechnique*, Vol. 29, No. 3, September, 1979, pp. 323-340.

5. Bazant, Z. P., and Krizek, R. J., "Endochronic Constitutive Law for Liquefaction of Sand," *Journal of the Engineering Mechanics Division*, ASCE, Vol. 102, No. EM2, April, 1976, pp. 225-238.

6. Bishop, A. W., "The Strength of Soils as Engineering Materials: Sixth Rankine Lecture," *Geotechnique*, Vol. 16, No. 2, June, 1966, pp. 91-128.

7. Carter, J. P., Randolph, M. F., and Wroth, C. P., "Stress and Pore Pressure Changes in Clay during and after Expansion of a Circular Cavity," *International Journal for Numerical and Analytical Methods in Geomechanics*, Vol. 3, No. 4, October-December, 1979, pp. 305-322.

8. Cavounidis, S., and Hoeg, K., "Consolidation during Construction of Earth Dams," *Journal of the Geotechnical Engineering Division*, ASCE, Vol. 103, No. GT10, October, 1977, pp. 1055-1067.

9. Chang, C. S., Adegoke, C. W., and Selig, E. T., "GEOTRACK Model for Railroad Track Performance," *Journal of the Geotechnical Engineering Division*, ASCE, Vol. 106, No. GT11, November, 1980, pp. 1201-1218.

10. Chang, C. S., Espinoza, J. M., and Selig, E. T., "Computer Analysis of Newtown Creek Culvert," *Journal of the Geotechnical Engineering Division*, ASCE, Vol. 106, No. GT5, May, 1980, pp. 531-556.

11. Chang, J. C., and Forsyth, R. A., "Finite Element Analysis of Reinforced Earth," *Journal of the Geotechnical Engineering Division*, ASCE, Vol. 103, No. GT7, July, 1977, pp. 711-724.

12. Christian, J. T., "Plane Strain Deformation Analysis of Soil," Ph. D. Thesis, Massachusetts Institute of Technology, 1966.

13. Christian, J. T., Hagmann, A. J., and Marr, W. A., "Incremental Plasticity Analysis of Frictional Soils," *International Journal for Numerical and Analytical Methods in Geomechanics*, Vol. 1, 1977, pp. 343-375.

14. Clough, G. W., "Stabilizing Berm Design for Temporary Walls in Clay," *Journal of the Geotechnical Engineering Division*, ASCE, Vol. 103, No. GT2, February, 1977, pp. 75-90.

15. Clough, G. W., and Duncan, J. M., "Finite Element Analyses of Retaining Wall Behavior," *Journal of the Soil Mechanics and Foundations Division*, ASCE, Vol. 97, No. SM12, December 1971, pp. 1657-1672.

16. Clough, G. W., and Tsui, Y., "Static Analysis of Earth Retaining Structures," Ch. 15 in *Numerical Methods in Geotechnical Engineering*, C. S. Desai and J. T. Christian, eds., New York, McGraw-Hill, 1977, pp. 506-527.

17. Constantopoulos, I. V., Roesset, J. M., and Christian, J. T., "A Comparison of Linear and Exact Nonlinear Analysis of Soil Amplification," *Proceedings, Fifth World Conference on Earthquake Engineering*, Rome, 1973, paper 225.

18. Coulomb, C. A., "Essai sur une application des regles de maximis et minimis a quelques problemes de statique, relatifs a l'architecture," *Memoires de mathematique et de physique, presentes a l'academie royale des sciences*, Paris, 1776.

19. Dafalias, Y. F., and Herrmann, L. R., "A Generalized Bounding Surface Constitutive Model For Clays," presented at NSF/NSERC symposium, Montreal, May 1980.

20. Domaschuk, L., and Valliappan, P., "Nonlinear Settlement Analysis by Finite Elements," *Journal of the Geotechnical Engineering Division*, ASCE, Vol. 101, No. GT7, July, 1975, pp. 601-614.

21. Drnevich, V. P., "Constrained and Shear Moduli for Finite Elements," *Journal of the Geotechnical Engineering Division*, *ASCE*, Vol. 101, No. GT5, May, 1975, pp. 459-473.

22. Drnevich, V. P., and Massarch, K. R., "Sample Disturbance and Stress-Strain Behavior," *Journal of the Geotechnical Engineering Division*, *ASCE*, Vol. 105, No. GT9, September, 1979, pp. 1001-1016.

23. Drucker, D. C., "Some Implications of Work Hardening and Ideal Plasticity," *Quarterly of Applied Mathematics*, Vol. 7, No. 4, 1950, pp. 411-418.

24. Drucker, D. C., and Prager, W., "Soil Mechanics and Plastic Analysis or Limit Design," *Quarterly of Applied Mathematics*, Vol. 10, No. 2, 1952, pp. 157-165.

25. Drucker, D. C., "Coulomb Friction, Plasticity and Limit Loads," *Journal of Applied Mechanics*, Vol. 21, 1954, pp. 71-74.

26. Drucker, D. C., Gibson, R. E., and Henkel, D. J., "Soil Mechanics and Work Hardening Theories of Plasticity," *Proceedings*, *ASCE*, Separate 798, 1955.

27. Dumas, F., and Lee, K. L., "Cyclic Movements of Offshore Structures on Clay," *Journal of the Geotechnical Engineering Division*, *ASCE*, Vol. 106, No. GT8, August, 1980, pp. 877-897.

28. Duncan, J. M., "Behavior and Design of Long Span Metal Culverts," *Journal of the Geotechnical Engineering Division*, *ASCE*, Vol. 105, No. GT3, March, 1970, pp. 399-418.

29. Duncan, J. M., and Chang, C. Y., "Nonlinear Analysis of Stress and Strain in Soils," *Journal of the Soil Mechanics and Foundations Division*, *ASCE*, Vol. 96, No. SM5, September, 1970, pp. 1629-1653.

30. Eisenstein, Z., and Law, S. T. C., "Analysis of Consolidation Behavior of Mica Dam," *Journal of the Geotechnical Engineering Division*, *ASCE*, Vol. 103, No. GT8, August, 1977, pp. 879-895.

31. Eisenstein, Z., and Thomson, S., "Geotechnical Performance of a Tunnel in Till," *Canadian Geotechnical Journal*, Vol. 15, No. 3, August, 1978, pp. 332-345.

32. Finn, W. D. L., Byrne, P. M., and Martin, G. R., "Seismic Response and Liquefaction of Sands," *Journal of the Geotechnical Engineering Division*, *ASCE*, Vol. 102, No. GT8, August, 1976, pp. 841-856.

33. Foott, R., and Ladd, C. C., " Behaviour of Atchafalaya Levees during Construction," *Geotechnique*, Vol. 27, No. 2, June, 1977, pp.137-160.

34. Hill, R., *The Mathematical Theory of Plasticity*, Oxford, Clarendon Press, 1950.

35. Hoeg, K., "Finite Element Analysis of Strain-Softening Clay," *Journal of the Soil Mechanics and Foundations Division, ASCE*, Vol. 98, No. SM1, January, 1971, pp. 43-58.

36. Hoeg, K., Christian, J. T., and Whitman, R. V., "Settlement of Strip Load on Elastic-Platic Soil," *Journal of the Soil Mechanics and Foundations Division, ASCE*, Vol. 94, No. SM2, March, 1968, pp. 431-445.

37. Hoeg, K., Andersland, O. B., and Rolfsen, E. N., "Undrained Behaviour of Thick Clay under Load Tests at Asrum," *Geotechnique*, Vol. 19, No. 1, March 1969, pp. 101-115.

38. Joyner, W. B., and Chen, A. T. F., "Calcualtion of Nonlinear Ground Responses in Earthquakes," *Bulletin of the Seismological Society of America*, Vol. 65, No. 5, October, 1975, pp. 1315-1336.

39. Kavazanjian, E., Jr., and Mitchell, J. K., "Time-Dependent Deformation Behavior of Clays," *Journal of the Geotechnical Engineering Division, ASCE*, Vol. 106, No. GT6, June 1980, pp. 611-630.

40. Kawamoto, T., and Okuzono, K., "Analysis of Ground Surface Settlement due to Shallow Shield Tunnels," *International Journal for Numerical and Analytical Methods in Geomechanics*, Vol. 1, No. 3, July-September, 1977, pp. 271-281.

41. Ko, H. Y., and Sture, S., "Data Reduction and Applications for Analytical Modeling," State of the Art paper presented at ASTM symposium, Chicago, June 1980.

42. Kondner, R. L., "Hyperbolic Stress-Strain Response: Cohesive Soils," *Journal of the Soil Mechanics and Foundations Division, ASCE*, Vol. 89, No. SM1, February 1963, pp. 115-143.

43. Krizek, R. J., and McQuade, P. V., "Behavior of Buried Concrete Pipe," *Journal of the Geotechnical Engineering Division, ASCE*, Vol. 104, No. GT7, July, 1978, pp. 815-836.

44. Kulhawy, F. H., Duncan, J. M., and Seed, H. B., "Finite Element Analyses of Stresses and Movements in Embankments during Construction," Report No. S-69-8, U. S. Army Engineer Waterways Experiment Station, November 1969.

45. Lacasse, S. M., and Ladd, C. C., "Undrained Behavior of Embankments on New Liskeard Varved Clay," *Canadian Geotechnical Journal*, Vol. 14, No. 3, August, 1977, pp. 367-388.

46. Ladd, C. C., and Foott, R., "New Design Procedure for Stability of Soft Clays," *Journal of the Geotechnical Engineering Division, ASCE*, Vol. 100, No. GT7, July 1974, pp. 763-786.

47. Lade, P. V., and Duncan, J. M., "Elasto-Plastic Stress-Strain Theory for Cohesionless Soil," *Journal of the Geotechnical Engineering Division*, ASCE, Vol. 101, No. GT10, October 1975, pp. 1037-1053.

48. Lade, P. V., "Elasto-Plastic Stress-Strain Model for Sand," presented at NSF/NSERC symposium, Montreal, May 1980.

49. Leathers, F. D., and Ladd, C. C., "Behavior of and Embankment on New York Varved Clay," *Canadian Geotechnical Journal*, Vol. 15, No. 2, May, 1978, pp. 250-268.

50. Lee, K. L., and Idriss, I. M., "Static Stresses by Linear and Nonlinear Methods," *Journal of the Geotechnical Engineering Division*, ASCE, Vol. 101, No. GT9, September, 1975, pp. 871-887.

51. Marr, W. A., Jr., and Christian, J. T., "Permanent Displacements Due to Cyclic Wave Loading," *Journal of the Geotechnical Engineering Division*, ASCE, Vol. 107, No. GT8, August 1981.

52. Martin, H. L., "A Three-Dimensional Deformation Analysis of the Storvass Dam," *International Journal for Numerical and Analytical Methods in Geomechanics*, Vol. 2, No. 1, January-March, 1978, pp. 1-17.

53. Mroz, Z., "On Hypoelasticity and Plasticity Approaches to Constitutive Modeling of Inelastic Behaviour of Soils," *International Journal for Numerical and Analytical Methods in Geomechanics*, Vol. 4, No. 1, January-March, 1980, pp. 45-55.

54. Mroz, Z., Norris, V. A., and Zienkiewicz, O. C., "An Anisotropic Hardening Model for Soils and Its Application to Cyclic Loading," *International Journal for Numerical and Analytical Methods in Geomechanics*, Vol. 2, 1978, pp. 203-221.

55. Mroz, Z., Norris, V. A., and Zienkiewicz, O. C., "Application of an Anisotropic Hardening Model in the Analysis of Elasto-Plastic Deformation of Soils," *Geotechnique*, Vol. 29, No. 1, March, 1979, pp. 1-34.

56. Murphy, D. J., Clough, G. W., and Woolworth, R. S., "Temporary Excavation in Varved Clay," *Journal of the Geotechnical Engineering Division*, ASCE, Vol. 101, No. GT3, March, 1975, pp. 279-295.

57. Osaimi, A. E., and Clough, G. W., "Pore Pressure Dissipation during Excavation," *Journal of the Geotechnical Engineering Division*, ASCE, Vol. 105, No. GT4, April, 1979, pp. 481-498.

58. Parry, R. G. H. (ed.), *Stress-Strain Behaviour of Soils, Proceedings of the Roscoe Memorial Symposium*, G. T. Foulis & Co., Henley-on-Thames, 1972.

59. Prevost, J. H., "Anisotropic Undrained Stress-Strain Behavior of Clays," Journal of the Geotechnical Engineering Division, ASCE, Vol. 104, No. GT8, August, 1978, pp. 1075-1090.

60. Prevost, J. H., "Plasticity Theory for Soil Stress-Strain Behavior," Journal of the Engineering Mechanics Division, ASCE, Vol. 104, No. EM5, October, 1978, pp. 1177-1194.

61. Prevost, J. H., and Hoeg, K., "Analysis of Pressuremeter in Strain Softening Soil," Journal of the Geotechnical Engineering Division, ASCE, Vol. 101, No. GT8, August, 1975, pp. 717-732.

62. Prevost, J. H., and Hughes, T. J. R., "Finite Element Solution of Boundary Value Problems in Soil Mechanics," Proceedings, International Symposium on Soils under Cyclic and Transient Loading, Swansea, U. K., January, 1980.

63. Prevost, J. H., Cuny, B., Hughes, T. J. R., and Scott, R. F., "Offshore Gravity Structures: Analysis," Journal of the Geotechnical Engineering Division, ASCE, Vol. 107, No. GT2, February, 1981, pp. 143-165.

64. Pyke, R. M., "Nonlinear Models for Irregular Cyclic Loadings," Journal of the Geotechnical Engineering Division, ASCE, Vol. 105, No. GT6, June, 1979, pp. 715-726.

65. Roscoe, K. H., Schofield, A. N., and Wroth, C. P., "On the Yielding of Soils," Geotechnique, Vol 8, No. 1, March, 1958, pp 25-53.

66. Sandler, I. S., DiMaggio, F. L., and Baladi, G. Y., "Generalized Cap Model for Geological Materials," Journal of the Geotechnical Engineering Division, ASCE, Vol. 102, No. GT7, July, 1976, pp. 683-699.

67. Schmidt, B., and Grantz, W. C., "Settlements of Immersed Tunnels," Journal of the Geotechnical Engineering Division, ASCE, Vol. 105, No. GT9, September, 1979, pp. 1031-1047.

68. Schofield, A. N., and Wroth, C. P., Critical State Soil Mechanics, London, McGraw-Hill, 1968.

69. Simon, R. N., Christian, J. T., and Ladd, C. C., "Analysis of Undrained Behavior of Loads on Clay," Proceedings, ASCE Specialty Conference on Analysis and Design in Geotechnical Engineering, Austin, TX, 1974, Vol. I, pp. 51-84, also Vol. II, pp. 139-140.

70. Simpson, B., O'Riordan, N. J., and Croft, D. D., "A Computer Model for the Analysis of Ground Movements in London Clay," Geotechnique, Vol. 29, No. 2, June, 1979, pp. 149-175.

71. Streeter, V. L., Wylie, E. B., and Richart, F. E., Jr., "Soil Motion Computation by Characteristic Methods," Journal of the Geotechnical Engineering Division, ASCE, Vol. 100, No. GT3, March, 1974, pp. 247-263.

72. Tan, D. Y., and Clough, G. W., "Ground Control for Tunnels by Soil Grouting," *Journal of the Geotechnical Engineering Division, ASCE*, Vol. 106, No. GT9, September, 1980, pp. 1037-1057.

73. Tavenas, F., Mieussens, C., and Bourges, F., "Lateral Displacements an Clay Foundations under Embankments," *Canadian Geotechnical Journal*, Vol. 16, No. 3, August, 1979, pp. 532-560.

74. Tavenas, F., and Leroueil, S., "The Behavior of Embankments on Clay Foundations," *Canadian Geotechnical Journal*, Vol. 17, No. 2, May, 1980, pp. 236-260.

75. Whang, B., "Elasto-Plastic Orthotropic Plates and Shells," *Proceedings, Symposium on Application of Finite Element Methods in Civil Engineering*, Nashville, TN, 1969, pp. 481-515.

DEFORMATION ANALYSIS FOR BRACED EXCAVATION IN CLAY

by

Ching S. Chang[1], A.M. ASCE and Mohd H.B. Abas[2]

ABSTRACT

In many practical cases of excavation in clay, the test data are neither extensive nor accurate enough for the determination of soil stress-strain parameters required for finite element analyses. For practical purposes, an empirical relationship is usually used to estimate the soil modulus, as suggested by many previous researchers. Utilizing one such empirical relationship, two case histories were analyzed with the finite element method. The calculated response of the excavation was compared with the measured field behavior. It was found that the empirical relationship was not applicable for sensitive or organic clays. Special consideration should be given to these conditions. The effects of increasing soil strength and modulus on strut loads, wall moments and deflections were compared.

INTRODUCTION

The primary quantities that are required for prediction of braced excavations are earth pressure loads, as well as lateral and vertical movements of the supports and of the soil. While prediction of loads can be made using the well known Terzaghi and Peck or the Tschebotarioff method, there is no conventional method for predicting support and soil movements. The complexities arise from the uncertainties regarding the actual earth pressure distributions at each stage of excavation, the location of the point of fixity of the embedded length of the piles or walls, and the degree of fixity. In order to model the excavation sequence and the complex boundary conditions of such problems, the finite element method is used and found satisfactory by many investigators (Refs. 4,8,12,19).

For clay foundations, the undrained compressive strength and soil modulus are the major parameters that govern the accuracy of the finite element analysis. Unfortunately, in order to have meaningful test data for determining these two parameters, a great deal of effort in carefully obtaining and testing undisturbed natural clay samples is required. However, for many cases such information is not sufficiently available. To supplement insufficient test data, an alternative method suggested by Bjerrum (Ref. 2), D'Appolonia, et al.

KEY WORDS: Soil Strength, Soil Modulus, Braced Excavation, Finite Element Method, Deformation

---

[1] Assistant Professor of Civil Engineering, University of Massachusetts, Amherst, Massachusetts 01003.

[2] Former Graduate Research Assistant, State University of New York at Buffalo, Buffalo, New York 14223.

(Ref. 7) and others is to use empirical relationships for determining soil modulus, based on undrained shear strength. In this paper, the results of finite element analyses of two case studies utilizing this established empirical relationship are presented in comparison with measured results. The validity of using this empirical relationship in excavation analysis is evaluated. The effects of soil strength and soil modulus on the wall deflection, wall bending moment, strut load, and earth pressures are also investigated.

Brief Review of Braced Excavation Analysis

One approach to the excavation analysis is to regard the soil and struts as spring restraints to the basic structural member, such as the wall or pile. Based on this approach, analyses have been made by Turabi and Balla (Ref. 20) and Haliburton (Ref. 11). Following the same approach, Rahut (Ref. 17) tried to incorporate friction between pile and soil in the model. The disadvantage of this approach is the model's incapability of handling the interaction between soil and the pile. In order to appropriately model the soil-structure interaction, another powerful approach, the finite element technique, is utilized. The finite element technique is not only capable of modeling soil-structure interaction, but is also capable of modeling the construction sequence. Among the works based on the finite element method are the efforts of Morgenstern and Einstein (Ref. 13), Wong (Ref. 22), Palmer and Kenney (Ref. 15), Clough and Duncan (Ref. 5), Tsui and Clough (Ref. 19) and Mana (Ref. 12). Effect of clay anisotropy has been considered in the finite element analysis by Clough and Hansen (Ref. 6).

Soil Parameters

The ability to simulate realistically the stress-strain characteristics of the soil determines the accuracy of the excavation analysis. The hyperbolic model developed by Duncan and Chang (Ref. 9) has been proven to be reasonably adequate for excavation analysis (Refs. 5,7). Thus, it was incorporated in the finite element program "EXCAV" (Ref. 4) for use in this study.

In the hyperbolic model, the initial tangent modulus, E_i, is a function of confining pressure, σ_3 as follows:

$$E_i = K P_a \left(\frac{\sigma_3}{P_a}\right)^n , \qquad (1)$$

where K is the elastic modulus value of the soil and P_a is the atmospheric pressure. For saturated clay soil, it is reasonable to assume that the elastic modulus exponent, n, is zero.

Determining values of E_i for natural clay requires a great deal of effort because a small amount of disturbance can significantly reduce the magnitude of the modulus. Therefore, one must carefully obtain and test a large number of undisturbed clay samples. In most practical cases, the test data are neither extensive nor accurate enough to adequately represent the stress-strain behavior of soil in the field. In the situation of insufficient available data, some empirical correlations would be useful guidelines in interpreting the available data. The empirical correlation

$$E_i = \eta S_u , \qquad (2)$$

between E_i and undrained compressive strength, S_u, is suggested by D'Appolonia, et al. (Ref. 7), Foott and Ladd (Ref. 10), Bjerrum (Ref. 2), and others. In Eq. 2, η is a function of plasticity index and is summarized in Fig. 1a. This chart was derived from empirical data on different types of normally consolidated clay. For over-consolidated clay, the correction factor, C_o, is given in Fig. 1b, and the value of E_i can be estimated by the following equation:

$$E_i = C_o \eta S_u . \tag{3}$$

Once E_i is determined, parameter K can be found from Eq. 1, assuming parameter n is zero. Besides parameter K and parameter n, the soil model requires the following parameters: R_f, c, ϕ, $\Delta\phi$ and ν, where the value of the failure ratio R_f for all soft clays ranges between 0.75 and 1.0, the cohesion parameter c is equal to $S_u/2$, ϕ and $\Delta\phi$ are zero, and the Poisson's ratio is set equal to 0.5 for saturated clays.

Case Study I

The excavation site chosen for the first case study is the Vaterland 1 site on the Oslo subway, which was the subject of one of a series of reports of measurements at braced excavations in Norway (Ref. 14). This recorded excavation was cited by Peck (Ref. 16) and was also the subject of a case study by Palmer and Kenney (Ref. 15). Figure 2 shows some of the details concerning the site and subsoil conditions.

The stratigraphy of the soils obtained at the site is shown in Fig. 3. The top-most layer consists of fill which is underlain by clay. Below this clay layer is a stratum of weathered clay containing some sand, followed by a layer of sensitive soft clay seated on bedrock. The depth to bedrock below ground surface is about 52 ft (15.9 m). The ground water table is close to the ground surface. Figure 3 also shows the properties of the soils on the site.

The excavation was carried out to a depth of 24 ft (10.5 m) and the width of the excavation was 36 ft (11.0 m). The earth retaining system used at the site consisted of Belval steel-sheet pile, section BZ-IVN-50, which was driven into bedrock. Wales were used and lateral restraints were provided by the installation of five cross-braces which were spaced at intervals of 10.5 ft (3.2 m) horizontally and with vertical spacing as shown in Fig. 2.

The finite element mesh used in the case study for the simulation of the Vaterland 1 excavation is shown in Fig. 4.

The section at the excavation site is symmetrical about the centerline; therefore, only half of the section needs to be considered. An artificial lateral boundary on the right side of the excavation is taken far enough away from the back of the sheet pile at a location where no disturbance is assumed in the soils. The bottom tip of the pile is driven to bedrock and a hinged joint is assigned to this end of the pile. The top of the pile being a free end, it is permitted movements in both directions and is also free to rotate. The connection between strut and wale and the pile is considered to be hinged.

The stages of excavation are indicated on Fig. 4. The time period to complete the excavation work was 73 days.

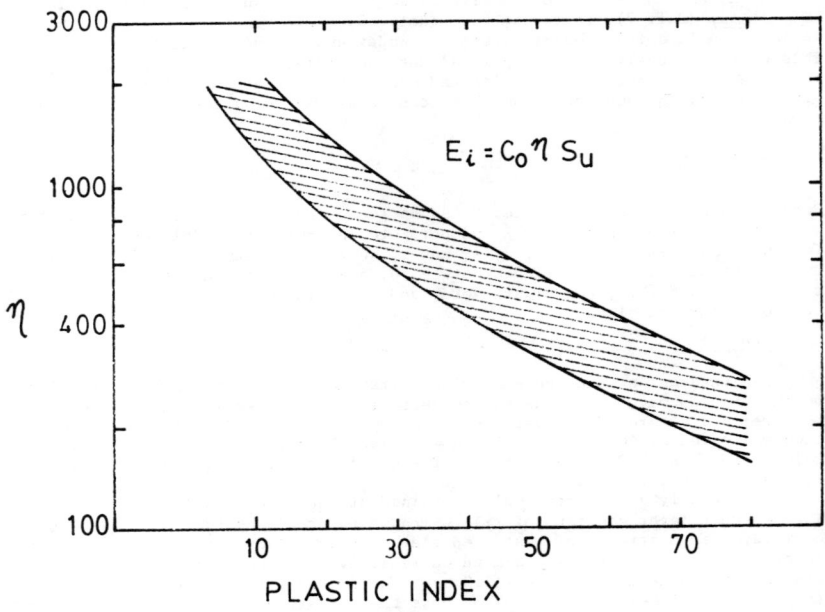

Fig. 1a. Values of η for Normally Consolidated Inorganic Clay

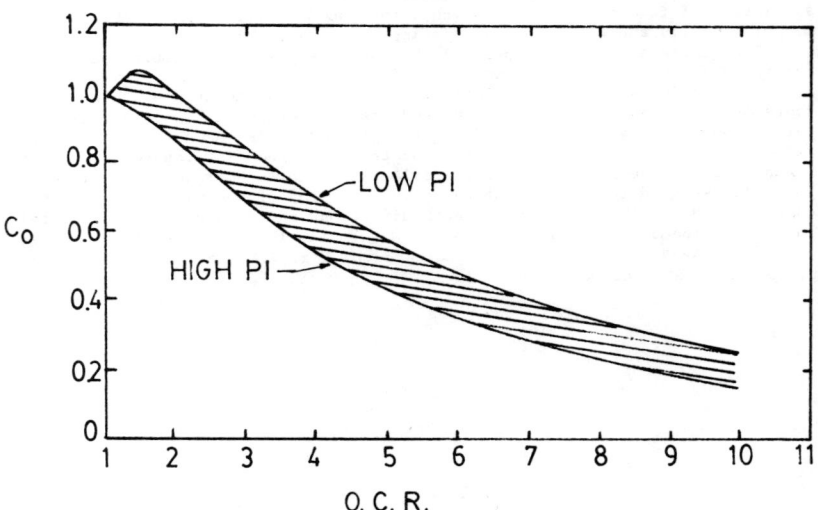

Fig. 1b. Values of Correction Factor for Overconsolidation Ratio

BRACED EXCAVATION IN CLAY

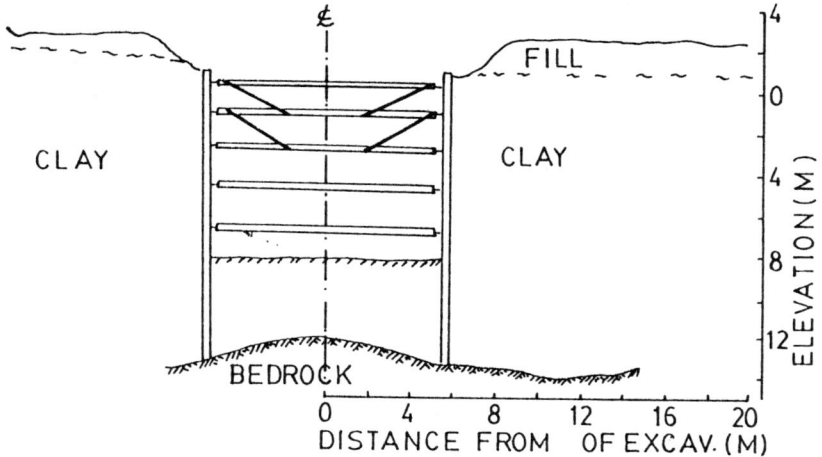

FIG. 2. Excavation Section - Vaterland 1 Site (Ref. 15)

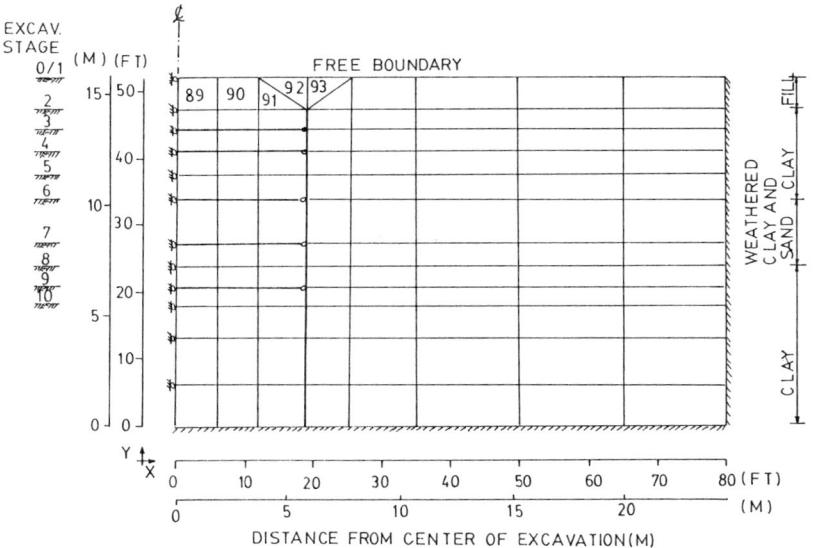

FIG. 4. Finite Element Mesh for Computer Simulation of Excavation - Vaterland 1

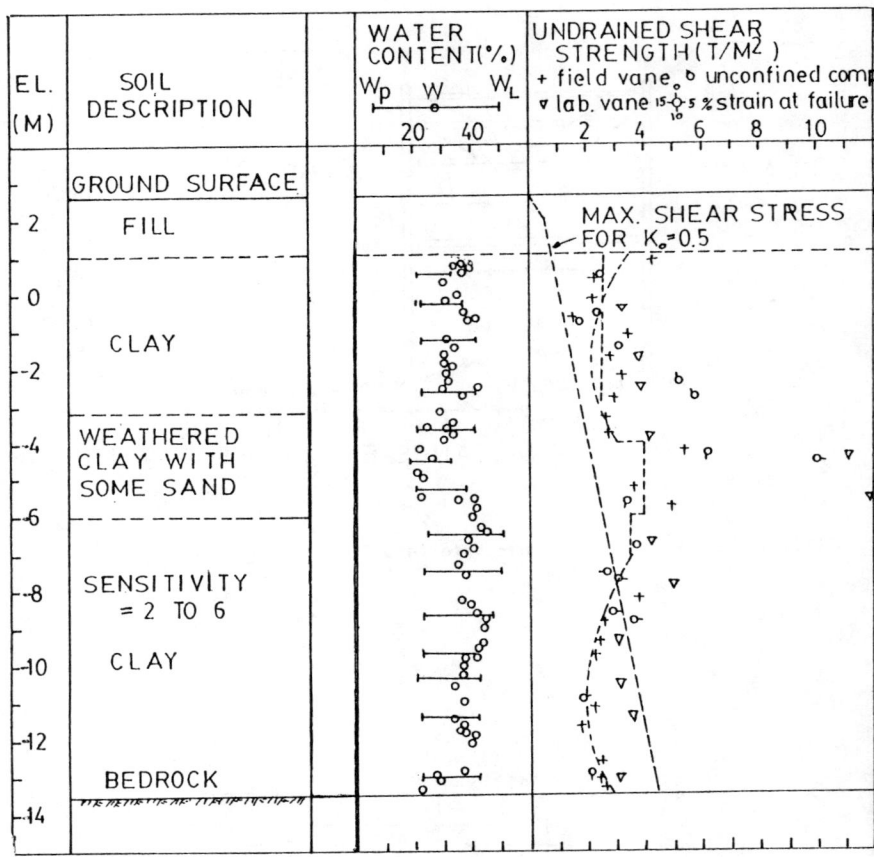

Fig. 3. Typical Soil Properties - Vaterland 1 Site (Ref. 15)

Properties of Soil and Sheet Piles

Results of both unconfined compression tests and vane shear tests for the soil layers are shown in Fig. 3. For the purpose of comparing results with those in Ref. 15 analyzed by Palmer and Kenney (Ref. 15), the values of undrained strength for each soil layer were chosen similar to those used in Ref. 15. An undrained strength S_u = 510 lb/ft^2 (24.5 kN/m^2) has been taken for the top clay layer. For the weathered clay, S_u is taken as 820 lb/ft^2 (39.2 kN/m^2). As no data are available for the shear strength of the fill, this layer is assumed to have the same shear strength as the underlying clay. An average shear strength of 720 lb/ft^2 (34.3 kN/m^2) is taken as uniform over the depth of bottom-most clay layer. With initial ground water table close to the ground surface, it is reasonable to assume that the soils had remained saturated throughout the duration of the excavation. The saturated unit weight, γ, equal to 125 lb/ft^3 (19.6 kN/m^3) is assumed for all soil layers.

The values of E_i were determined by using Eq. 2, where the values of η, based on soil plastic index, have been estimated from Fig. 1. The values of plastic index for the soil materials on site, shown in Fig. 3, range from 20 to 30. A complete tabulation of the soil parameters used in the analysis of the Vaterland 1 excavation are listed in Table 1. After E_i for each layer is determined, the values of K can then be determined by using Eq. 1.

The Belval steel sheet pile section BZ-IVN-50 has a moment of inertia, I = 217 in.^{4}/ft (29,600 cm^4/m), and modulus of elasticity E_s = 29.9 x 10^6 psi (2.1 x 10^5 KN/cm^2). The equivalent stiffness of the struts, wales and blocking varied from a low value of 1.34 x 10^6 lb/ft (196 kN/cm) to a high value of 3.36 x 10^6 lb/ft (491 kN/cm), based on actual field performance. The strut spacing was 10.5 ft (3.2 m). From the equivalent strut stiffness, it was calculated that the equivalent strut area varied from 1.44 in.^{2}/ft (30.5 cm^2/m) to 3.60 in.^{2}/ft (76.2 cm^2/m). As a simplification, an equivalent strut area of 2.25 in.^{2}/ft (47.6 cm^2/m) was assumed for the analysis.

Comparison of Measured and Computed Results

The measured maximum deflection of the wall, maximum ground settlement, and maximum strut load, compared with the computed results, are listed in Table 2. The results analyzed by Palmer and Kenney (Ref. 15), using a bilinear elastic model, are also included in Table 2 for comparison. Palmer and Kenney indicated that the values of soil moduli for this site should be 90 S_u or lower in order to have reasonable agreement in measured and predicted deformations. Whereas the values of E_i estimated from Fig. 1 are approximately 700 S_u, it is obvious that the soil moduli estimated from Fig. 1 represent much stiffer soils than they really are in the field. The results also indicate that the empirical relationship shown in Fig. 1 is not applicable to the estimated E_i for sensitive clays. Since Fig. 1 is a summary of data from inorganic clays, it is recommended that, besides sensitive clays, Fig. 1 is also not used for organic highly plastic clays.

In Table 2, besides the deformations, all the computed values of maximum strut loads were also underestimated. From the computed maximum strut loads, the apparent earth pressures are plotted and compared with Peck's design pressure diagram with m = 0.4 and 1.0 in Fig. 5.

Peck's design pressure diagrams shown in the figure have included the effect of the 5 ft (1.5 m) thickness of soil above the sheet pile as a surcharge effect, thus defining the magnitude of the earth pressure as

TABLE 1 - Excavated Soil Parameters for the Vaterland 1 Excavation

Soil Parameters	TYPE OF SOIL			Unit
	Fill and Underlying Clay Layer	Weathered Clay and Some Sand	Bottom-most Clay layer	
S_u	510 (24.5)	820 (39.2)	720 (34.3)	lb/ft^2 (kN/m^2)
γ	125 (19.6)	125 (19.6)	125 (19.6)	lb/ft^3 (kN/m^3)
K	168	271	238	
n	0	0	0	
R_f	0.95	0.95	0.95	
ν	0.49	0.49	0.49	
c	255 (12.2)	410 (19.6)	360 (17.1)	lb/ft^2 (kN/m^2)
ϕ	0	0	0	

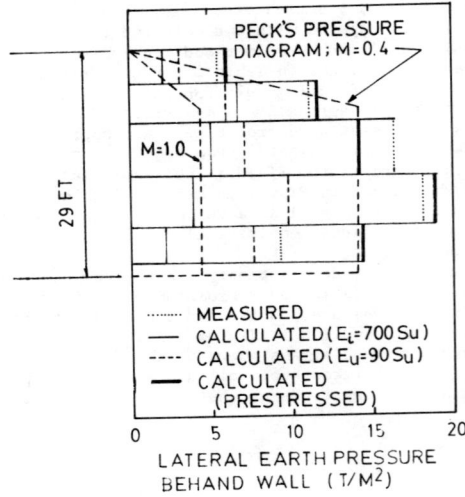

Fig. 5. Comparison of Computed Apparent Earth Pressures and Peck's Design Earth Pressure

Table 2. Comparison Between Computed and Measured Results

Maximum Deflection of the Wall	Measured	23 cm
	Hyperbolic $E_i = 700\ S_u$	6 cm
	Bilinear Elastic* $E_i = 150\ S_u$	13 cm
	Bilinear Elastic* $E_i = 90\ S_u$	18 cm
Maximum Settlement of the Ground	Measured	17 cm
	Hyperbolic $E_i = 700\ S_u$	4.2 cm
	Bilinear Elastic* $E_i = 150\ S_u$	10 cm
	Bilinear Elastic* $E_i = 90\ S_u$	11 cm
Maximum Strut Load	Measured	40 t/m
	Hyperbolic $E_i = 700\ S_u$	14 t/m
	Bilinear Elastic* $E_i = 150\ S_u$	20 t/m
	Bilinear Elastic* $E_i = 90\ S_u$	20 t/m

1 ton = 9.8 kN

*From Palmer and Kenney (Ref. 15)

$$(\gamma H + \gamma h) \; (1 - \frac{4mS_u}{\gamma H}) \; , \tag{4}$$

where γ is the unit weight of soil mass, H is the depth of excavation, h is the thickness of the soil surcharge, S_u is the unconfined shear strength of clay, and m is a coefficient varying between 0.4 and 1.0. The value of S_u assumed is that of the bottom clay layer.

From Fig. 5, it can be seen that the measured maximum apparent earth pressure agrees quite closely with Peck's design diagram for the value of m = 0.4, which is typical of Oslo clay. All the computed apparent earth pressures fall within the envelope, except the results computed based on 90 S_u, and assuming the struts were 100% pre-stressed.

Case Study II

The subway station "Onoue-Cho," on which the second case study is based, is located in Yokohama, Japan. This station is a three-floor underground structure, as shown in Fig. 6. This case study has previously been analyzed in Ref. 1. In the present paper, the soil moduli, based on two different estimated profiles of soil strength as shown in Fig. 7, were determined by the previously described empirical relationship. The computed results are compared to investigate the effects of soil modulus and soil strength on the deformation and the bending moment of the wall, and on the earth pressures behind the wall.

The station is located on a clay deposit which extends downwards 30 to 40 meters. On top of the clay deposit is a 5 to 7 meter thick clay fill. The water content, unit weight and undrained strength of the clay are plotted against the depth in Fig. 7. Other properties are shown in Table 3. Underlying the clay is a layer of very firm silty soil into which large capacity foundation piles are penetrated. The ground water table is about 1.5 meters below the surface of the ground, and the distribution of pore pressure was reported as almost hydrostatic.

The station is located under a road with heavy traffic in a highly developed business area of high-rise buildings and utilities. A slurry wall was chosen among many alternatives, because in the construction of a slurry wall, the noise, vibration and ground settlement are reduced to a minimum. The slurry wall was designed to have enough strength not only for the retaining wall during excavation, but also for the permanent structure of the station.

In order to reduce the inclination of the vertical deep trench to a minimum, a Kelly Excavator with a hydraulic grab and a 40-m-high Kelly guide were employed. The Kelly Excavator is efficient for deep excavation, and therefore excavation of the trench could be quickly done at night, when the traffic was relatively light. The excavation of soil between the slurry walls was done mostly by manpower, because the clay was too soft for heavy equipment.

There were eight struts on the vertical section. During the excavation, these struts were placed and preloaded to support the walls. Measurements of earth pressure, stress in the concrete and stress in the reinforcement were made during the excavation. The measurements were made at two panels on both sides of the station. The gages for those measurements were installed in the steel reinforcement cages. The gages for measuring earth pressure were installed on the surface of the slurry wall by using hydraulic jacks to prevent the gages from being covered with mortar paste. Strut loads were also measured. Details of these measurements can be found in Ref. 1. The station was completed successfully in 1974.

The finite element mesh employed in this excavation is shown in Fig. 8,

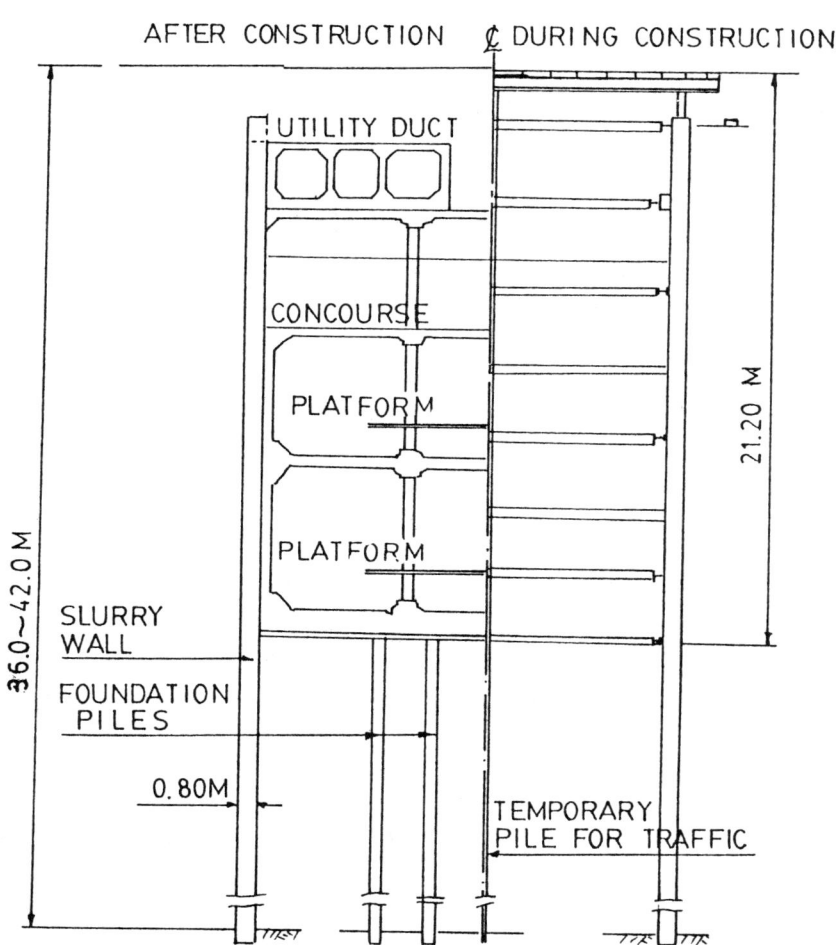

Fig. 6. Typical Section of the Excavation Plan

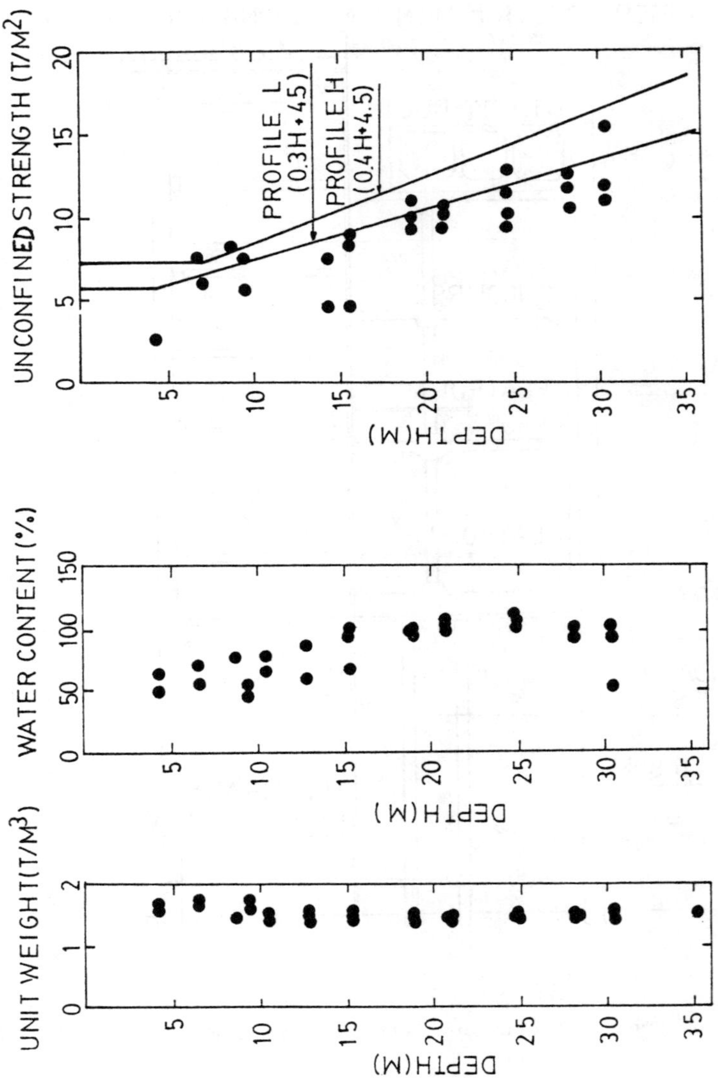

216 SOIL STRESS STRAIN APPLICATIONS

FIG. 7. Soil Properties of the Soft Clay in Excavation Site

TABLE 3. Soil Properties of the Yokohama Clay

Soil Properties	Unit	Depth		
		10m - 15m	15m - 25m	25m - 35m
Unit Weight	t/m^3	1.5 - 1.7	1.5	1.5
Water Content	%	50 - 100	100	100
Plastic Index	%	? - 60	60	55
Liquid Limit	%	? - 120	115	110
S_u	t/m^2	5 - 9	9 - 13	10 - 16
OCR	--	1.0 - 2.0	1.0 - 1.8	1.0 - 1.2

* 1 ton = 9.8 kN

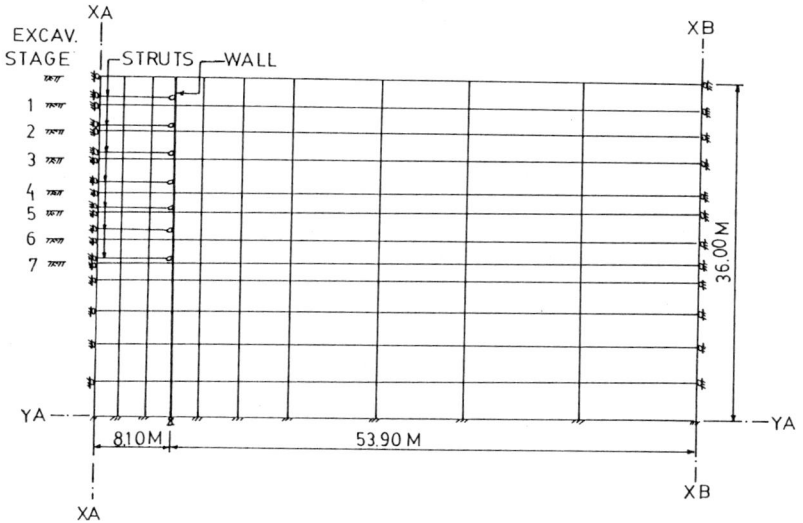

FIG. 8. Finite Element Mesh

where only a half-section of the excavation is shown, because of the symmetrical assumption. The distance from the wall to boundary line XA-XA is half the width of the excavation.

The depth of the excavation is 19.4 meters and the distance from wall to boundary line XB-XB is 53.9 meters. The nodal point at the bottom of the wall is fixed in the x and y directions, and allows only rotation. The other nodal points on the bottom boundary line YA-YA in Fig. 8 are fixed in all directions. The nodal points on the vertical boundary line, XA-XA or XB-XB, are fixed in x direction and in rotation, and movable in the y direction. The soil between the walls is excavated layer by layer in seven stages, as shown in Fig. 8.

Properties of Soil and Wall

Laboratory S_u values were determined from small diameter specimens which were obtained in the field with a thin-walled sampler. In-situ field S_u values should be, in general, higher than those obtained from the laboratory, due to unavoidable sample disturbance. To account for this effect, in Fig. 7, profile H and profile L were chosen to give the possible range of S_u values. Both profiles were used in the analysis to investigate the effects of soil strength on the excavation behavior.

E_i can be estimated from the chart shown in Fig. 1 using PI = 60 and OCR = 1.5. The values of E_i are

$$E_i = 200 \ S_u \sim 500 \ S_u \quad . \tag{5}$$

Both 200 S_u and 500 S_u were used in the analysis to investigate the effects of soil modulus on the excavation behavior. The density γ = 1.5 t/m^3 (14.7 kN/m^3) is estimated from the test results shown in Fig. 7. Failure ratio, R_f, is assumed to be 0.95 for the clay in this excavation. The modulus exponent n is assumed to be zero. Thus, the value of K is computed from Eq. 1.

The coefficient of earth pressure at rest for effective stress analysis is estimated based on values of plastic index and overconsolidation ratio from the chart by Brooker and Ireland (Ref. 3). For this excavation site, PI is about 55 and OCR is 1 to 2; K_o is approximated to be 0.65 from the chart.

For total stress analysis, the coefficient of earth pressure at rest, K_o^t, can be expressed as follows:

$$K_o^t = K_o + \frac{\gamma_w}{\gamma} (1-K_o) \quad , \tag{6}$$

where γ is the total unit weight of soil and γ_w is the unit weight of water. K_o^t is 0.88 for the case that K_o is 0.65, γ is 1.5 t/m^3 (14.7 kN/m^3), and γ_w is 1.0 t/m^3 (4.8 kN/m^3).

The slurry walls were made of reinforced concrete with a thickness of 0.8 meters. The moment of inertia is 0.0427 m^4/m. The average Young's modulus of the reinforced concrete is 2.8 x 10^6 t/m^2 (27.5 kN/m^2). The types of H-beams used as struts and preloads applied on struts are shown in Table 4.

Comparisons of Measured and Computed Results

Figures 9 and 10 show the comparison of calculated and measured earth pressures after the 4th and the 7th stages of excavation, respectively. As excavation depth increases, both the active and passive earth pressure decrease. Although the calculated values of earth pressures are overestimated in both cases,

TABLE 4. Strut Properties in the Yokohama Subway Excavation

Strut No.	Depth Below Ground (m)	H Beam	Section Area (m^2)	Preload (tons)
1	-1.8	H - 300 × 300 × 10 × 15	0.0118	15
2	-4.6	H - 388 × 402 × 15 × 15	0.0178	15
3	-7.85	H - 300 × 300 × 10 × 15	0.0118	15
4	-10.85	H - 388 × 402 × 10 × 15	0.0178	15
5	-13.3	H - 388 × 402 × 10 × 15	0.0178	15
6	-16.25	H - 388 × 402 × 10 × 15	0.0178	30
7	-18.60	H - 388 × 402 × 10 × 15	0.0178	20

*1 ton = 9.8 kN

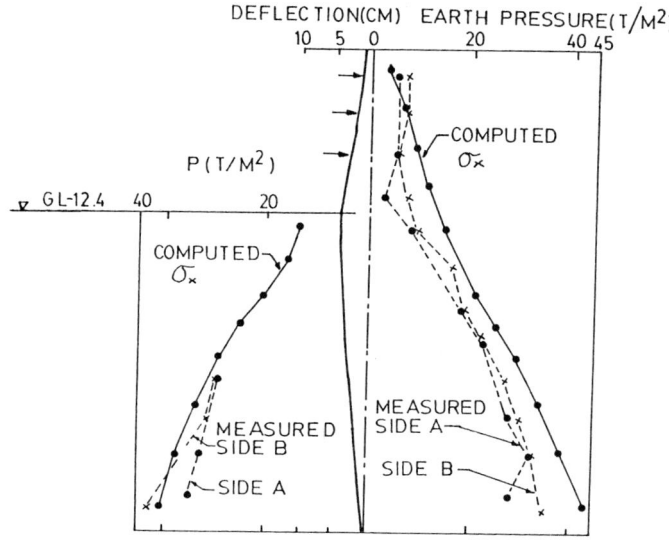

FIG. 9. Comparison of Measured and Computed Lateral Earth Pressure After 4th Excavation Stage

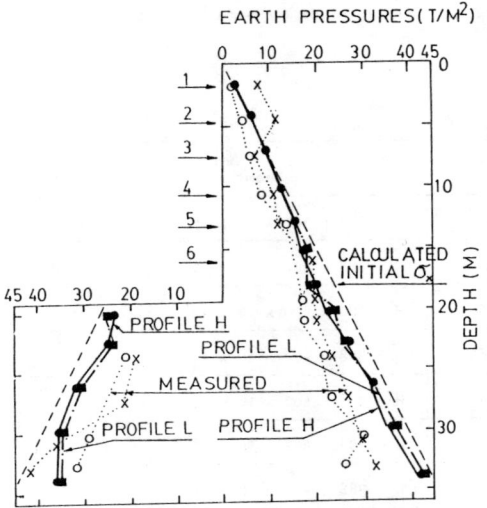

FIG. 10. Comparison of Computed and Measured Lateral Earth Pressures After the Final Stage of Excavation

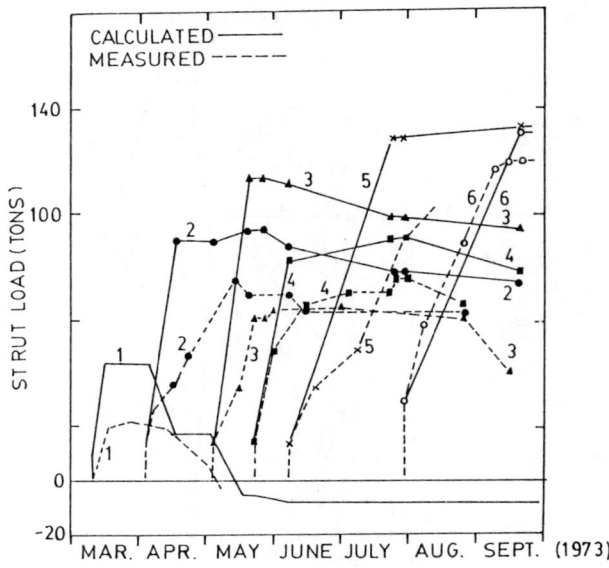

FIG. 11. Comparison of Measured and Computed Strut Loads (High Strength Profile)

the trends of changes in earth pressures during excavation are similar to those measured. The effect of soil strength on earth pressures is also shown in Fig. 10. Very little difference between the calculated earth pressures occurs as a result of employing different soil strengths.

Figure 11 shows the computed and measured strut loads. The behavior of changes in strut loads during excavation in the field agreed well with FEM results. However, the calculated magnitude of strut loads for the upper layers of struts are larger than those measured in the field. The comparison between apparent earth pressures calculated from the analysis and the apparent earth pressure envelope suggested by Terzaghi and Peck is shown in Fig. 12.

Figure 13 shows the computed bending moments for the two cases and the measured bending moments which were computed from the measured stresses in concrete and reinforcements. The measured maximum bending moments after the 7th excavation was 113 t-m/m (1107 kN-m/m). The computed maximum bending moment after the 7th (final) excavation for low strength profile is 129 t-m/m (1260 kN-m/m), and for high strength profile, 107 t-m/m (1050 kN-m/m). The shapes and magnitude of the computed bending moments agree with those measured in the field.

The results shown in Figs. 9 through 12 were computed using $E_i = 200\ S_u$. Similar analysis was performed for $E_i = 500\ S_u$, and the comparisons are summarized in Table 5. The effects of both soil strength and soil modulus on deflection, bending moment of the wall, strut loads, and earth pressures are listed in Table 6.

This case study indicated that the values of soil modulus E_i estimated from Fig. 1 provide a reasonable predicted behavior of wall during excavation.

Summary and Conclusions

1. Deformation analysis of excavations in clay often requires two principal parameters: undrained compressive strength, and a soil modulus. Unfortunately, due to sample disturbance, the values determined from the unconfined compression test or the unconsolidated-undrained test usually result in considerable scatter and are subject to variable interpretation. Data from field vane tests or consolidated undrained tests appear to be more reliable for estimating S_u values. For practical purposes, it is suggested that the soil modulus E_i be determined from the empirical relationship between E_i and S_u, shown in Fig. 1.

2. The case studies presented herein show that values of soil modulus E_i, estimated from Fig. 1, seem to yield reasonable solutions in the finite element excavation analysis. However, for sensitive clay, organic clay, or highly plastic clay, the values of soil modulus estimated from Fig. 1 may be significantly higher than those of the soil in the field. Future research is needed for better understanding of these types of soils.

3. From the scatter shown in the empirical relationship of Fig. 1, one can easily overestimate or underestimate the soil modulus E_i by 150%. Scatter also appears in the values of S_u. The uncertainty of the values is around 20-30% from the second case study. Table 5 shows changes of the amount in bending moment, deflection, strut loads and earth pressures, due to an increase in soil strength by 20%, and an increase in soil modulus by 150%. It can be seen that an overestimation of both soil strength and soil modulus would result in an underestimation of apparent earth pressure by 25% in the finite element analysis. On the other hand, by underestimating both soil strength and soil modulus, one can overestimate by 50% the wall deflections, and by 25% the apparent earth pressure. However, there are equal chances that one of the parameters is overestimated and the other is underestimated, which results in a solution with

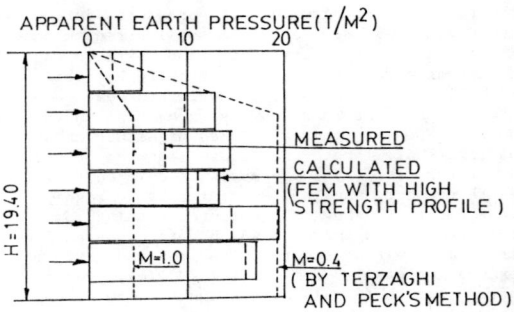

FIG. 12. Comparison of Computed Apparent Earth Pressure and Peck's Design Earth Pressure

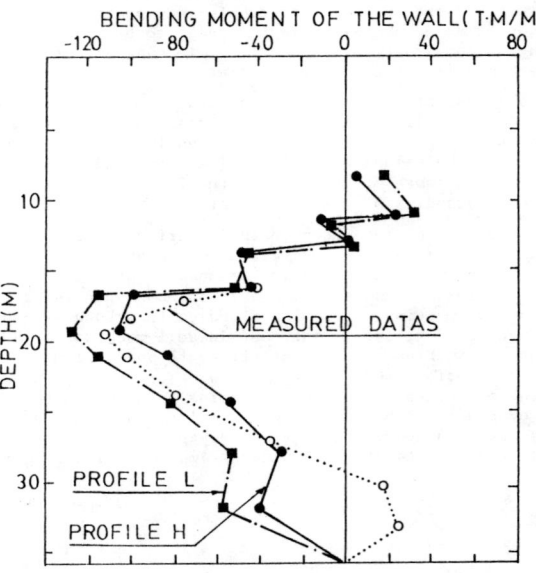

FIG. 13. Computed and Measured Bending Moment After the Final Stage of Excavation

BRACED EXCAVATION IN CLAY   223

Table 5. Comparison of Cases with Different Soil Strength and Modulus

	Measured	$E_i = 200 S_u$		$E_i = 500 S_u$	
		Low Strength Profile	High Strength Profile	High Strength	Low Strength
Maximum Bending Moment of the Wall (t-m/m)	11.3	129	107	74.9	92
Maximum Apparent Earth Pressure (t/m^2)	16	20.8	19.5	17.1	18.9
Maximum Strut Load (tons)	130	145	136	119	131
Maximum Delfection of the Wall (cm)	---	8.22	6.59	4.35	5.49

1 ton = 9.8 kN

Table 6. Effects of Soil Strength and Modulus on Bending Moment, Deflection, Strut Load and Earth Pressures

	Effect of Increased Strength (Avg. 20%)	Effect of Increased E_i/S_u 150%
Maximum Bending Moment	Decrease 20%	Decrease 30%
Maximum Apparent Earth Pressure (Maximum Strut Load)	Decrease 10% Decrease 10%	Decrease 15% Decrease 15%
Maximum Deflection	Decrease 25%	Decrease 35%
Active Earth Pressure	Decrease 0.5-1%	Decrease 0.5-1%
Passive Earth Pressure	Decrease 0-4%	Decrease 0-8%

compensating errors.

4. The constitutive model used in the analysis is a very simplified model, which cannot appropriately account for the different stress paths exhibited in the passive and the active soil zones. Besides the unreliability of the accuracy in soil models, there are still other uncertainties in excavation analysis, such as construction sequence, work performance, and ground water conditions. With this many uncertain factors about which we know very little, engineering judgment still plays the most important role in predicting deformation behavior, until a better understanding of the unknown factors is obtained.

REFERENCES

1. Aihara, I., Duncan, J.M. and Chang, C.S., "Finite Element Analysis of Sheathing for Soft Ground," Journal of the Japan Society of Civil Engineering, Vol. 64, No. 12, November, 1979, p. 41 (in Japanese).

2. Bjerrum, L., "Problems of Soil Mechanics and Construction on Soft Clays and Structurally Unstable Soils," State-of-the-Art Report, Proceedings, 8th International Conference on Soil Mechanics and Foundation Engineering, Vol. 3, Moscow, U.S.S.R., 1973, pp. 111-159.

3. Brooker, E.W. and Ireland, H.O., "Earth Pressure at Rest Related to Stress History," Canadian Geotechnical Journal, Ontario, Vol. 2, No. 1, 1965.

4. Chang, C.S. and Duncan, J.M., "EXCAV - A Finite Element Computer Program for Excavation Analysis," Department of Civil Engineering, University of California at Berkeley, June, 1977.

5. Clough, G.W. and Duncan, J.M., "Finite Element Analyses of Retaining Wall Behavior," Journal of the Soil Mechanics and Foundations Division, ASCE, Vol. 97, No. SM12, Proceedings Paper 8583, December, 1971, pp. 1657-1673.

6. Clough, G.W. and Hansen, L.A., "Clay Anisotropy and Braced Wall Behavior," Journal of the Geotechnical Engineering Division, ASCE, No. GT7, July, 1981, pp. 893-914.

7. D'Appolonia, D.J., Poulos, H.G. and Ladd, C.C., "Initial Settlement of Structures on Clays," Journal of the Soil Mechanics and Foundations Division, ASCE, Vol. 97, No. SM10, October, 1971, pp. 1359-1377.

8. Duncan, J.M. and Buchignani, A.L., "An Engineering Manual for Settlement Analysis," Department of Civil Engineering, University of California at Berkeley, June, 1976.

9. Duncan, J.M. and Chang, C.Y., "Nonlinear Analysis of Stress and Strain in Soils," Journal of the Soil Mechanics and Foundations Division, ASCE, No. SM5, September, 1970, pp. 1629-1652.

10. Foott, R. and Ladd, C.C., "Prediction of End of Construction Undrained Deformations of Atchafalaya Levee Foundation Clays," Massachusetts Institute of Technology, Soil Publication No. 305, June, 1972.

11. Haliburton, T.A., "Numerical Analysis of Flexible Retaining Structures," Journal of the Soil Mechanics and Foundations Division, ASCE, Vol. 94, No. SM6, 1968, pp. 1233-1235.

12. Mana, A.Z., "Finite Element Analysis of Deep Excavation Behavior in Soft Clay," Ph.D. Dissertation, Department of Civil Engineering, Stanford University, Stanford, California.

13. Morgenstern, N.R. and Einstein, Z., "Methods of Estimating Lateral Loads and Deformation," ASCE Specialty Conference, Cornell University, <u>Lateral Stresses in the Ground and Design of Earth-Retaining Structures</u>, 1970, pp. 51-102.
14. Norwegian Geotechnical Institute, "Measurements at a Strutted Excavation," Reports 1-9, 1962.
15. Palmer, J.H.L. and Kenney, T.C., "Analytical Study of a Braced Excavation in Weak Clay," <u>Canadian Geotechnical Journal</u>, No. 145, 1972, pp. 145-164.
16. Peck, R.B., "Deep Excavations and Tunneling in Soft Ground," <u>Proceedings</u>, 7th International Conference on Soil Mechanics and Foundation Engineering, State-of-the-Art Volume, Mexico City, 1969, pp. 225-290.
17. Rahut, J.B., "Discussion of Numerical Analysis of Flexible Retaining Structures by T.A. Haliburton," <u>Journal of the Soil Mechanics and Foundations Division</u>, ASCE, Vol. 95, No. SM6, 1968, pp. 1553-1564.
18. Tschebotarioff, G.P., <u>Foundations, Retaining and Earth Structures</u>, McGraw Hill Book Company, New York, 1973.
19. Tsui, Y. and Clough, G.W., "Plane Strain Approximations in Finite Element Analyses of Temporary Walls," <u>Proceedings</u>, ASCE Geotechnical Engineering Division Specialty Conference on Analysis and Design in Geotechnical Engineering, Vol. 1, Austin, Texas, 1974, pp. 1973-1998.
20. Turabi, D.A. and Balla, A., "Sheet-Pile Analysis by Distribution Theory," <u>Journal of the Soil Mechanics and Foundation Engineering Division</u>, ASCE, Vol. 94, No. SM1, 1968, pp. 291-322.
21. Turabi, D.A. and Balla, A., "Distribution of Earth Pressure on Sheet-Pile Walls," <u>Journal of the Soil Mechanics and Foundation Engineering Division</u>, ASCE, Vol. 94, No. SM6, 1968, pp. 1271-1301.
22. Wong, I.H., "Analysis of Braced Excavations," Ph.D. Dissertation, Department of Civil Engineering, Massachusetts Institute of Technology, 1971.

ACKNOWLEDGMENT

The writers wish to express their gratitutde to I. Aihara, who did some of the computer runs in this study, J.M. Duncan, who provided valuable input during the development of computer program EXCAV, and E.T. Selig, who reviewed the paper.

RE-EVALUATION OF WORK HARDENING MODEL

By Erman Evgin[1] and Zdenek Eisenstein[2], M.ASCE

ABSTRACT

The elasto-plastic work hardening constitutive model for cohesionless soils as proposed by Lade has been chosen for application in a finite element analysis. Previous applications of the model attempted by others have not been fully satisfactory despite their proposed modifications. In the present study the model is examined on the basis of analytical predictions of laboratory testing results. The model has been re-evaluated and a new approach is proposed. The re-evaluation centers around the work hardening law which controls the magnitude of plastic strain increments. Although the original formulation of the law is not changed a new interpretation is offered regarding its validity in a truly three-dimensional stress space. Predictions made with the re-evaluated constitutive model show markedly improved capabilities compared with the previous attempts.

KEY WORDS: Cohesionless soils; Elasticity; Finite element analysis; Plasticity; Soil mechanics; Strains; Stresses; Stress-strain curves.

[1] Research Asst., University of Alberta, Edmonton, Alberta, Canada.
[2] Prof. of Civil Engineering, University of Alberta, Edmonton, Alberta, Canada.

2

INTRODUCTION

The progress in stress and displacement analysis of soil structures using elastic stress-strain models has been seriously limited by difficulties in modelling several important features of soil behavior. While general stress-strain nonlinearity could be incorporated into elastic models in a satisfactory manner, the features associated with stress path dependency and dilatancy are much more difficult to deal with. The dependency of soil behavior on the stress path taken during loading has been long recognized (9). Attempts to employ stress path dependent soil models in elastic analysis have been reasonably successful and certainly can provide solutions for a number of practical problems (e.g., 3, 10). These solutions, however, lack generality and require special, nonconventional testing to determine soil deformation properties.

Many researchers have recently turned their attention to the development of constitutive models which would incorporate all the important features of soil behavior. However, due to their complexity not all the proposed models are readily adaptable for analyses of engineering problems by the finite element technique. Yet the applicability of a model to a practical problem is obviously the most important criterion in determining its value.

From this point of view the elasto-plastic stress-strain model as proposed by Lade (6) has been shown most promising by Evgin (4). However, its application in finite element analyses has not been without difficulties.

In the following, after a brief review of the model, the problems so far encountered during its application will be discussed. An attempt will then be made to restore its practical value by re-evaluating the model.

REVIEW OF THE MODEL

The work-hardening model proposed by Lade (6) accounts for several aspects of the stress-strain and strength characteristics of cohesionless soils under general three-dimensional stress conditions. To a certain extent, it has the capability of modelling nonlinearity, shear dilatancy, yielding, stress path dependency and influence of intermediate principal stress.

The theory is based on experimental data from cubical triaxial tests in sand and uses the concepts of the flow theory of plasticity. The total strain increments $\{d\varepsilon_{ij}\}$ are divided into an elastic part $\{d\varepsilon^e_{ij}\}$ and a plastic part $\{d\varepsilon^p_{ij}\}$ as

$$\{d\varepsilon_{ij}\} = \{d\varepsilon^e_{ij}\} + \{d\varepsilon^p_{ij}\} \qquad (1)$$

Each part is then calculated separately. The elastic strain increments are calculated from Hooke's Law, using the unloading-reloading modulus defined by Duncan and Chang (1). The value of Poisson's ratio is assumed to be equal to zero. This assumption is based on the triaxial test data where the axial and volumetric strains are very nearly equal for the first increment of load.

For the calculation of plastic strain increments, the model follows the basic requirements of plasticity theory as summarized by Lade and Duncan (7).

The yield surface is assumed to have the shape of a cone with the apex at the coordinate center of the principal stress space. It is expressed as a function of the first and third stress invariants:

$$f = \frac{I_1^3}{I_3} \tag{2}$$

Function f denotes the stress level and its value varies from 27 for hydrostatic stress conditions up to a value K_1 at failure. With increasing values of f, the yield surface expands continuously, and becomes identical with the failure surface at its outermost shape.

A non-associated flow rule is used for this model in which the plastic potential surface is no longer assumed to be the same as the yield surface. The plastic potential function incorporated in the theory is expressed by Lade and Duncan (7) in a form:

$$g = I_1^3 - K_2 I_3 \tag{3}$$

where:
  g is the value of the plastic potential and
  K_2 is a constant for a given value of f. K_2 is calculated from

$$K_2 = Af + 27(1 - A) \tag{4}$$

where:
  A is a material constant.

Eq. 3 describes a series of surfaces which are normal to the plastic strain increment directions.

The work hardening law adopted for the model is an experimentally determined relationship between the plastic work W_p and the stress level f. Test data by Lade (6) have indicated that the plastic work was very small for a range of f starting with 27 at hydrostatic state of stress up to a certain value which was called the threshold stress level f_t. As a convenience in fitting curves to the experimental data, it has been assumed that, for values of f between 27 and f_t, no plastic strains occur and no plastic work is done. Therefore, the relations between W_p and $(f-f_t)$ have been approximated by hyperbolae for which Eq. 5 has been proposed.

$$f - f_t = \frac{W_p}{a + bW_p} \tag{5}$$

The initial slope of a curve representing the W_p versus $(f-f_t)$ relationship is the reciprocal of the parameter a. The value of a increases with confining pressure, and this variation has been expressed as in Eq. 6.

$$a = Mp_a \left(\frac{\sigma_3}{p_a}\right)^\ell \tag{6}$$

where:
  p_a is atmospheric pressure expressed in the same units as a and σ_3.
  M and ℓ are dimensionless numbers.

The parameter b in Eq. 5 is the reciprocal of the ultimate value of $(f-f_t)$ which the hyperbola approaches asymptotically with increasing values of W_p. This relationship is given by Eq. 7.

$$b = \frac{1}{(f - f_t)_{ult}} \tag{7}$$

Since the value $(f-f_t)_{ult}$, determined from a curve fitting procedure, is always larger than the value of $(f-f_t)$ at failure for all finite values of W_p, a new parameter called r_f was introduced to relate the asymptotic value of $(f-f_t)$ to its value at failure as given in the following expression:

$$r_f = \frac{K_1 - f_t}{(f - f_t)_{ult}} \tag{8}$$

In the theory of plasticity, the relation between the plastic strain increments and the plastic potential function is given by an expression as follows:

$$\Delta \varepsilon_{ij}^p = \Delta\lambda \frac{\partial g}{\partial \sigma_{ij}} \tag{9}$$

The determination of proportionality constant $\Delta\lambda$ follows the development outlined by Hill (5) and is given as:

$$\Delta\lambda = \frac{dW_p}{3g} \tag{10}$$

in which the increment in plastic work is expressed as:

$$dW_p = \frac{a\,df}{(1 - r_f \frac{(f - f_t)}{(K_1 - f_t)})^2} \tag{11}$$

The plastic strain increments expressed in suffix notation in Eq. 9 can now be written in matrix form:

$$\begin{Bmatrix} \Delta\varepsilon_x^p \\ \Delta\varepsilon_y^p \\ \Delta\varepsilon_z^p \\ \Delta\varepsilon_{xy}^p \\ \Delta\varepsilon_{yz}^p \\ \Delta\varepsilon_{zx}^p \end{Bmatrix} = \frac{a \cdot df}{3g(1 - r_f \frac{f-f_t}{K_1-f_t})^2} \begin{Bmatrix} 3I_1^2 - K_2(\sigma_y\sigma_z - \tau_{yz}^2) \\ 3I_1^2 - K_2(\sigma_z\sigma_x - \tau_{zx}^2) \\ 3I_1^2 - K_2(\sigma_x\sigma_y - \tau_{xy}^2) \\ 2K_2(\sigma_x\tau_{yz} - \tau_{xy}\tau_{zx}) \\ 2K_2(\sigma_y\tau_{zx} - \tau_{xy}\tau_{yz}) \\ 2K_2(\sigma_z\tau_{xy} - \tau_{yz}\tau_{zx}) \end{Bmatrix} \tag{12}$$

Therefore, the elastic and plastic parts of strain increments are fully described.

APPLICATION OF THE MODEL

Applications of the model to laboratory tests and engineering problems are reviewed next.

1. Predictions of laboratory testing results:

To assess the ability of the model to predict soil behavior, a variety of tests with different stress paths were conducted and the comparative results were published by Lade and Duncan (7,8). They concluded that the theory was reasonably accurate for cohesionless soils for the conditions of primary loading, unloading and reloading. The theory was less satisfactory for proportional loading with increasing stresses and for unloading and reloading at constant confining pressure.

However, since the strains observed in these cases were relatively small, the calculated strains were considered accurate enough for many purposes. The effectiveness of the theory in modelling stress strain behavior along several different stress paths was also confirmed by Medeiros (10).

2. Engineering problems:

Ozawa (11) was the first to use the model for the analysis of stresses and movements in earth masses. His results of finite element analysis of passive earth pressure problems were published later on by Ozawa and Duncan (12). Due to errors in the analysis of this published data and the problems related to the formulation of the model, (13), the results were not suitable to determine the usefulness of the model.

The second attempt to find the effectiveness of the model for engineering problems was made by Wong (14). Using the modifications summarized in the following section, the predictions of a finite element analysis were compared with experimental results from passive pressure tests on sand. The calculated values of average stresses on the wall exceeded those measured by a considerable amount. Attention was then drawn to the limitations of the model as possible reasons for the discrepancies. The limitations which were considered to be significant are as follows:
1) According to the theory, no plastic straining takes place as a result of proportional loading.
2) Strain softening cannot be handled.
3) The failure envelope in the triaxial plane is linear.
4) The value of Poisson's ratio was assumed to be equal to zero. Consequently, no straining should be expected in any other direction but on the stress increment direction in elastic behavior.

PREVIOUS MODIFICATIONS

Modifications and corrections for the model were first proposed by Ozawa and Duncan (13) as follows:
1. The plastic potential function was found to be mathematically inconsistent with the expressions for plastic strain increments and a new equation was proposed as:

$$g = I_3 (f - 27)^\alpha \qquad (13)$$

where:
α is a constant for a particular material and it is defined as:

$$\alpha = \frac{1}{1 - \overline{A}} \qquad (14)$$

where A is the same as in Eq. 4.
2. Due to the error in the calculation of derivatives of plastic potential function with respect to shear stresses, the shear strain increments were in error by a factor of 2.0. Therefore, on the right hand side of Eq. 12, the multiplier 2.0 of each term related to plastic shear strain increments has to be removed.
3. The parameter, a, contained in Eq. 5 varies with σ_3 according to Eq. 6. This variation was neglected in calculating $\partial f/\partial\{\sigma\}$ which is used in finite element formulation to form the elastoplastic constitutive matrix $[C^{ep}]$.

Having made these corrections the authors then presented the elastoplastic constitutive relation in matrix notation, suitable for use in a finite element analysis. Wong's predictions (14) are based on this form of the model and therefore utilize all the previous modifications.

RE-EVALUATION OF THE MODEL

The discouraging results of finite element analyses of earth pressure problems (14) as opposed to the satisfactory predictions of laboratory tests (6,7,8,10) suggested that either the limitations of the model were too severe or the modifications proposed by Ozawa and Duncan (13) have not been effective enough for the model to become useful to the engineering practice.

In the following, certain parts of the original formulation of the model and the previous modifications are examined. The significance of a correct interpretation of the laws of Lade's original theory is demonstrated by comparing the test results of Lade (6) and the predictions by the previously modified and presently re-evaluated forms of the model.

First, to find the effect of modifications on the existing abilities of the model, the predictions of laboratory test measurements, obtained by using previous modifications, are presented with the predictions made by Lade and Duncan (7,8).

The variation of parameter a with respect to σ_3 for calculating $\partial f/\partial\{\sigma\}$ in finite element formulation, as proposed in (13), has equivalent effect on the results of stress analysis, as the parameter a be used as a variable in calculating the plastic work increments. Therefore, realizing that the parameter a of Eq. 5 has to be treated as a variable rather than a constant and using the new plastic potential function given by Eq. 13, the constitutive relationship in modified form can be written as:

$$\begin{Bmatrix} \Delta\varepsilon^p_x \\ \Delta\varepsilon^p_y \\ \Delta\varepsilon^p_z \\ \Delta\varepsilon^p_{xy} \\ \Delta\varepsilon^p_{yz} \\ \Delta\varepsilon^p_{zx} \end{Bmatrix} = \left[\frac{(M+bh)^2(\sigma_3^\ell)(\Delta f) + M\ell\,\sigma_3^{\ell-1}h(\Delta\sigma_3)}{3\,M\,g} \right] (\alpha(f-27)^{\alpha-1}) \begin{Bmatrix} 3I_1^2 - K_2(\sigma_y\sigma_z - \tau_{yz}^2) \\ 3I_1^2 - K_2(\sigma_z\sigma_x - \tau_{zx}^2) \\ 3I_1^2 - K_2(\sigma_x\sigma_y - \tau_{xy}^2) \\ -K_2(\tau_{yz}\tau_{zx} - \tau_{xy}\sigma_z) \\ -K_2(\tau_{zx}\tau_{xy} - \tau_{yz}\sigma_x) \\ -K_2(\tau_{xy}\tau_{yz} - \tau_{zx}\sigma_y) \end{Bmatrix} \quad (15)$$

where: $K_2 = A_1 f + A_2$ $\alpha = 1/(1 - A)$

$A_1 = 1 - 1/\alpha$ $h = W_p/(\sigma_3^\ell)$

$A_2 = 27/\alpha$

Notes: 1) W_p has to be updated at each loading step

2) The column vector on the right hand side of Eq. (15) and (16) will be denoted by {COLUMN} and the left hand side by $\{\Delta\varepsilon^p\}$ in the following as required.

Several stress-strain curves following different stress paths were reproduced by substituting Eq. 15 into Eq. 1, and are presented with the test results, if available, in Figures 1 to 5. The predictions by using Lade's approach (6) are also plotted in the same figures. Although the predicted curves of the previously modified approach are as good as the curves produced by using Lade's approach for Figure 1 and 2, the gap between these curves rapidly opens up for large changes in σ_3 but relatively small changes in f. Figures 3 to 5 demonstrate that the modified approach may produce significantly different results from Lade's approach, even

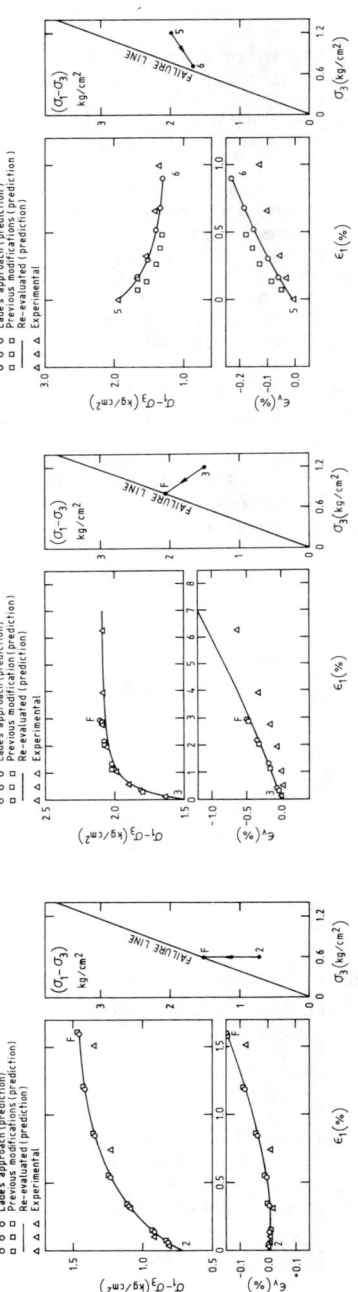

Figure 1. Stress-path increasing deviator stress and constant confining pressure. Monterey No. 0 Sand. [Experimental results from Lade and Duncan (8)]

Figure 2. Stress-path with increasing deviator stress and decreasing confining pressure. Monterey No. 0 Sand. [Experimental results from Lade and Duncan (8)]

Figure 3. Stress-path with decreasing deviator stress and decreasing confining pressure. Monterey No. 0 Sand. [Experimental results from Lade and Duncan (8)]

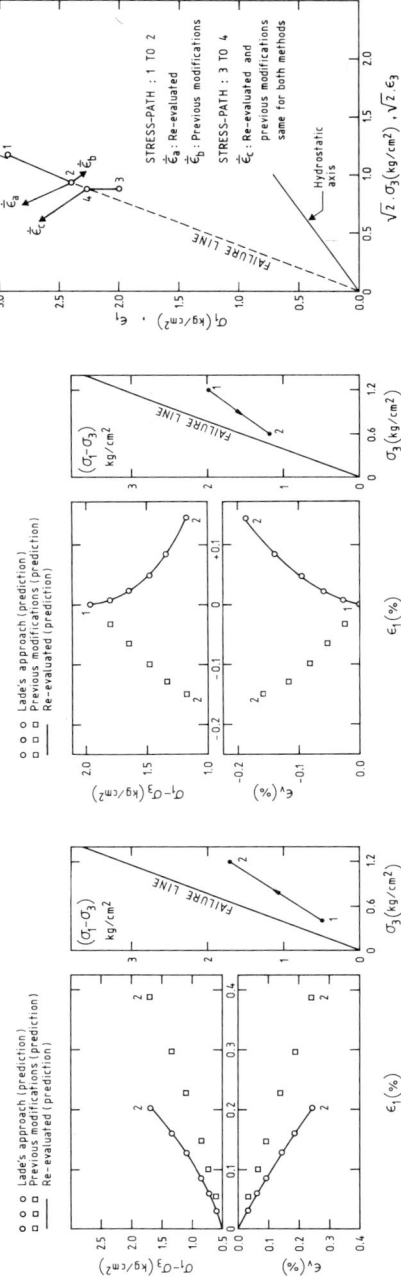

Figure 4. Stress-path with increasing deviator stress and increasing confining pressure. Monterey No. 0 Sand.

Figure 5. Stress-path with decreasing deviator stress and decreasing confining pressure. Monterey No. 0 Sand.

Figure 6. Predicted strain increment vectors.

resulting in different signs for strains as shown in Figure 5.
Although there are no measured test results to be plotted in Figures 4 and 5 to support either one of the predictions, it can be concluded that the use of Eq. 16 alters the abilities of the model for certain stress paths.

Fig. 6 shows the strain increment vectors for two different stress paths in triaxial plane as predicted by the previously modified form and the presently re-evaluated approach which is described later in this chapter.

For similar stress paths, Lade and Duncan (8) provide the measured test results, which are reproduced on Fig. 7. A comparison of measured and predicted strain increment vectors shows that the re-evaluated procedure predicts the increments of strain reasonably well for both stress paths, while the modified approach fails to do so for the stress path with decreasing deviator stress and decreasing confining pressure.

To find out whether any of the previous modifications adversely affects the results, the effect of each deviation from the original formulation is investigated separately. The modification related to the plastic potential function is treated first while holding the parameter a as a constant in calculating the plastic work increment. Rather than reproducing the stress-strain curves, the following proof is provided to show that the modified plastic potential function (Eq.13) does not change the stress-strain relation in incremental form.

To differentiate between the original and modified plastic potential functions, g of Eq. 3 and Eq. 13 are renamed here as g_o and g_m respectively.

Substituting Eq. 10 into Eq. 9 and using Eq. 3 for plastic potential function results in the following plastic strain increment values:

$$\{\Delta\epsilon^p\} = (\frac{dW_p}{3g_o}) \times \{COLUMN\} \qquad (16)$$

where: $\{\Delta\epsilon^p\}$ and $\{COLUMN\}$ are defined in the notes for Eq. 15. If Eq. 13 is used rather than Eq. 3, the set of equations to calculate the plastic strain increments becomes:

$$\{\Delta\epsilon^p\} = (\frac{dW_p}{3g_m}) \times \alpha \ (f - 27)^{\alpha-1}) \ \{COLUMN\} \qquad (17)$$

If the multipliers on the righthand side of Eq. 16 and Eq. 17 have the same values, then $\Delta\epsilon^p_{ij}$ calculated by either way are the same. That is in fact the case as shown next.

Ignoring the common multiplier ($dW_p/3.0$), and writing f, g_o and g_m in terms of stress invariants gives:

Reduced multiplier for Eq. 16:

$$\frac{1}{g_o} = \frac{1}{I_1^3 - K_2 I_3} = \frac{1}{I_1^3 - [A\frac{I_1^3}{I_3} + 27(1-A)]I_3} = \frac{1}{I_1^3 - AI_1^3 - 27I_3 + 27AI_3}$$

Reduced multiplier for Eq. 17:

$$\frac{1}{g_m} \times \alpha \ (f-27)^{\alpha-1} = \frac{1}{1-A} \times \frac{(f-27)^{\alpha-1}}{I_3(f-27)^\alpha} = \frac{1}{(1-A)(f-27)I_3} = \frac{1}{I_1^3 - AI_1^3 - 27I_3 + 27AI_3}$$

where $\alpha = 1/(1-A)$

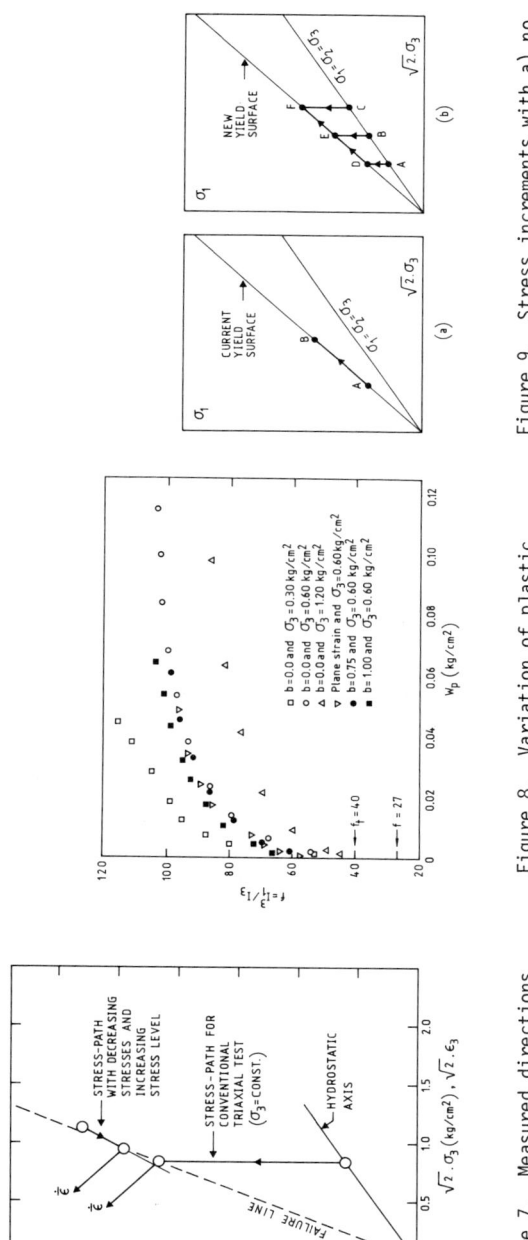

Figure 7. Measured directions of strain increment vectors for different stress-paths in triaxial plane [After Lade and Duncan (8)]

Figure 8. Variation of plastic work for dense Monterey No. 0 sand [After Lade and Duncan (7)]

Figure 9. Stress increments with a) no change in stress level b) conventional triaxial test path followed by no change in stress level.

Eq. 18 and Eq. 19 have equal values, therefore, the use of modified plastic potential function will produce results identical to those obtained from Lade's original formulation. If Eq. 13 is used, it improves the formulation of the model in a sense of mathematical integrity. The explanation for this statement is as follows: when the derivatives of the plastic potential function are needed for flow rule, the use of Eq. 3 requires that K_2 be held constant in the derivation process. Yet K_2 is a function of f which in turn is a function of stresses. The use of Eq. 13 does not require such an assumption.

In the light of the above observations with regard to the plastic potential function, the cause for large discrepancies between the predictions by Lade's approach and the results of modified formulation as plotted in Fig. 3 to Fig. 6 seems to be related to the parameter a, or in other words, to the work hardening law of the theory.

PROPOSED INTERPRETATION OF WORK HARDENING LAW

The work hardening law and its modification are re-evaluated in this section, to establish a correct method for its use.

Lade's test results from which the work hardening law was obtained are reproduced in Fig. 8. The results show that there is a unique relationship between f and W_p for tests where σ_3 is held constant. This relationship is expressed by Lade and Duncan (7) in mathematical form (see Eq. 5). Fig. 8 also shows some data points for various b values [b = $(\sigma_2-\sigma_3)/(\sigma_1-\sigma_3)$] obtained from tests where σ_3 is still kept constant. These results allowed Lade and Duncan (7) to state that "the relationship between W_p and f (...) depends on the confining pressure, σ_3, but it is essentially the same for all values of b". There are no other test results with plastic work calculations to support the validity of Eq. 5 for stress paths with varying σ_3. Therefore, this equation is applicable strictly for stress changes for which σ_3 remains constant.

In order to use Eq. 5 in the analysis of a general three dimensional problem where σ_3 generally changes during loading, an assumption has to be made. The theory which was first published by Lade and Duncan (7) does not elaborate on this point, but simply treats the parameter a as a constant when Eq. 5 is used to derive an expression for the increment of plastic work. However, Lade (6) employs special procedures in an attempt to use Eq. 5 in its true sense. Some of his predictions of laboratory tests can be found in (8). But the importance of the non-generality of Eq. 5 for changing σ_3 values has never been emphasized nor has the basic assumption which allows its use in such problems been clearly stated. In the publications by Ozawa and Duncan (12,13), Eq. 5 remained as a three dimensional representation of the work hardening law. Subsequently, as a deviation from Lade's original work, Duncan et al. (2) and Wong (14) treated the parameter a as a variable when a differentiation was needed on Eq. 5.

The consequences of assuming that Eq. 5 is unique for general 3-D problems are as follows:

Case I:
Referring to Fig. 9a, the initial state of stress is assumed to be on the current yield surface as indicated by point A with a total plastic work value equal to W_p. An increment of stress is applied such that the stress path follows the route from A to B without travelling outside the

current yield surface. The limits of the elastic region are not exceeded, so there are no plastic deformations, and no increment in plastic work. Therefore, the total plastic work done is the same for points A and B. But since the value of σ_3 has changed, and with it, the value of parameter a, the Eq. 5 is no longer satisfied for point B.

Case II:

Figure 9b illustrates three different stress paths, all of which start from the hydrostatic axis. All three follow the stress path of the conventional triaxial test up to the new yield surface, then, without leaving that surface end at point F. The change in f value in going from the hydrostatic axis to the new yield surface is the same for all points A, B, and C. But the parameter a calculated for point A will be the smallest of all, and the plastic work increment for path A to D will be smaller than for path B to E or C to F. Therefore, depending on which path is followed, the total plastic work at point F will have a different value. Although all these work values as well as an infinite number of others are possible for point F, only one of them satisfies Eq. 5. Subsequent increments of stress from point F with increasing stress level will create problems for any total plastic work value other than the one which satisfies Eq. 5. The reason for this is that Eq. 5 will become an inequality and cannot be used any further to calculate the increments of plastic work, which are needed for Eq. 10.

Case III:

This case is related to the use of Eq. 5 in its modified form, in which the parameter a is treated as a function of σ_3 in calculating the increment of plastic work, as formulated below:

$$\Delta W_p = \frac{1}{M} (M + bh)^2 (\sigma_3^\ell) (\Delta f) + M \ell \sigma_3^{\ell-1} h (\Delta \sigma_3) \tag{20}$$

The problem does not surface with this formulation until the term with $\Delta \sigma_3$ becomes larger in absolute value with a negative sign than the term with Δf on the right hand side of the equation as in stress paths with decreasing σ_3 and increasing f values. As a result of this, ΔW_p will have a negative value. As long as the plastic deformation is an irreversible process from which the energy cannot be recovered, a negative ΔW_p, will contradict the laws of the theory of plasticity.

Due to the lack of generality as demonstrated by Case I and II, Eq. 5 is not assumed here as a work hardening law applicable for all problems. Instead the following procedure is followed.

Rather than searching for a general relationship between f and W_p, and then differentiating that function to find a relation between df and dW_p, the following assumption is made to establish a criterion about how the yield surface will expand as the plastic work is done:

Using Fig. 8, the tangent to the related experimental curve specified by σ_3, at the point corresponding to known f defines the relation between the increments of f and W_p without any regard for the actual value of W_p. As it is shown earlier, depending on the loading history of the soil, the accumulated value of W_p may vary for the same values of f and σ_3. However, the increments of plastic work will always be unique for a given state of stress and stress increment. In mathematical form this statement is equivalent to the expression given by Eq. 11, which is Lade's original equation for plastic work increments. Therefore, it can be concluded that the re-evaluated approach makes full use of the experimental data

and Eq. 5 without influencing adversely what has been formulated so far.
To find out the results of such an approach the stress-strain curves of Figures 1 to 6 were reproduced and plotted as "re-evaluated" predictions on the same figures.

CONCLUSIONS

As the predictions made in this paper indicate, the measured stress-strain curves can be successfully predicted by the newly proposed interpretation without making the previous modifications of Ozawa and Duncan (13) or using special procedures given by Lade (6). At the same time, the problems that are created in finite element analysis by using Eq. 5 as the work hardening law can be avoided as shown by Evgin (4).

ACKNOWLEDGMENT

This study was carried out with the assistance of the Natural Sciences and Engineering Research Council of Canada.

REFERENCES

1. Duncan, J.M., and Chang, C.-Y., "Nonlinear Analysis of Stress and Strain in Soils", Journal of the Soil Mechanics and Foundation Division, ASCE, Vol. 96, No. SM5, Proc. Paper 7513, Sept., 1970, pp. 1629-1653.
2. Duncan, J.M., Ozawa, Y., Lade, P.V. and Booker, J.R., "An Elasto-Plastic Stress-Strain Relationship for Cohesionless Soil", Specialty Session No. 9, Ninth International Conference on Soil Mechanics and Foundation Engineering, Tokyo, Japan, 1977, pp. 45-50.
3. Eisenstein, A. and Law, S.T.C., "The Role of Constitutive Laws in Analysis of Embankments", Proceedings of the 3rd International Conference on Numerical Methods in Geomechanics, Aachen, April, 1979, pp. 1413-1430.
4. Evgin, E., "Application of Elasto-Plastic Constitutive Law", thesis presented to the University of Alberta, at Edmonton, Alberta, in 1980, in partial fulfillment of the requirements for the degree of Doctor of Philosophy.
5. Hill, R., "The Mathematical Theory of Plasticity", Oxford University Press, London, 1950.
6. Lade, P.V., "The Stress-Strain and Strength Characteristics of Cohesionless Soils", thesis presented to the University of California, at Berkeley, Calif., in 1972, in partial fulfillment of the requirements for the degree of Doctor of Philosophy.
7. Lade, P.V., and Duncan, J.M., "Elastoplastic Stress-Strain Theory for Cohesionless Soil", Journal of the Geotechnical Engineering Division, ASCE, Vol. 101, No. GT10, Proc. Paper 11670, Oct., 1975, pp. 1037-1053.
8. Lade, P.V., and Duncan, J.M., "Stress-Path Dependent Behavior of Cohesionless Soil", Journal of the Geotechnical Engineering Division, ASCE, Vol. 102, No. GT1, Proc. Paper 11841, January, 1976, pp. 51-68.
9. Lambe, T.W., "The Stress Path Method", Journal of the Soil Mechanics and Foundations Divisions, ASCE, Vol. 93, SM6, Proc. Paper 5613, Nov. 1967, pp. 309-331.
10. Medeiros, L., "Deep Excavations in Stiff Soils", thesis presented to the University of Alberta, at Edmonton, Alberta, in 1979, in partial fulfillment of the requirements for the degree of Doctor of Philosophy.

11. Ozawa, Y., "Elasto-Plastic Finite Element Analysis of Soil Deformation", thesis presented to the University of California, at Berkeley, Calif., in 1972, in partial fulfillment of the requirements for the degree of Doctor of Philosophy.
12. Ozawa, Y. and Duncan, J.M., "Elasto-Plastic Finite Element Analyses of Sand Deformation", Proceedings of the 2nd International Conference on Numerical Methods in Geomechanics, Blacksburg, Virginia, June, 1976, pp. 243-263.
13. Ozawa, Y. and Duncan, J.M., "Correction to Elasto-Plastic Finite Element Analyses of Sand Deformations", Proceedings of the 2nd International Conference on Numerical Methods in Geomechanics, Blacksburg, Virginia, June, 1976, pp III-1477, III-1479.
14. Wong, K.S., "Elasto-Plastic Finite Element Analyses of Passive Earth Pressure Tests", thesis presented to the University of California, at Berkeley, Calif., in 1978, in partial fulfillment of the requirements for the degree of Doctor of Philosophy.

VERIFICATION OF NON-LINEAR EFFECTIVE STRESS MODEL IN SIMPLE SHEAR

W.D. Liam Finn[1], M.ASCE and Shobha K. Bhatia[2]

INTRODUCTION

In 1975 Martin, Finn and Seed (9) presented a model of the generation of pore-water pressures in sands under cyclic loading. The model is based on the physical properties of the soil skeleton and the pore-water and the plastic and elastic strains that occur in the sand skeleton and the water. Equipment and test procedures have since been developed that allow the individual assumptions underlying this model to be tested directly. The predictive capability of the model for complex loading and drainage patterns can also be tested. In this paper, an extensive verification testing of the model is reported.

ASSUMPTIONS

In its simplest form, the model assumes that the water is an order of magnitude stiffer than the soil skeleton and therefore that the net volumetric strain during undrained cyclic loading is negligible. This is the usual assumption that undrained tests are constant volume tests. The consequence of this assumption is that the plastic and elastic volumetric strains that occur during cyclic loading must be equal and of opposite sense. This, for example, is a fundamental assumption of the critical state theory as applied to undrained loading (11). Since residual (or permanent as opposed to transient) pore-water pressures are generated during undrained cyclic loading of sands, there must be a tendency for plastic volumetric strains to occur. These strains are absorbed by elastic rebound in the soil skeleton due to the reduction in effective stresses and "constant" volume is preserved, but they are recoverable on allowing the sample to drain. The increment in pore-water pressure, Δu, during a short time interval, Δt, of cyclic loading is then assumed to be given by

$$\Delta u = \bar{E}_r \Delta \varepsilon_{vd} \qquad (1)$$

in which $\Delta \varepsilon_{vd}$ is the elastic (or plastic) volumetric strain increment that occurs due to the increment in pore-water pressure and $\bar{E}_r$ is the rebound modulus appropriate for the effective stress state in the sand at the beginning of the time increment, Δt.

The pore-water pressure model given by Eqn. (1) posed immediately the important question; how was $\Delta \varepsilon_{vd}$ to be obtained? It was assumed that the plastic volumetric strain, $\Delta \varepsilon_{vd}$, which developed during one cycle of uniform shear strain, γ, in an undrained simple shear test would be the same as the volumetric strain in a drained simple shear test. Therefore, a fundamental assumption of the pore-pressure model is that there is a unique relationship between volumetric strains in drained tests and pore-water pressures in undrained tests on a given sand at corresponding strain histories. A study of the relationship between $\Delta \varepsilon_{vd}$ and γ in drained

---

[1] Professor, Soil Dynamics Group, Faculty of Graduate Studies and Dept. of Civil Engineering, Faculty of Applied Science, University of British Columbia, Vancouver, B.C., Canada, V6T 1W5.

[2] Doctoral Student, Dept. of Civil Engineering, Faculty of Applied Science, University of British Columbia, Vancouver, B.C., Canada, V6T 1W5.

cyclic simple shear tests showed that the increment in volumetric strain for any γ decreased with increasing number of cycles of loading. As the number of cycles increased, so did the accumulated volumetric strain, ε_{vd}. The following relationship, independent of effective stress level, was found between $\Delta\varepsilon_{vd}$ and γ.

$$\Delta\varepsilon_{vd} = C_1(\gamma - C_2\varepsilon_{vd}) + C_3\varepsilon_{vd}^2/(\gamma + C_4\varepsilon_{vd}) \qquad (2)$$

in which C_i, $i=1,4$ are experimentally determined constants.

The measurement of $\bar{E}_r$ posed difficult problems. Rebound occurs during undrained cyclic loading but direct measurement of $\bar{E}_r$ under such conditions was not possible. Therefore, $\bar{E}_r$ was measured during static rebound in a consolidation ring. Test data obtained at the time indicated that $\bar{E}_r$ measured in this way resulted in satisfactory predictions of liquefaction strength curves. As will be seen later, $\bar{E}_r$ measured in this way is too stiff. It gave reasonable results for earlier test data primarily because compliance in the early simple shear equipment absorbed the excess pore-water pressure generated by the excessive stiffness of $\bar{E}_r$.

In 1976 and 1977, Finn, Lee and Martin (3,4) incorporated the pore-water pressure model into a method for dynamic effective stress analysis of saturated sands during earthquakes. Since earthquakes do not apply uniform cycles of shear strain an additional assumption is required that the pore-water pressure model based on uniform cyclic shear strain data can be used also to predict pore-water pressure generation during an irregular shear strain history.

It was noticed during cyclic drained tests that the maximum shear modulus G_m and the maximum shear strength τ_m for any load cycle increased with increasing volumetric strain (2). This is the phenomenon of strain-hardening. The parameters G_m and τ_m at any stage of cyclic loading corresponding to an accumulated volumetric strain ε_{vd} are given in terms of the initial values G_{mo} and τ_{mo} by

$$G_m = G_{mo}\left[1 + \frac{\varepsilon_{vd}}{H_1 + H_2\varepsilon_{vd}}\right] \qquad (3)$$

$$\tau_m = \tau_{mo}\left[1 + \frac{\varepsilon_{vd}}{H_3 + H_4\varepsilon_{vd}}\right] \qquad (4)$$

in which H_i, $i=1,4$ are experimentally determined constants. The parameters G_m and τ_m were used to define the Masing type stress-strain curve given by Eqn. (5)

$$\frac{\tau - \tau_r}{2} = \frac{G_m(\gamma - \gamma_r)}{2} / \left[1 + \frac{|G_m(\gamma - \gamma_r)|}{2\tau_m}\right] \qquad (5)$$

in which τ_r and γ_r are the stress and strain values at the last stress reversal. This stress-strain relationship, which was incorporated in the dynamic effective stress method (3,4), is used later in the analysis of constant stress cyclic loading tests.

Can the strain-hardening effect occur during undrained tests in which the plastic strains are absorbed by rebound or is it postponed until the strains develop after drainage? In critical state theory it is considered that strain-hardening does occur in clays in undrained shear and, therefore, the area enclosed by the yield surface increases (11). Until this study, the authors have been uncertain whether strain-hardening occurs during undrained cyclic shear in sands, although, in applying their method in engineering practice, they have not included hardening effects for undrained conditions. This matter is discussed in detail later.

Finn, Bransby and Pickering (1) have noted that previous levels of cyclic loading which did not lead to liquefaction and disruption of initial sand structure increased the resistance to liquefaction and reduced considerably the rate of pore-water pressure increase during future cyclic loading if drainage occurred between the two loadings. This effect of previous strain history was confirmed by Seed, Mori and Chan (13) who used the slower generation rates to predict successfully the effects of previous cyclic loading. A necessary and additional assumption for effective stress calculations using the model is that all the relevant effects of previous cyclic loading are included in the strain-hardening described by Eqns. (3) and (4).

The verification of the basic assumptions of the model consists of providing adequate and experimentally based answers to these questions:

(i) In constant strain cyclic loading tests, is there a unique relationship between volume changes in drained tests and pore-water pressures in undrained tests?

(ii) Can $\bar{E}_r$ be measured statically?

(iii) Can the model accurately predict the history of development of pore-water pressures in constant stress tests and not just the liquefaction strength curve?

(iv) Does strain-hardening occur during undrained tests?

(v) Can the model predict the effects of strain history under previous loadings?

Finally, can the model make useful predictions of pore-water pressures under general loading and drainage conditions in simple shear?

APPARATUS, EXPERIMENTAL PROCEDURE AND TEST MATERIAL

The University of British Columbia simple shear apparatus has been described by Finn, Pickering and Bransby (5). This apparatus was later modified (6,7) for carrying out constant volume cyclic loading tests as an alternative means of measuring liquefaction potential. The new method of testing was developed in order to eliminate some of the inherent difficulties and errors associated with cyclic loading tests on saturated undrained samples. The new apparatus has a virtual independence from system compliance, which Martin, Finn and Seed (10) have shown to result in an overestimation of the resistance to liquefaction, and reduced rates of pore-water pressure generation. Maximum gross volume change introduced at the onset of liquefaction in the "constant" volume test is very small and arises as a result of the recovery of deformation of the vertical loading components. For liquefaction tests with initial σ'_{vo} = 2 kg/cm^2 (200 kN/m^2) this movement amounted to a maximum of 2.10^{-4} inches (7.9×10^{-6}m) which was only 5% of the movement of the floating head due to system compliance in conventional liquefaction tests on saturated undrained samples in the same equipment and is equivalent to a strain of the order of 0.01%.

The constant volume test uses dry sand which greatly simplifies sample preparation with the result that it is possible to prepare, consistently, very uniform samples. Cyclic shear loads, with a sinusoidal wave form and a frequency of 2 Hz, were applied by an MTS, servo-controlled electro-hydraulic piston. During each test, vertical pressure, cyclic shear strain and cyclic shear stress were continuously monitored with electronic transducers and records obtained on a light beam chart recorder.

Most tests were performed on normally consolidated Ottawa sand, ASTM designation C-109. This is a natural silica sand consisting of rounded particles with grain sizes between 0.15 and 0.59 mm and $D_{50} = 0.40$ mm. The maximum and minimum void ratios are 0.82 and 0.50, respectively. Cyclic loading tests were carried out at two relative densities $D_r = 45\%$ and $D_r = 60\%$. All samples were consolidated under an initial vertical confining pressure $\sigma'_{vo} = 2$ kg/cm^2 (200 kN/m^2). The method of sample preparation has been described by Vaid and Finn (16).

EXPERIMENTAL VERIFICATION

Volumetric strains were measured in drained constant strain cyclic simple shear tests on Ottawa sand at relative densities $D_r = 45\%$ and $D_r = 60\%$. Pore-water pressures were also measured in undrained cyclic tests at the same relative densities and initial effective confining pressures. Silver and Seed (15) have shown that volumetric strains are independent of effective vertical confining pressure so the requirement of equal confining pressures in both kinds of tests was not required. However, since uniformity in procedure is conducive to uniformity in samples, the initial effective confining pressures were kept identical for all tests. Volumetric strains are shown plotted against pore-water pressures in Fig. 1 for $D_r = 45\%$. Ideally, each point should represent corresponding values of these variables for a given number of cycles with equal cyclic strain amplitudes. It will be noticed that there are slight deviations from equality in the applied shear strain amplitudes. The deviations are not considered to be important. The data indicates an apparently unique relationship between the volumetric strains and the pore-water pressures. A unique relationship was also found for $D_r = 60\%$, (Fig. 2).

The slope of the curve representing the relationship between pore-water pressure normalized with respect to confining pressure and volumetric strains is the dynamic rebound modulus K_d, normalized with respect to the initial confining pressure. This value of the rebound modulus is less than that measured under static conditions as shown in Fig. 3. This agrees with the observations of Seed, Pyke and Martin (14) that the recoverable strains after cyclic loading are greater than under static conditions. The reasons underlying this difference in response to unloading are being researched further.

For convenience in computation the rebound modulus $\bar{E}_r$ measured in static rebound tests is expressed as

$$\bar{E}_r = (\sigma'_v)^{1-m}/mK_2(\sigma'_{vo})^{n-m} \tag{6}$$

in which σ'_{vo} is the initial value of the vertical effective stress, σ'_v the current value and K_2, m and n are experimental constants for the given sand.

Thus, in applying the effective stress method in undrained response analyses 7 constants are measured, the 4 C_i and the 3 constants specifying the rebound characteristics of the sand. In the application of the predictive method to practical problems, these constants are always finally adjusted so that the model predicts the liquefaction resistance curve selected as representative of field conditions. This representative field curve is either determined by undrained cyclic loading tests on undisturbed samples or by modifying laboratory curves for reconstituted samples in the manner recommended by Seed (12). Adjustments in the constants to fit the selected field strength curve may be made in either the volume change or rebound constants. If the required adjustments (usually in K_2) are made to the rebound constants the effect is to scale the static rebound modulus towards the dynamic modulus. Thus, except for very special problems in which the precise pattern of development of pore-water pressure is considered important, it is not necessary to measure the rebound modulus dynamically in practice.

It is crucial to verification of the model to test its predictive capability under test conditions other than those under which the model parameters are derived. Since constant __strain__ cyclic tests were used to determine the model parameters, the predictive capability of the model was first tested under constant __stress__ cyclic test conditions. In constant stress tests the shear strains needed for use with the model Eqns. (1) and (2), must be measured or calculated. Calculation of the shear strains requires a knowledge of the initial shear modulus (at very low strains). This cannot be measured accurately in the simple shear apparatus due to small tolerances in linkages and joints so the recorded shear strains were used. Using these shear strains, the pore-water pressures were computed using the pore-water pressure model, Eqn. (1), with $\bar{E}_r = K_d$. The comparisons between predicted and measured pore-water pressures are shown in Fig. 4 for $D_r = 45\%$ and in Fig. 5 for $D_r = 60\%$. The predictive capability of the model appears to be satisfactory.

The pore-water pressures were next computed for two tests using strains calculated from the stresses by Eqn. (5). Since G_{mo} could not be measured, both it and τ_{mo} were computed for the void-ratios, $e = .676$ ($D_r = 45\%$) and $e = .628$ ($D_r = 60\%$) using the well-known Hardin-Drnevich equations (8). The magnitudes of G_m and τ_m in Eqn. (5) were modified during the strain calculations for the effect of increasing pore-water pressure by the equations

$$G_m/G_{mo} = (\sigma_v'/\sigma_{vo}')^{\frac{1}{2}} \qquad (6)$$

and

$$\tau_m/\tau_{mo} = \sigma_v'/\sigma_{vo}' \qquad (7)$$

These expressions neglect the strain-hardening functions in Eqns. (3) and (4). Neglecting strain-hardening (strain history) effects for undrained tests appears justified on the evidence in (1,13) that increased resistance to pore-pressure development as a result of previous loading or strain history is achieved only if dissipation of the pore-water pressures caused by the previous loading is allowed. Later on the correctness of this approach will be demonstrated. Comparisons between predicted and measured pore-water pressures are shown in Fig. 6 for $D_r = 45\%$ and in Fig. 7 for $D_r = 60\%$. The comparisons are quite good, although not as good as when measured strains were used in the pore-water pressure model. This is not unexpected since the actual initial in-situ moduli were not measured but estimated by the Hardin-Drnevich equations (8).

A more severe test of the predictive capability of the model is provided by the irregular strain history shown in Fig. 8a. The computed pore-water response to this strain pattern is shown in Fig. 8b and is satisfactory. Some of the scatter is probably due to the fact that the irregular strain pattern was imposed by manual control and not by a programmed automatic control. Thus, the variation between peaks is not as uniform as Fig. 8a suggests.

A final verification consisted of subjecting the sample to a loading-drainage-loading sequence. The purpose of this test was to check the strain-hardening feature of the model and its ability to take into account the effects of previous cyclic loading. Sample A, at $D_r = 45\%$, was first subjected to 70 cycles of a stress-ratio $\tau/\sigma_{vo}' = 0.066$ and then allowed to drain. The rate of development of pore-water pressure under this loading is shown in Fig. 9 by curve A. Sample B, also at $D_r = 45\%$, was subjected to a cyclic stress ratio of $\tau/\sigma_{vo}' = 0.104$. The rate of development of pore-water pressure in this sample is shown in Fig. 9 by curve B. Sample A is now subjected to the cyclic stress ratio $\tau/\sigma_{vo}' = 0.104$ and the rate of development of pore-water pressure is given by curve C in Fig. 7. This rate of increase in pore-water pressure is considerably less than that generated in sample B which had no previous stress history by the same cyclic stress ratio. This phenomenon was first reported by Finn, Bransby and Pickering (1) and was attributed to the effects of strain history by them.

To predict the pore-water pressure response given by curve C, the strain-hardening effects of the volumetric strains which developed during drainage after the first cyclic loading on G_m and τ_m were calculated by Eqns. (3) and (4). The new values of G_m and τ_m were used as G_{mo} and τ_{mo} for the second application of cyclic loading. The softening effect of the increasing pore-water pressure was also included using Eqns. (6) and (7). The predicted pore-water pressure is given by curve D in Fig. 9. The comparison between predicted and measured pore-water pressures is good and may reasonably be viewed as indicating that strain-hardening due to plastic volumetric strains in sands occurs only after drainage.

RESPONSE OF OVERCONSOLIDATED SANDS

The original pore-pressure model was based entirely on tests on normally consolidated sands. It is well-known that overconsolidated sands have a greater resistance to liquefaction, the resistance increasing with overconsolidation ratio, OCR (Seed, (12)). Both drained and undrained constant strain cyclic simple shear tests were conducted on Ottawa sand at OCR = 2,3 and 4 to check the effects of overconsolidation on the model parameters and to investigate whether the model could successfully predict the porewater pressure response for overconsolidated sands. In terms of the model parameters, the chief effects of overconsolidation were increased values of the initial shear modulus G_{mo} and the initial maximum shear stress in the horizontal plane, τ_{mo} and reduced volume changes due to a given drained cyclic loading compared with those in normally consolidated Ottawa sand. The reduced volumetric strain potential of the overconsolidated samples was reflected in lower values of the volume change constants C_i (i=1,4).

The unique relationship between volumetric strains in drained cyclic simple shear tests and pore-water pressures in undrained tests after similar strain histories established for normally consolidated sands was found to hold also for the overconsolidated sands. Pore-water pressures and corresponding volumetric strains for Ottawa sand at D_r = 47% and OCR = 4 shown in Fig. 10. The data fall very closely on a single curve. The corresponding curve for OCR = 1 is shown for comparison. Similar relationships between pore-water pressures and corresponding volumetric strains were also found for OCR = 2 and 3. On the basis of this evidence it may be claimed that the fundamental assumption of the pore-water pressure model that there is a direct connection between pore-water pressures in undrained cyclic simple shear tests and volumetric strains in drained cyclic simple shear tests holds for both normally and overconsolidated sands.

The ability of the model to predict pore-water pressures in overconsolidated sands was tested by predicting the liquefaction strength curves for various OCR and comparing the results with experimental curves. The strength curves are plots of the cyclic stress ratio τ/σ_{vo}' versus the number of cycles to liquefaction N; τ being the applied cyclic shear stress. Figure 11 shows liquefaction strength curves for OCR = 1,2,3 and 4 computed using Eqns. (1) through (5) for Ottawa sand at D_r = 45%. The points are experimental data from undrained constant volume cyclic simple shear tests. The initial effective vertical pressure σ_{vo}' in all tests after the OCR was established was σ_{vo}' = 200 kN/m².

The comparison between the computed and measured liquefaction strengths is very good. It seems that the pore-water pressure model originally developed for normally consolidated sands, may also be used for overconsolidated sands at least up to the OCR = 4 for which direct verification is available.

CONCLUSIONS

The pore-water pressure model developed by Martin, Finn and Seed (9) has been tested under a variety of drained and undrained loading conditions. The basic assumptions of the model appear to be well-founded. There is strong verification of a unique relationship between volumetric strains in drained tests and pore-water changes in undrained tests on a given sand at similar strain histories. An important point to emerge from the study is that the rebound modulus used to convert volumetric strains to pore-water pressures (Eqn. (1)) must be measured under cyclic loading conditions. Moduli measured under static rebound conditions are too stiff and for a given volumetric strain generate too much pore-water pressure. In practice, a viable alternative to measuring the rebound modulus under cyclic loading conditions is to adjust the static rebound modulus so that the computed liquefaction resistance curve, defined by the stress-ratios τ/σ_{vo}' to cause liquefaction in N cycles, agrees within acceptable limits with the selected field strength curve.

The model predicts successfully the pore-water pressure response under drained conditions for uniform loading and for irregular cyclic loading histories representative of earthquake loading. When combined with the effective stress model proposed by Finn, Lee and Martin (3,4) it predicts successfully the effects of previous cyclic loading history on pore-water pressure response. It appears from the test data that in undrained tests on sand strain-hardening or strain history effects do not occur. However, at the conclusion of such tests if drainage is allowed to take place plastic strains are recovered and the sands strain-harden.

If proper values for the rebound modulus are used and the effects of strain-hardening are included whenever drainage occurs, it appears that

MODEL VERIFICATION

the effective stress pore-pressure model can make good predictions of the development of pore-water pressures under fairly general loading patterns and drainage conditions in simple shear.

It must be stressed that the model was developed and verified on the basis of simple shear data. Therefore, it may be used with confidence only for the analysis of cyclic loading problems in the field for conditions which can be approximated by simple shear conditions. This implies more or less level ground underlain by horizontal soil layers and excited by shear waves propagating vertically.

ACKNOWLEDGEMENTS

The support of the National Science and Engineering Research Council of Canada under grant no. 1498 and the Canadian Commonwealth Scholarship are gratefully acknowledged.

APPENDIX I - REFERENCES

1. Finn, W.D. Liam, Bransby, P.L. and Pickering, D.J. (1970), "Effect of Strain History on Liquefaction of Sand," Jour. of the Soil Mech. and Found. Div., ASCE, Vol. 97, No. SM6, Proc. Paper 7670, November, pp. 1917-1934.

2. Finn, W.D. Liam, Lee, K.W. and Martin, G.R. (1975), "Stress-Strain Relations for Sand in Simple Shear," ASCE National Convention, Denver, Colorado, Nov. 1975, Meeting Preprint 2517. Also in Soil Mechanics Series No. 26, Dept. of Civil Engineering, University of British Columbia, Vancouver, B.C.

3. Finn, W.D. Liam, Lee, K.W. and Martin, G.R. (1976), "An Effective Stress Model for Liquefaction," ASCE National Convention, Philadelphia, Specialty Session, Liquefaction Problem in Geotechnical Engineering, September, pp. 169-198.

4. Finn, W.D. Liam, Lee, Kwok W. and Martin, Geoffrey R. (1977), "An Effective Stress Model for Liquefaction," Jour. of the Geotech. Eng. Div., ASCE, Vol. 103, No. GT6, Proc. Paper 13008, June, pp. 517-533.

5. Finn, W.D. Liam, Pickering, D.J. and Bransby, P.L. (1971), "Sand Liquefaction in Triaxial and Simple Shear Tests," Jour. of the Soil Mech. and Found. Div., ASCE, Vol. 97, SM4, pp. 639-59.

6. Finn, W.D. Liam and Vaid, Y.P. (1977), "Liquefaction Potential from Drained Constant Volume Cyclic Simple Shear Tests," Proceedings, 6th World Conf. on Earthquake Engineering, New Delhi, Session 6, pp. 7-12.

7. Finn, W.D. Liam, Vaid, Y.P. and Bhatia, S.K. (1978), "Constant Volume Cyclic Simple Shear Testing," 2nd Int. Conf. on Microzonation, San Francisco, Calif., Nov. 26-Dec. 1.

8. Hardin, B.O. and Drnevich, V.P. (1972), "Shear Modulus and Damping in Soils: Design Equations and Curves," Jour. of the Soil Mech. and Found. Div., ASCE, Vol. 98, No. SM7, Proc. Paper 9006, July, pp. 667-692.

9. Martin, G.R., Finn, W.D. Liam and Seed, H.B. (1975), "Fundamentals of Liquefaction Under Cyclic Loading," Jour. of the Geotech. Eng. Div., ASCE, Vol. 101, No. GT5, Proc. Paper 11284, May, pp. 423-438.

10. Martin, G.R., Finn, W.D. Liam and Seed, H.B. (1978), "Effect of System Compliance on Liquefaction Tests," Jour. of the Geotech. Eng. Div., ASCE, Vol. 104, No. GT4, Proc. Paper 13667, April, pp. 463-479.

11. Schofield, A.N. and Wroth, C.P. (1968), "Critical State Soil Mechanics," McGraw Hill, London.

12. Seed, H.B. (1976), "Evaluation of Soil Liquefaction Effects on Level Ground During Earthquake," ASCE National Convention, Philadelphia, Specialty Session, Liquefaction Problems in Geotechnical Engineering, September, pp. 1-104.

13. Seed, H.B., Mori, K. and Chan, C.K. (1977), "Influence of Seismic History on Liquefaction of Seeds," Jour. of the Geotech. Eng. Div., ASCE, Vol. 103, No. GT4, Proc. Paper 12841, April, pp. 246-270.

14. Seed, H.B., Pyke, R.M. and Martin, G.R. (1979), "Effect of Multi-Directional Shaking on Pore-Pressure Development in Sand," Jour. of the Geotech. Eng. Div., ASCE, Vol. 104, No. GT1, Proc. Paper 13485, January, pp. 27-44.

15. Silver, M.L. and Seed, H.B. (1971), "Volume Changes in Sands During Cyclic Loading," Jour. of the Soil Mech. and Found Div., ASCE, Vol. 97, No. SM9, Proc. Paper 8354, September, pp. 1171-1182.

16. Vaid, Y.P. and Finn, W.D. Liam (1979), "Static Shear and Liquefaction Potential," Jour. of the Geotech. Eng. Div., ASCE, Vol. 105, No. GT10, Proc. Paper 14909, October, pp. 1233-1246.

APPENDIX II - NOTATION

C_1, C_2, C_3, C_4 = constants;

D_r = relative density;

$\bar{E}_r$ = tangent modulus of one-dimensional unloading curve;

e = void ratio;

G_{mo} = initial maximum tangent shear modulus;

G_m = maximum (initial) tangent shear modulus of the nth cycle;

H_1, H_2, H_3, H_4 = constants;

K_d = dynamic rebound modulus;

γ = shear strain;

γ_r = shear strain value at which reversal occurs;

Δt = time increment;

Δu = increment of pore-water pressure;

$\Delta \varepsilon_{vd}$ = increment of volumetric strain occurring during cyclic loading;

ε_{vd} = accumulated volumetric slip strain from cyclic loading sequence;

σ'_v = vertical effective stress;

σ'_{vo} = initial vertical effective stress;

τ = shear stress;

τ_m = cyclic shear stress amplitude;

τ_{mo} = maximum shear stress (initial);

τ_r = shear stress at reversal; and

ϕ = effective angle of shearing resistance.

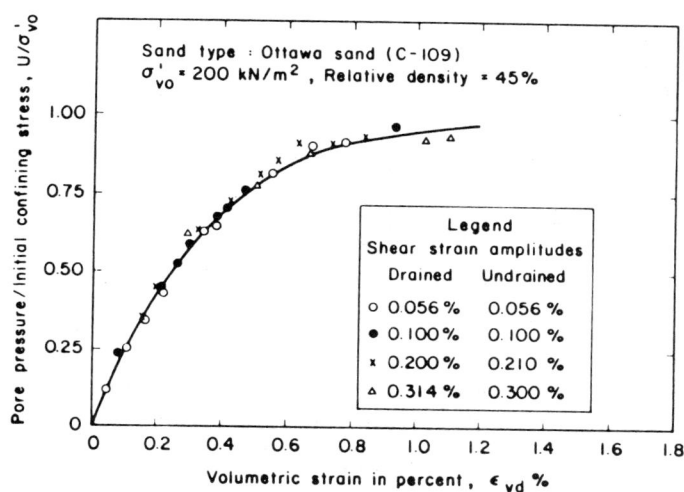

FIG. 1 : RELATIONSHIP BETWEEN VOLUMETRIC STRAINS AND PORE-WATER PRESSURES IN CONSTANT STRAIN CYCLIC SIMPLE SHEAR TESTS

SOIL STRESS STRAIN APPLICATIONS

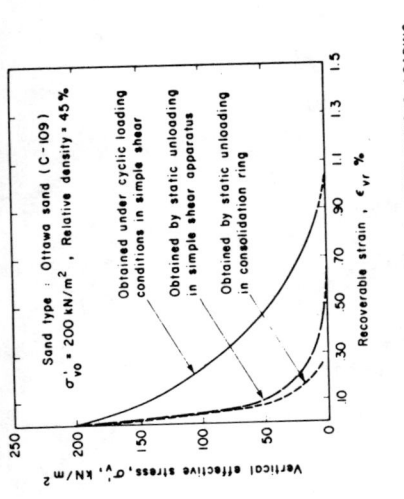

FIG. 3 : REBOUND OF OTTAWA SAND UNDER VARIOUS LOADING CONDITIONS

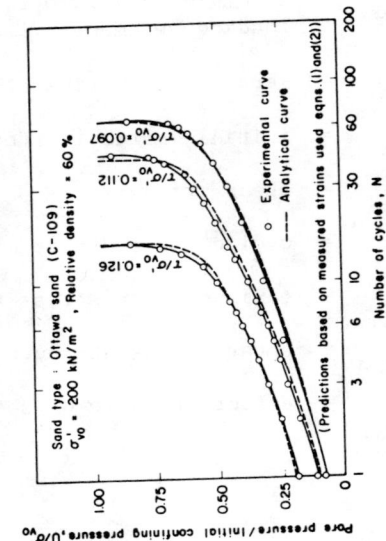

FIG. 5 : PREDICTED AND MEASURED PORE-WATER PRESSURE IN CONSTANT STRESS CYCLIC SIMPLE SHEAR TESTS, Dr = 60%

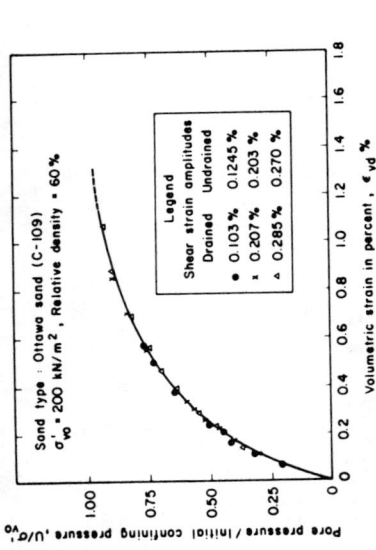

FIG. 2 : RELATIONSHIP BETWEEN VOLUMETRIC STRAINS AND PORE-WATER PRESSURES IN CONSTANT STRAIN CYCLIC SIMPLE SHEAR TESTS

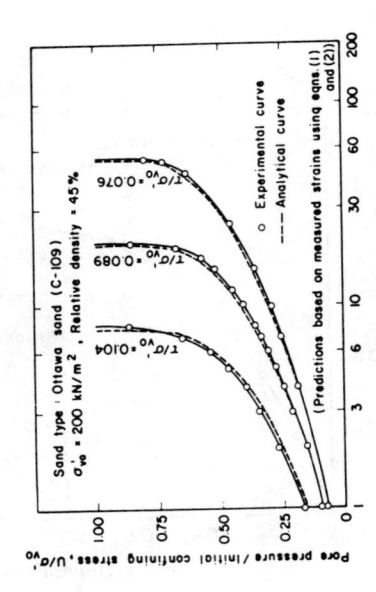

FIG. 4 : PREDICTED AND MEASURED PORE-WATER PRESSURE IN CONSTANT

MODEL VERIFICATION 251

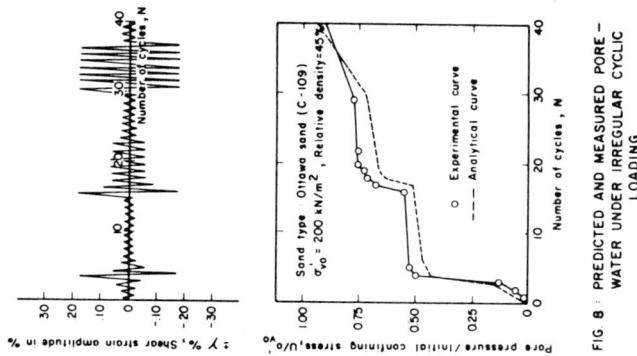

FIG. 8 : PREDICTED AND MEASURED PORE-WATER UNDER IRREGULAR CYCLIC LOADING

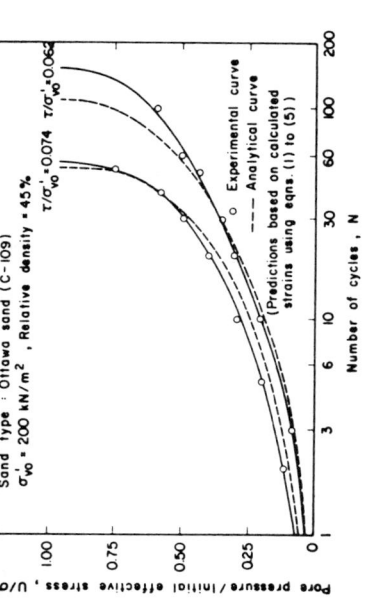

FIG. 6 : PREDICTED AND MEASURED PORE-WATER PRESSURES IN CONSTANT STRESS CYCLIC SIMPLE SHEAR TESTS, Dr = 45 %

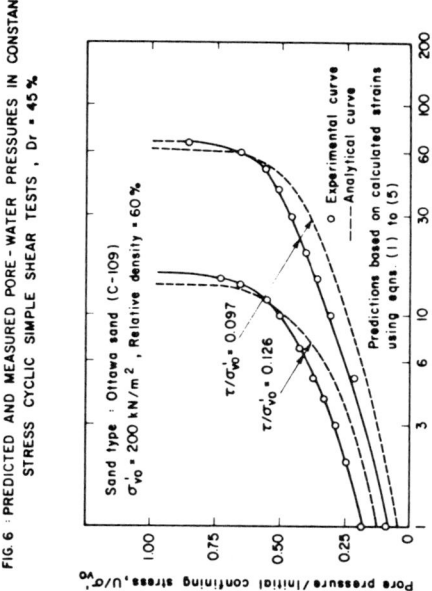

FIG. 7 : PREDICTED AND MEASURED PORE-WATER PRESSURES IN CONSTANT STRESS CYCLIC SIMPLE SHEAR TESTS, Dr = 60 %

252 SOIL STRESS STRAIN APPLICATIONS

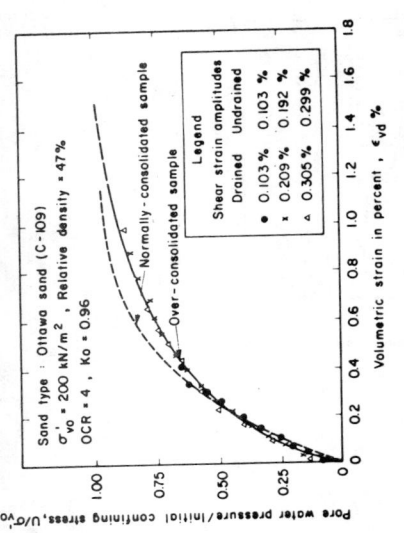

FIG. 10: RELATIONSHIP BETWEEN VOLUMETRIC STRAINS AND PORE-WATER PRESSURES IN CONSTANT STRAIN CYCLIC SIMPLE SHEAR TESTS

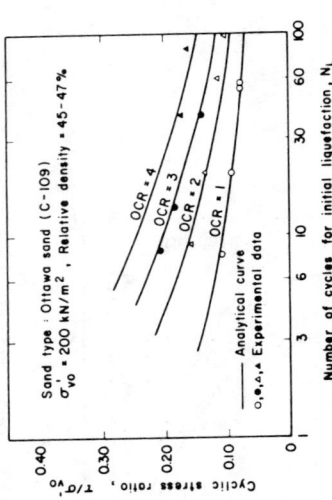

FIG. 11 CYCLIC STRESS RATIO vs. NUMBER OF CYCLES FOR

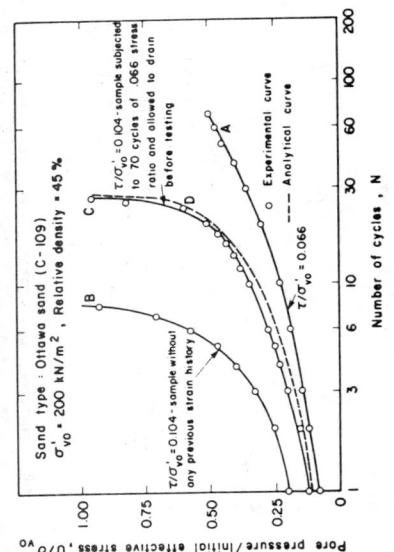

FIG. 9: PREDICTED AND MEASURED PORE-WATER PRESSURES IN A SAND WITH PREVIOUS LOADING HISTORY

CHARACTERIZATION OF THE UNDRAINED ANISOTROPY OF CLAYS

By Lawrence A. Hansen,[1] A.M., ASCE
and G. Wayne Clough,[2] M., ASCE

ABSTRACT

Undrained strength and stress-strain laboratory test data for eight normally consolidated clays are reviewed. Using the loading response trends determined, a utilitarian, nonlinear elastic model for undrained analysis is developed. The model allows stress-strain data to be numerically characterized for input to a finite element code. The response of a cantilever wall founded in clay to excavation in front of the wall is described, illustrating the effect of strength anisotopy.

INTRODUCTION

The focus of this paper is the development of a model for characterization of the undrained anisotropic behavior of clays, based on a review of available laboratory test results. Anisotropy in clays has received considerable attention in the last 15 years. This research has emphasized strength anisotropy, with particular attention to slope stability and bearing capacity problems; less attention has been paid to the equally important question of stress-strain response anisotropy. This paper reviews published and unpublished experimental data for eight normally consolidated clays, in terms of both strength and stress-strain behavior.

Using loading response trends determined for these clays, a utilitarian, nonlinear elastic model is developed. The model is an extended form of hyperbolic models presented previously by Duncan and Chang (8) and Simon (19). It allows stress-strain data to be numerically characterized, and thus it can be readily programmed into a finite element code for analysis of a wide variety of undrained loading problems in clays. The influence of strength anisotropy is illustrated in this paper by finite element analyses of the response of excavation in anisotropic clay supported by cantilever walls.

KEY WORDS: Clays; Anisotropy; Shear strength; Stress-strain response; Retaining walls; Finite element method; Earth pressure; Deformation.

---

[1]Assistant Professor of Engineering, Arizona State University, Tempe, Arizona.
[2]Professor of Civil Engineering, Stanford University, Stanford, California.

SUMMARY OF SELECTED LOAD TEST RESULTS

A comprehensive listing of all available load test data pertaining to clay anisotropy is beyond the scope of this paper. Instead, detailed test results for a group of eight normally consolidated clays are presented. These results were selected from a large body of data representing 47 laboratory prepared and naturally occurring clays summarized in the first author's dissertation (12). Index properties and a summary of testing procedures and undrained test results for each of the clays are listed in Tables 1 and 2. These clays were selected primarily because reliable stress-strain data from high quality tests measuring anisotropic variations in response was available for each of them. The purpose of this review is to establish trends in such response which can be implemented in a model.

Sources of Strength and Stress-Strain Anisotropy

It is useful to briefly consider the source of anisotropy in clays. During deposition and subsequent one-dimensional consolidation of naturally occurring clays, the fabric of the clay tends to become oriented in the horizontal direction, with the vertical direction an axis of rotational symmetry. The effect of this natural bedding is termed inherent or material anisotropy. It is characterized by variations in friction angle (ϕ'), cohesion (c') and pore pressure parameter ($\bar{A}$) and, therefore, undrained strength for loading at different orientations to the direction of deposition (or vertical, as considered in this paper).

A second, equally important component of anisotropy can only be determined if undrained shear is initiated from an anisotropic state of stress. This component, termed "stress system induced anisotropy" by Ladd et al. (16), is induced by rotating the principal stresses during loading from their orientations at the end of consolidation, and is exhibited whenever k_o is not equal to unity. It is evident that if induced anisotropy is included in the test procedure, then the combined effect of both induced and inherent anisotropy will actually be measured. A comprehensive discussion of both components of anisotropy, and in particular induced anisotropy, is included in Duncan and Seed (9).

Effect of Combined Anisotropy on Undrained Strength

In order to measure the combined effect of both components, anisotropically consolidated triaxial and plane strain tests (TX-ACU and PS-ACU) are typically used. Each of the clays included in Table 1 were tested using one or both of these procedures. Vertically oriented samples are used (note the exception for San Francisco Bay Mud in Table 1) and principal stress directions are either not rotated ($\beta=0°$) or rotated such that the direction of the major principal change in stress is horizontal ($\beta=90°$), corresponding to the direction of the major principal stress at failure. Intermediate values of β are not possible. The value of the

TABLE 1 - INDEX PROPERTIES AND SUMMARY OF UNDRAINED
TESTS FOR SELECTED NORMALLY CONSOLIDATED CLAYS

Soil Description (1)	w_l % (2)	w_p % (3)	PI % (4)	w_n % (5)	Sensitivity (6)	K_s (7)	K_{45} (8)	Test Type (9)	Undrained Stress Path[a] (10)	Notes (11)	Reference (12)
1. Haney Clay (British Columbia)	44	26	18		6-10	0.70 0.63		PS-ACU TX-ACU	LU,UC,LE,UE LU,UC,LE,UE	b b	21
2. Vaterland	47	27	20	40	5	0.37	0.81	TX-ACU & DSS	LC,UE	c,d	1,18
3. Drammen Lean Clay	33	23	10	31	8	0.22	0.65	TX-ACU & DSS	LC,UE	c,d	1,18
4. Drammen Plastic Clay	60	30	30	52	8	0.39	0.75	TX-ACU & DSS	LC,UE	c,d	1,18
5. Postgirobygget (Norway)						0.35		TX-ACU	LC,UE	c	18
6. San Francisco Bay Mud	88	43	45	90	8	0.76	0.67	PS-ACU & SS		e	10,11
7. Kalix Clay (Sweden)	170	65	105	140-170	11	0.54		TX-ACU	LC,UE	c	20
8. Boston Blue Clay	35-43	15-22	18-24		5-10	0.58 0.48	0.59 0.61	PS-ACU & DSS TX-ACU & DSS	LC,LE LC,LE	d,f d,f	6,13 15,16

Notes -

a. Indicates total stress path to failure after consolidation for vertically oriented sample:
    LC - loading compression - axial stress increased, lateral stress held constant
    UC - unloading compression - axial stress held constant, lateral stress decreased
    LE - loading extension - axial stress held constant, lateral stress increased
    UE - unloading extension - axial stress decreased, lateral stress held constant.
    Nomenclature is from Vaid and Campanella (21).
    $\beta = 0°$ for compression tests; $\beta = 90°$ for extension tests.

b. Specimens tested in k_o consolidation triaxial and plane strain devices. K_s is ratio of strengths from extension (LE,UE) and compression (UC,LC) tests.

c. Specimens reconsolidated in triaxial cell or simple shear device to field stresses. K_s is ratio of strengths from extension (UE) and compression (LC) tests.

d. K_{45} is ratio of $(\tau_h)_{max}/p_o'$ from DSS test to $1/2\ (\sigma_1-\sigma_3)/p_o'$ from triaxial compression (LC) test.

e. Specimens trimmed in vertical and horizontal orientations. Vertical samples tested in compression (LC). Horizontal samples were placed in vertical orientation in cell, then tested by reducing lateral stress until isotropic stress condition, then by increasing vertical stress. K_s is equal to the ratio of strengths for horizontal and vertical samples, K_{45} is the ratio of $(\tau_{xy})_{max}/p_o'$ from SS test to $1/2\ (\sigma_1-\sigma_3)/p_o'$ from tests on vertical samples.

f. Specimens trimmed from resedimented samples then reconsolidated to insitu stresses. K_s is ratio of strengths from extension (LE) and compression (LC) tests.

TABLE 2 - SUMMARY OF NORMALIZED PS-ACU OR TX-ACU TEST RESULTS FOR SELECTED NORMALLY CONSOLIDATED CLAYS

Soil identi-fication (1) [a]	Test Type (2)	σ'_{c_2} kN/m^2 (3)	$\dfrac{\sigma'_{ic}}{\sigma'_{ic}}$ (4)	s_u/σ'_{ic} Comp. (5)	s_u/σ'_{ic} Ext. (6)	E_i/σ'_{ic} Comp. (7)	E_i/σ'_{ic} Ext. (8)	K_s (9)	K_E (10)	Axial Strain, ϵ_{af}, % Comp. (11)	Axial Strain, ϵ_{af}, % Ext. (12)
1	PS	590	0.5	0.30	0.21	75	120	0.70	1.6	1.0	6.0
2	TX	58-104	0.5	0.27	0.10	400	485	0.37	1.2	0.6	3.5
3	TX	102	0.5	0.37	0.08	430	290	0.22	0.69	0.3	2.5
4	TX	68	0.6	0.41	0.16	260	345	0.39	1.3	1.0	2.4
5	TX	155-195	0.6	0.35	0.12	305	385	0.35	1.3	1.0	4.0
6	PS	60-400	0.4-0.5	0.37	0.28	85	170	0.76	2.0	2.5	8.0
7	TX	20	0.5	0.83	0.41	280	250	0.54	0.87	0.87	6.0
8	PS	-	0.6	0.34	0.19	310	525	0.58	1.7	0.2	1.2

[a] Refer to Table 1 for soil descriptions.

TABLE 3 - SUMMARY OF NORMALIZED HYPERBOLIC PARAMETERS FOR SELECTED NORMALLY CONSOLIDATED CLAYS

Soil identi-fication (1) [a]	$a = (\sigma'_{ic}/E_i) \times 10^{-2}$ Comp. (2)	$a = (\sigma'_{ic}/E_i) \times 10^{-2}$ Ext. (3)	$b = \sigma'_{ic}/2\Delta q_f$ Comp. (4)	$b = \sigma'_{ic}/2\Delta q_f$ Ext. (5)	$R_f = b(2\,q_f)/\sigma'_{ic}$ Comp. (6)	$R_f = b(2\,q_f)/\sigma'_{ic}$ Ext. (7)
1	1.3	0.82	5.3	0.97	0.75	0.85
2	0.25	0.21	3.8	1.3	0.93	0.92
3	0.23	0.34	5.0	1.4	0.85	0.90
4	0.38	0.29	2.1	1.3	0.85	0.92
5	0.33	0.26	3.8	1.5	0.93	0.92
6	1.2	0.59	4.8	0.83	0.92	0.90
7	0.35	0.40	0.75	0.71	0.87	0.94
8	0.32	0.19	4.6	1.1	0.83	0.93

[a] Refer to Table 1 for soil descriptions

TABLE 4 - SOIL PARAMETERS ASSUMED FOR CANTILEVER WALL ANALYSES

Parameter	Symbol	Value
Total unit weight	γ	20.0 kN/m^3
At rest earth pressure coefficient	k_o	0.60
Poisson's ratio	ν	0.49
Undrained shear strength ($\beta = 0°$)	s_{uo}	28.4+0.2σ'_v kN/m^2
Anisotropic strength ratio	K_s	0.50
Anisotropic modulus ratio	K_E	4.0
Parameters for anisotropic strength variation	A_s B_s	0.50 0.00
Parameter for anisotropic modulus variation	A_E	-0.25
Initial tangent modulus ($\beta = 0°$)	E_{io}	600 s_{uo}
Hyperbolic correlation parameter	R_f	0.9

anisotropic strength ratio, K_s, is equal to the ratio of strengths for $\beta=90°$ and $\beta=0°$.

As listed in Table 1, values of K_s vary from 0.22 to 0.76 for the eight clays. Ladd, et al (16) have previously indicated that normally consolidated clays with a low plasticity index are more anisotropic than clays with a high plasticity index.[3] This is supported in Figure 1, where K_s is plotted versus PI. Figure 1 is an updated version of a similar plot from Ladd, et al. (16). Figure 1 also indicates that tests in which both inherent and induced anisotropy are included (PS-ACU and TX-ACU) result in lower values of K_s than tests in which only inherent anisotropy is included (UC,UU,TX-ICU). Both groups of results, however, illustrate similar trends in K_s with PI. Results for two cases shown in Figure 1 in which the same clay was tested in both plane strain and triaxial shear indicates that values of K_s based on triaxial tests are somewhat lower than from plane strain tests. Ladd, et al. (16) gives data for six clays which can be used to calculate an average ratio of K_s for triaxial to K_s for plane strain tests of 0.90.

Points representing data reported by Berre and Bjerrum (1) are singled out in Figure 1 because of the significantly lower values of K_s reported. The points represent Norwegian clays of marine origin having highly developed structures. The undrained strength for $\beta=90°$ for each was determined by unloading extension (UE). The lower values of K_s may result from a combination of structure and testing procedure, or they may represent realistic values for these clays

Estimated Strength Variations Due to Combined Anisotropy

Bjerrum (3) and Ladd (14) have suggested that the undrained strength measured by a direct simple shear (DSS) test can be used to represent an intermediate value of rotation angle β, which would allow estimation of a relative strength variation with β. Berre and Bjerrum (1) and Bjerrum and Kenney (4) report results of DSS-ACU tests on several clays, in which vertical samples were reconsolidated to field stresses prior to undrained shear. Assuming the maximum shear stress on the horizontal plane, τ_h, for this test is equal to the undrained strength (or $q_f = 1/2\ (\sigma_1-\sigma_3)_f = s_u$), and that the principal stresses are rotated $\beta=45°$, DSS-ACU strengths can be used in conjunction with TX-ACU or PS-ACU test results to estimate undrained strength variations.

Estimated variations for four clays, included in Table 1, determined in this manner are plotted in Figure 2. The variations in Figure 2 indicate that the four clays considered have the same general form: if the principal stresses are rotated from their insitu orientations during shear

---

[3] Bjerrum (3) first reported this trend, based on the results of field vane shear tests.

258 SOIL STRESS STRAIN APPLICATIONS

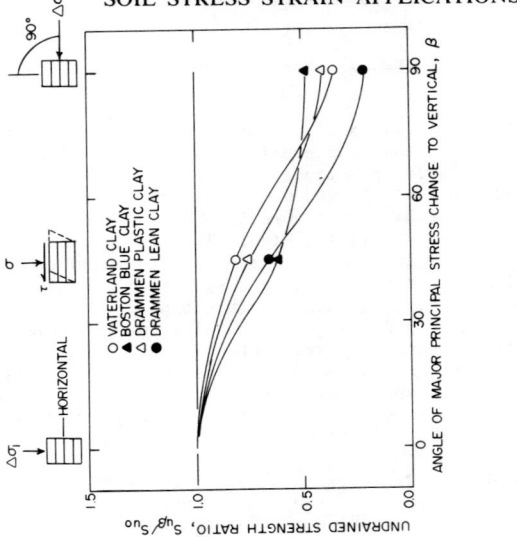

Figure 2. DIRECTIONAL VARIATION OF UNDRAINED SHEAR STRENGTH FROM DSS-ACU AND TX-ACU TESTS.

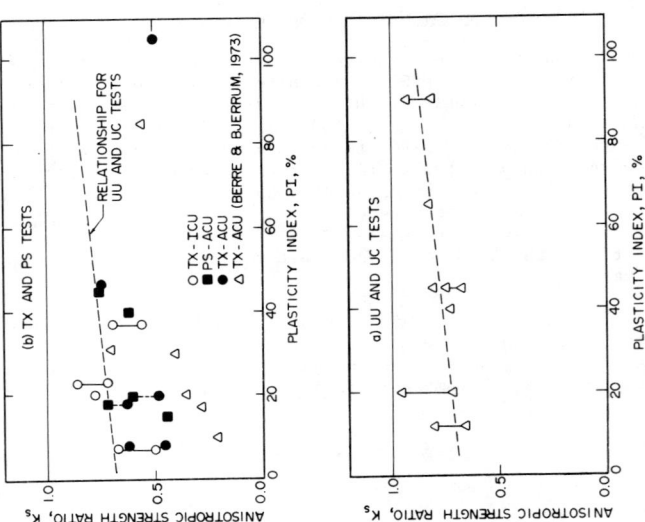

FIGURE 1. K_S vs. PI FOR NORMALLY CONSOLIDATED CLAYS.

a lower strength will be measured. The ratio of DSS-ACU strength ($\beta=45°$) to plane strain compressive strength ($\beta=0°$) varies from 0.59 to 0.81 for the clays included in Figure 2. This ratio is listed under the heading K_{45} in Table 1. Since K_s for these clays varies between 0.22 and 0.76, the range in strength for DSS-ACU tests varies approximately between the strength in extension and the average of the strengths in compression and extension.

The variations plotted in Figure 2 are to some extent a function of the tests used to measure the clay strength. It is recognized that the stress conditions in the DSS test are not well defined. In particular, the direction of the principal stresses continuously rotate during shear and do not exactly correspond to the assumption of $\beta=45°$. However, the results shown in Figure 2 represent the present state of the art and, as will be shown later, the authors' model can be easily modified to consider alternative schemes should future results prove this to be necessary.

Effect of Combined Anisotropy on Initial Modulus

Normalized stress-strain response data from PS-ACU and TX-ACU tests on the eight clays considered are summarized in Table 2. Initial modulus values, E_i, were calculated using best fit lines to transformed hyperbolic stress-strain plots; the initial modulus values correspond to initial anisotropic stresses. The ratio of minor and major principal consolidation stresses varied from 0.5 to 0.6, except for San Francisco Bay Mud where generally lower ratios of 0.4 were used for extension tests. Major principal consolidation stresses for all clays varied from 20 to 400 kN/m^2.

The ratio of initial modulus for extension tests ($\beta=90°$) to that for compression tests ($\beta=0°$), termed the anisotropic modulus ratio, is listed under the heading K_E in Table 2. Values of K_E vary from 0.69 to 2.0, with K_E for six of the eight clays being 1.2 or larger. This illustrates a reversal in the trend established for anisotropic strength ratio: apparently clays which have a greater strength in compression, are also less stiff in compression. The difference in stiffness is related to the reversal in direction of the incremental shear stress during extension from the direction during deposition. Values of K_E for six of the clays indicate that K_E decreases with decreasing PI; however, the value of $K_E = 0.89$ for Kalix clay with PI equal to 105% does not fit this trend. This may be related to the organic content or structure of this particular clay.

In order to get modulus values for other than $\beta=0°$ and $\beta=90°$, the DSS test could be used, as it was for intermediate values of S_{uo}. However, few published data gave consistent results. An exception is the data of Ladd and Edgers (15). They calculated modulus values for DSS-ACU tests on Boston Blue Clay, assuming the shear modulus can be calculated from $G = \tau_h/\gamma$, where τ_h is the shear stress on the horizontal plane and γ is the shear strain. Extrapolating their data an initial normalized Young's modulus value, E_i/σ'_{vc}, of 300 is calculated. This value falls between Young's

modulus values of 240 and 390 computed from the deformation modulus values
listed in Table 2. Ladd and Edgers (15) indicate that for the same strain
level, the undrained secant modulus, calculated based on the same assumptions, increases with increasing rotation of the principal stress directions. In sum, then, it may be tantatively concluded that the relative
initial modulus value, or ratio of E_i for any value of β to E_i for $\beta=0°$,
increases with increasing β.

Hyperbolic Representation of Stress-Strain Data

In characterizing the effect of combined anisotropy on overall stress-strain response, it is necessary to determine the modulus values for other
than initial conditions. Values of tangent modulus, E_t, were determined
for numerous points on the stress-strain curve for seven of the clays
listed in Table 2 using hyperbolic models of the curves. The general procedure is illustrated by Figure 3. For either compression or extension
tests the normalized axial stress is computed relative to the anisotropic
state of stress existing at the end of consolidation, or $2|q-q_o|/\sigma'_{1c}$
where $2q_o$ is the principal stress difference at the end of consolidation,
$(\sigma_{ac}-\sigma_{1c})$, and $2q$ is the principal stress difference at a given value of
strain, $(\sigma_a-\sigma_1)$. For ease of presentation, axial, a, and lateral, 1,
stress designations are used. Similarly, absolute values of axial strain
$|\varepsilon_a|$ and relative stress $|q-q_o|$ are used to avoid confusion.

The hyperbolic transformation is made, as indicated in Figure 3 by
plotting transformed stress, $|\varepsilon_a|\sigma'_{1c}/2|q-q_o|$, versus axial strain, $|\varepsilon_a|$.
This is analogous to the procedure of plotting $\varepsilon_a/(\sigma_1-\sigma_3)$ versus ε_a
described by Duncan and Chang (8) for isotropic test results. The equation
for the transformed linear relation is then written as

$$\frac{2|q-q_o|}{\sigma'_{1c}} = \frac{|\varepsilon_a|}{a+b|\varepsilon_a|} \qquad (1)$$

where a and b are the intercept and slope of the relationship. It can be
shown that $a = \sigma'_{1c}/E_i$, or the inverse of the normalized initial tangent
modulus, and that $b = \sigma'_{1c}/(2|q-q_o|)_{ult}$, where $(2|q-q_o|)_{ult}$ is the asymptotic value of principal stress difference. Analogous to the isotropic
procedure, a term R_f can be introduced, relating the asymptotic hyperbolic
principal stress difference to the actual principal stress difference at
failure, $2|q-q_o|)_f$. Thus,

$$R_f = \frac{\Delta q_f}{\Delta q_{ult}} = \frac{|q-q_o|_f}{|q-q_o|_{ult}} \qquad (2)$$

Hyperbolic curves were fitted to normalized stress-strain data for
each of the clays listed in Table 2. A representative curve is shown in
Figure 4 for Vaterland clay. (Note in Figure 4 that extension strains are
plotted as negative, and that the stress difference for the extension test
is plotted as negative after the isotropic state is passed.) Values of

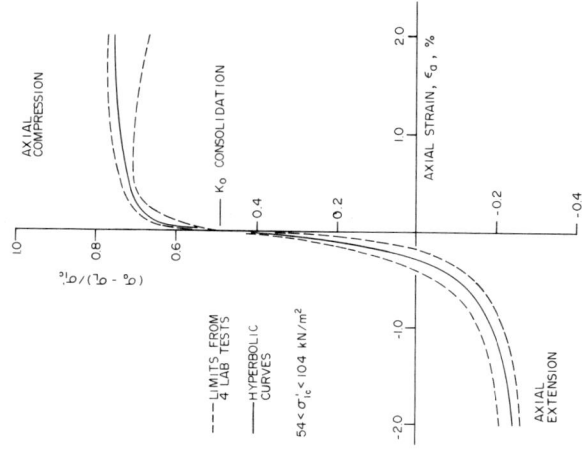

Figure 4. NORMALIZED STRESS-STRAIN RESPONSE OF VATERLAND CLAY IN PLANE STRAIN AXIAL COMPRESSION AND EXTENSION.

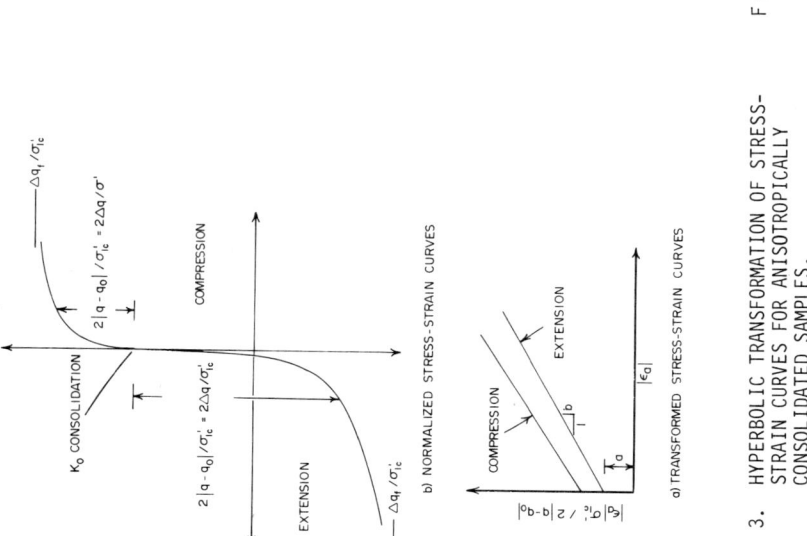

Figure 3. HYPERBOLIC TRANSFORMATION OF STRESS-STRAIN CURVES FOR ANISOTROPICALLY CONSOLIDATED SAMPLES.

hyperbolic fitting parameters a and b, and the term R_f, are listed in Table 3 for each clay. The lack of variation in R_f is striking: R_f does not vary significantly for extension or compression tests for the same clay, nor does it vary significantly for the clays considered. The average value of R_f based on compression tests is 0.87, compared to 0.90 for extension tests. Since the stress difference is measured relative to the initial stress state, the value of the parameter b is three to five times as large in compression than extension.

Effect of Combined Anisotropy on Tangent Modulus

The relationship for normalized tangent modulus is determined by taking the derivative of Equation 1 with respect to axial strain:

$$\frac{E_t}{\sigma'_{1c}} = \frac{a}{(a+b|\varepsilon_a|)^2} \qquad (3)$$

Values for tangent modulus for both extension and compression tests for seven clays, expressed as a fraction of the initial modulus, E_i, are plotted in Figure 5 versus axial strain. The relationships for both extension and compression tests plot as smooth curves; for either test, relationships for three or more soils can be grouped into a single representative band. Though the bands for the different tests do not always include the same soils, the fact that the curves are parallel suggests that for a given clay the ratio of tangent modulus in extension to that in compression is relatively constant with increasing strain. This is particularly the case for tangent modulus ratios, E_t/E_i, greater than 0.25 and axial strains less than 0.20%, since the curves are nearly linear in this region.

The same trend is indicated in Figure 6, where the ratio of tangent modulus values for extension and compression tests are plotted versus the shear stress ratio, $\Delta q/\Delta q_f$. For stress ratios less than 0.6 to 0.8 the value of the ratio is constant for a particular clay, indicating the effect of combined anisotropy characterized by the anisotropic modulus ratio, K_E, is consistent. Because this ratio is constant, it can be assumed that the same variation in tangent modulus value with β holds for almost all stress or strain levels. At higher stress levels the variation is no longer applicable, but at these stress levels tangent modulus values have been greatly reduced, and any differences are not nearly as significant.

Summary

This review has isolated several trends in the anisotropic response of normally consolidated, soft to medium clays that are pertinent to a model characterizing this behavior:

UNDRAINED ANISOTROPY OF CLAYS

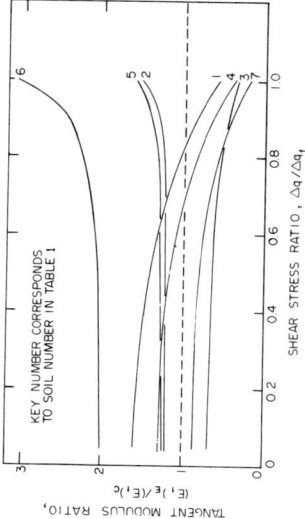

Figure 6. RATIO OF TANGENT MODULUS IN EXTENSION TO COMPRESSION VS. SHEAR STRESS RATIO FOR SEVEN CLAYS.

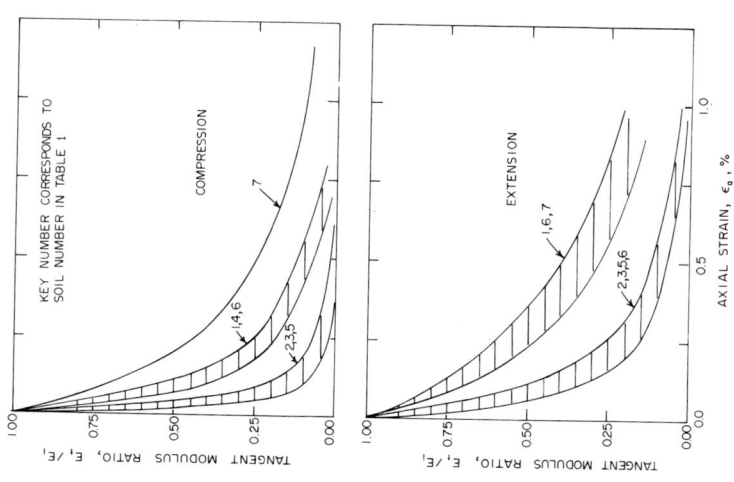

Figure 5. TANGENT MODULUS RATIO VS. AXIAL STRAIN FOR SEVEN CLAYS.

(1) the variation in undrained strength determined by DSS-ACU and TX-ACU or PS-ACU tests defines a smooth relationship, with the undrained strength decreasing as the angle of rotation of principal stresses increases;
(2) the variation in modulus is apparently the reverse of this, with the modulus increasing with rotation;
(3) the variation in modulus is applicable to both initial modulus and tangent modulus for high levels of either stress or strain.

UNDRAINED ANISOTROPIC MODEL

The undrained anisotropic model includes two basic elements: (1) the soil response to undrained loading is represented by hyperbolic stress strain curves, with a different curve being calculated for each principal stress rotation angle, β, based on the key parameters of undrained strength, s_u, and initial tangent modulus, E_i; (2) anisotropic variations for s_u and E_i are assumed, based on available laboratory test data. The hyperbolic stress-strain curves facilitate incremental analysis of the nonlinear response, because the tangent modulus is readily obtained. Each element of the model is considered in the following sections.

Anisotropic Variation in Undrained Strength and Modulus

Davis and Christian (7) and Hansen (12) have reviewed several relationships proposed for defining the anisotropic variation in undrained strength. The equation proposed by Bishop (2),

$$S_{u\beta} = S_{uo}(1-A_s \sin^2\beta)(1-B_s \sin^2 2\beta), \qquad (4)$$

is used in this model, where A_s and B_s are parameters for fitting the equation to experimental data. Use of Equation 4 requires strengths for three values of β be known; these can be provided by results of compression and extension PS-ACU tests and the DSS-ACU test. The value of β is taken as the angle of the major principal stress change, $\Delta\sigma_1$, to the vertical. This corresponds to the definition of β given previously for the special cases of laboratory tests, and is applicable to the more general case where both shear and normal stress changes are applied to horizontal and vertical planes of finite elements. Equation 4 calculates a value of $s_{u\beta}/s_{uo}$ of one for $\beta = 0°$, and allows a minimum value to be calculated at any value of β depending on the parameters A_s and B_s.

A generalized form of the expression developed by Dunlop, et al. (11) is used in the model to define the anisotropic modulus variation. The initial or tangent modulus for a rotation β, or $E\beta$, is determined using appropriate values for $\beta = 0°$ and $\beta = 90°$ from

$$E_\beta = E_o + (E_{90} - E_o)(\sin^2\beta + A_E \sin 2\beta) \qquad (5)$$

where A_E is a shape parameter and β is defined in the preceding paragraph. Since only two values of modulus are required, a value of A_E may be assumed or used to fit a curve to experimental data.

If Equation 4 is rewritten in the form

$$\frac{E_\beta - E_o}{E_{90} - E_o} = \sin^2\beta + A_E \sin 2\beta \tag{6}$$

a normalized modulus term, dependent on A_E and β, is defined. Values of the left hand side of Equation 6 for various values of A_E are plotted versus β in Figure 7. For $A_E = 0.0$, as shown in Figure 7 the modulus increases almost linearly with increasing β; increasing or decreasing A_E from 0.0 shifts the curve to the left or right, respectively. As indicated in Figure 7, Stille, et al. (20) assumed a modulus variation in the form of Equations 5 or 6, with $A_E = 0.0$, for use in a finite element analysis of the response of an embankment. Dunlop, et al. (11) reported Equations 5 or 6 with $A_E = -0.2$ reasonably described the modulus variation of San Francisco Bay Mud. This was determined by a trial and error procedure in which different forms of relationships were used in finite element analyses of simple shear tests. The relationship depicted in Figure 7 allowed computation of a stress-strain curve that was in good agreement with actual test curves. Young's modulus values computed for PS-ACU extension and compression tests and simple shear tests (interpreted as pure shear with $\beta = 45°$) on San Francisco Bay Mud, based on stress-strain curves in Dunlop, et al. (11), can be used in Equations 5 or 6 to calculate a value of $A_E = -0.25$ for this clay. This is in good agreement with the value determined by finite element analyses, which suggests this simplified procedure can be used to calculate a realistic anisotropic modulus variation.

Generalized Hyperbolic Model

To accommodate any degree of principal stress reorientation a transformed hyperbolic relationship can be written in generalized vectorial form as described by Hansen (12). Complete development of this model is beyond the scope of this paper. Since the purpose of the hyperbolic representation is to allow computation of tangent modulus values for use in incremental, nonlinear finite element analyses, only the necessary computational steps serving this purpose are outlined herein. In general, the model uses Equation 5 in conjunction with scalar hyperbolic representations for $\beta = 0°$ and $\beta = 90°$ to compute tangent modulus values.

The stress-strain curves for rotation angles $\beta = 0°$ and $\beta = 90°$ depicted in Figure 7 are defined by the single expression

$$2|q-q_o|_\beta = \varepsilon_\beta \left[\frac{1}{E_{i\beta}} + \frac{\varepsilon_\beta R_f}{(2|q-q_o|_f)_\beta} \right]^{-1} \tag{7}$$

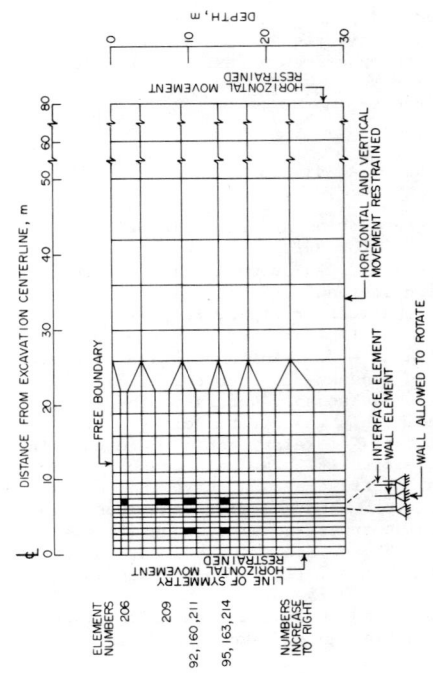

Figure 7. COMPARISON OF POSSIBLE MODULUS VARIATIONS COMPUTED USING EQUATION 6.

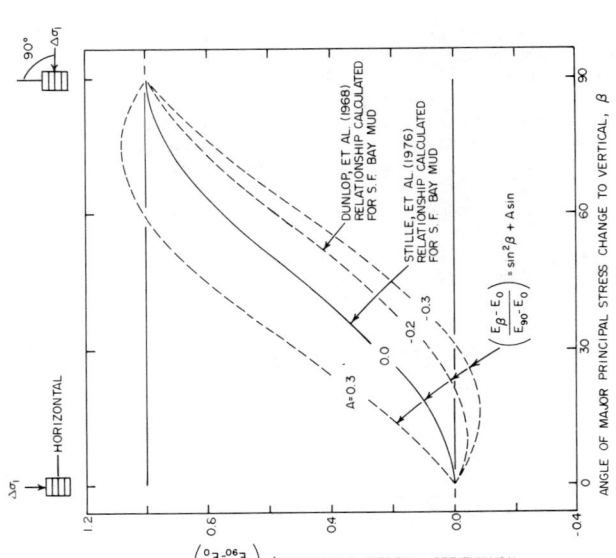

Figure 8. FINITE ELEMENT MESH FOR CANTILEVERED WALL ANALYSES.

where β is either $0°$ or $90°$, ε_β is the axial strain for angle β, $E_{i\beta}$ is the initial modulus for angle β, and $2|q-q_0|_\beta$ is the principal stress difference for angle β (subscript f indicates the value at failure). Differentiating Equation 7, using particular values of β, and then replacing axial strain with maximum shear strain, γ_{max}, results in the following expressions for tangent modulus for $\beta = 0°$ and $\beta = 90°$, or E_{to} and E_{t90}:

$$E_{to} = E_{io} \left[\frac{1}{E_{io}} + \frac{R_f \gamma_{max}/2}{2(s_{uo}-q_o)} \right]^{-2} \quad (8)$$

$$E_{t90} = E_{i90} \left[\frac{1}{E_{i90}} + \frac{R_f \gamma_{max}/2}{2(s_{u90} + q_o)} \right]^{-2} \quad (9)$$

All terms in Equations 8 and 9 have been defined previously.

Equations 8 and 9 relate tangent modulus values to maximum shear strain, hyperbolic parameter R_f, and relative stress difference. Each term is deserving of comment. First, as suggested by Dunlop, et al. (11), maximum shear strain is used as the strain parameter because it affords a measure of the total amount of shear distortion, independent of the stress or strain conditions. Thus, it provides a consistent parameter for any degree of principal stress rotation imposed by any change in stresses. The particular form of Equations 8 and 9 is dependent on the relationship $1/2 \gamma_{max} = \varepsilon_a$, and is valid only for undrained plane strain conditions. Second, R_f is assumed independent of rotation angle β, which eliminates the need for an equation similar to Equations 4 or 5 relating R_f to β. Values of R_f listed in Table 3 for eight clays indicate little difference in R_f for extension and compression tests. Finally, the value of principal stress difference is referenced to the initial anisotropic state of stress, but undrained strength is referenced to a value of principal stress difference of zero. Thus the relative principal stress difference at failure, $2|q-q_0|_f$, is equal to $(s_{uo}-q_o)$ for $\beta = 0°$ and $(s_{u90} + q_o)$ for $\beta = 90°$.

Incremental Analysis

Description of the use of Equations 4,5,8 and 9 in an iterative, incremental finite element analysis procedure is beyond the scope of this paper. The interested reader is referred to Hansen (12). A general outline of an incremental procedure would include the following steps:

(1) using changes in stress relative to the initial anisotropic state of stress, the direction, β, of the major principal change in stress, $\Delta\sigma_1$, is calculated;

(2) using the same relative changes in stress, the value of maximum shear strain, γ_{max}, is calculated;

(3) using the computed value of γ_{max} and given values of E_{i0}, E_{i90}, R_f, s_{uo}, s_{u90} and q_o, tangent modulus values E_{to} and E_{t90} are calculated from Equations 8 and 9, respectively:

(4) using computed values of E_{to} and E_{t90} a value of tangent modulus corresponding to the computed value of β, or $E_{t\beta}$, is calculated using Equation 5.

At each incremental step, the undrained strength for rotation angle β, or $s_{u\beta}$, would also be calculated by using values of s_{uo} and s_{u90} in Equation 4. The value of the stress level is then calculated.

FINITE ELEMENT ANALYSES

Finite element analyses of an excavation supported by a cantilever wall, using the anisotropic soil model developed herein and an isotropic version of the same model, are described in this section. This type of support was selected for the analysis because the absence of lateral bracing was expected to allow the development of well defined active and passive yielded zones, and the distribution of earth pressure acting on the wall, as well as the deformation conditions within the soil, can be compared to theoretical and empirical evidence allowing verification of the model. The effect of soil anisotropy can be assessed by comparing the results of the isotropic and anisotropic analyses.

The analyses model the cantilever wall problem by simulating the soil-structure system and excavation in front of the wall. At each stage of excavation, strain levels and degree of stress reorientation for each element in the finite element mesh are calculated and used to determine a new modulus, shear strength and stress level. Thus, the properties of each element are allowed to change so as to reflect nonlinear and anisotropic soil response. An iterative scheme is used for each stage of excavation, allowing the program to search for proper parameter values.

Finite Element Model

The finite element representation of the problem is shown in Figure 8. Only one half of the excavation is depicted because of symmetry; its total width is 12m (40 ft) and the maximum depth is 9m (30 ft). The soil is assumed to be homogeneous clay to a depth of 30m (100 ft), at which point a rigid layer is encountered. Values of soil parameters used are listed in Table 4. The undrained strength increases with depth in a manner consistent with the known behavior of soft to medium clays. Initial modulus values for $\beta = 0°$ are equal to 600 times the undrained strength. The wall, having the stiffness of a PZ 38 sheetpile, is assumed to penetrate the full depth of the clay; a pinned connection is assumed at the level of the rigid base. Soil and wall elements are separated by interface elements which are

UNDRAINED ANISOTROPY OF CLAYS 269

frictionless and allow no transfer of shear stresses. Initial conditions consist of a level ground surface with the wall in place.

Analysis Results

Predicted wall displacements for each excavation depth in the anisotropic soil are plotted in Figure 9. The wall rotated progressively inward toward the excavated area as excavation proceeded. The displacements at the top of the wall in the anisotropic soil are 0.024H and 0.084H for excavation depth, H, of 5.5 and 9m (17 and 30 ft), respectively. As shown in Figure 9, wall displacements due to excavation in the isotropic soil are significantly less. For an excavation depth of 9 m (30 ft), the top of the wall in the isotropic soil moved inward an amount equal to 0.052H, or approximately 60% of the displacement of the wall in the anisotropic soil.

These magnitudes of movement were sufficient to lower lateral wall pressures acting on the unexcavated or active side of the wall from at rest to approximately Rankine active values, as shown for excavation depths of 5.5 and 9m (17 and 30 ft) in Figure 10. For both anisotropic and isotropic soils, the zone of lowered earth pressures on the active side of the wall extends to a depth of 4 to 10m (12 to 33 ft) below the depth of excavation. Nearly zero or slightly tensile stresses are predicted for the upper zone where the Rankine theory predicts tensile stresses. In the anisotropic soil the active state is essentially reached when the excavation depth is 5.5m (17 ft), compared to 9m (30 ft) for the isotropic soil; this difference reflects the difference in wall displacements. On the excavated or passive side of the wall the effect of stress reorientation on shear strength results in much lower mobilized wall pressures for the anisotropic case.

Total stress paths, normalized with respect to vertical overburden stress, for selected soil elements (see Figure 8) are plotted for both anisotropic and isotropic analyses in Figure 11. Cartesian stresses, σ_x and σ_y, are used to avoid the confusion associated with using principal stresses, σ_1 and σ_3. Loading and unloading compression ($\beta = 0°$ for both) and unloading extension ($\beta = 90°$) stress paths are plotted for reference; these stress paths represent the condition of no shear stress development on the horizontal or vertical planes of the element.

As shown in Figure 11, elements located next to the wall on the unexcavated or active side (only results for element 206 are shown) followed the unloading compression stress path, corresponding to the lateral stress relief due to excavation. The rotation angle β calculated for elements 206, 209, 211 and 214 is less than 2°, regardless of the type of analysis. Elements located next to the wall on the excavated or passive side of the wall (element 160 is shown), in both the anisotropic and isotropic cases, generally followed the unloading extension stress path up to failure. The close adherence to this stress path is due in part to the frictionless

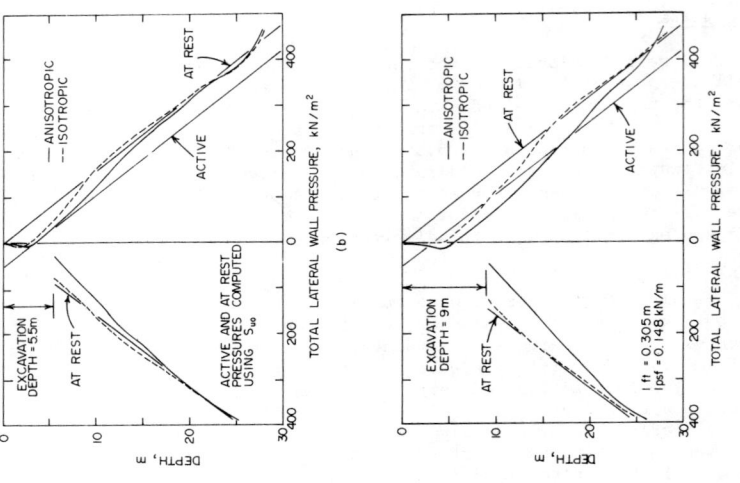

Figure 10. TOTAL LATERAL WALL PRESSURE DISTRIBUTIONS FOR CANTILEVERED WALL IN ISOTROPIC AND ANISOTROPIC SOIL.

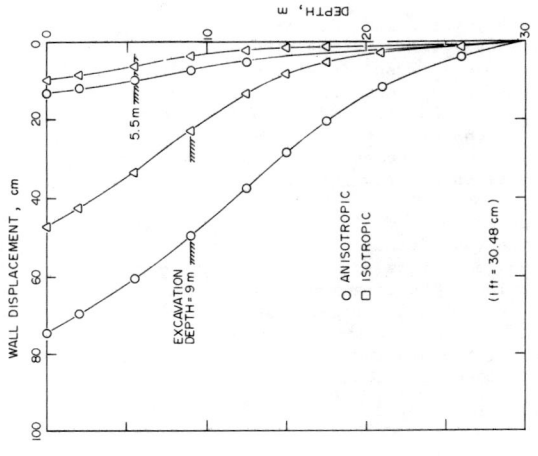

Figure 9. DISPLACEMENT OF CANTILEVERED WALL IN ISOTROPIC AND ANISOTROPIC SOILS.

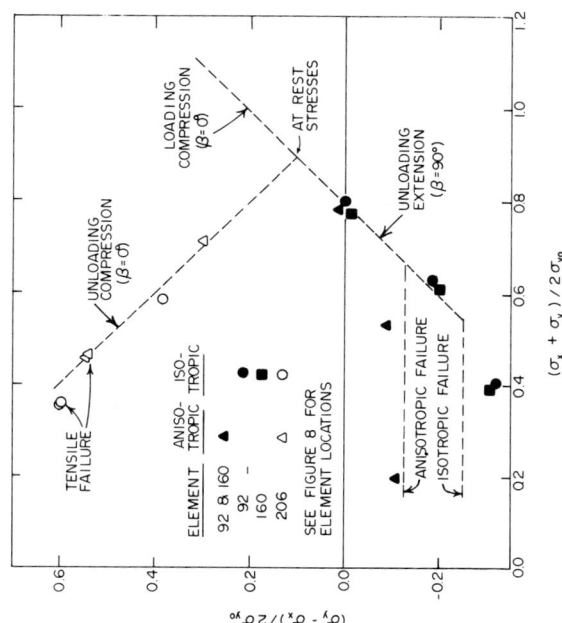

Figure 11. TOTAL STRESS PATHS FOR ELEMENTS 92, 160 AND 206 FOR CANTILEVERED WALL ANALYSES.

soil-wall interface. Since the undrained strength for these elements is less for the anisotropic case, failure is reached earlier in the anisotropic analysis. Element 92, located midway between the excavation centerline and the wall, followed a stress path nearly identical to that of element 160, indicating that the effect of excavation approximates a condition of one-dimensional heave on the excavated side of the wall. Values of β calculated for elements 92, 95, 160 and 163 were in all cases approximately equal to 90°.

SUMMARY AND CONCLUSIONS

Most naturally occurring soft to medium clays are to some degree anisotropic with respect to undrained strength and stress-strain response. Recognition of this fact is expressed in a trend toward inclusion of anisotropic behavior in analysis methods, and in particular the finite element method. In this paper, available load test results are reviewed in order to establish a means of characterizing the undrained anisotropy of clay for inclusion in such analyses.

Summarized test results indicate that anisotropic variations in undrained strength and modulus can be related to the angle of rotation of principal stresses during loading. Typically, undrained strength decreases and undrained modulus increases with increasing rotation of the major principal stress from its initial vertical orientation. Reductions in strength of as much as 80%, and increases in modulus of as much as 200% have been measured for 90° rotations.

The variations in response with rotation of principal stresses can be expressed using simple equations. Data required for input in these equations includes a minimum of two test results, if linear or nearly linear variations are assumed. These should include the effects of combined anisotropy, thus extension ($\beta = 90°$) and compression ($\beta = 0°$) PS-ACU or TX-ACU tests are suggested for use. A DSS-ACU or simple shear test provides an intermediate point, but a simplistic interpretation of the stress conditions in these tests is required.

A method for modeling the anisotropic stress-strain response of clays, using the variation equations and hyperbolic representations of response curves for tests in which 0° and 90° rotations occur, is presented. The model is used in a finite element code to analyze the behavior of an excavation supported by a cantilever wall. Analysis results, including wall pressure distributions and element stress paths, are in close agreement with theoretical predictions, indicating that the anisotropic model developed can provide realistic predictions of anisotropic behavior.

ACKNOWLEDGEMENTS

The authors wish to express their appreciation to Professor A. I. Mana, of the University of Petroleum and Minerals, Saudi Arabia, who assisted in developing the anisotropic finite element code. Dr. Kaare Hoeg, Director of the Norwegian Geotechnical Institute, graciously provided materials pertaining to the anisotropy of Norwegian clays. Financial support for the research was provided by NSF Grant No. 24308.

LIST OF SYMBOLS

a	curve fitting parameter for hyperbolic representation of stress-strain curve (intercept on transformed stress axis)
A_E	curve fitting parameter for characterization of anisotropic modulus variation with β
A_s, B_s	curve fitting parameters for characterization of anisotropic undrained strength variation with β
b	curve fitting parameter for hyperbolic representation of stress-strain curve (slope of transformed linear relationship)
ε_a	axial strain
E_o, E_{90}, E_β	modulus for major principal stress reorientation angles of $0°$, $90°$ and $\beta°$
E_i	initial tangent modulus
$E_{io}, E_{i90}, E_{i\beta}$	initial tangent modulus for major principal stress reorientation angles of $0°$, $90°$ and $\beta°$
E_t	tangent modulus
$E_{to}, E_{t90}, E_{t\beta}$	tangent modulus for major principal stress reorientation angles of $0°$, $90°$ and $\beta°$
K_E	anisotropic modulus ratio ($K_E = E_{90}/E_o$)
K_s, K_{45}	anisotropic strength ratio ($K_s = s_{u90}/s_{uo}$, $K_{45} = s_{u45}/s_{uo}$)
R_f	failure ratio in hyperbolic representation of stress-strain curve
$s_{uo}, s_{u45}, s_{u90}, s_{u\beta}$	undrained strength for major principal stress reorientation angles of $0°$, $45°$, $90°$ and $\beta°$
β	angle of major principal stress reorientation
γ_{max}	maximum shear strain
σ_a, σ_1	normal stress in the axial and lateral directions
σ_x, σ_y	normal stress in the x and y directions
σ'_{ac}	axial effective consolidation stress
σ'_{1c}	major principal effective consolidation stress
ν	Poisson's ratio

REFERENCES CITED

1. Berre, T. and Bjerrum, L. "Shear Strength of Normally Consolidated Clay," *Proceedings of the 8th International Conference on Soil Mechanics and Foundation Engineering*, Moscow, Vol. 1, Pt. 1, pp. 39-49, 1973.

2. Bishop, A. W., "The Strength of Soils as Engineering Materials," Sixth Rankine Lecture, *Geotechnique*, Vol. 17, No. 2, pp. 91-130. Also published as Chapter 6 in Milestones in Soil Mechanics: *The First Ten Rankine Lectures*, ICE, London, 1975, pp. 131-171, 1966.

3. Bjerrum, L. "Problems of Soil Mechanics and Construction on Soft Clays and Structurally Unstable Soils (Collapsible, Expansive and Others)," *Proceedings of the 8th International Conference on Soil Mechanics and Foundation Engineering*, Moscow, Vol. 3, pp. 111-159, 1973.

4. Bjerrum, L. and T. C. Kenney, "Effect of Structure on the Shear Behavior of Normally Consolidated Quick Clay," *Proceedings of the Geotechnical Conference*, Oslo, Norway, Vol. 2, pp. 19-27, 1967.

5. Clough, G. W. and L. A. Hansen, "Effects of Clay Anisotropy on Braced Wall Behavior," paper submitted for presentation at the Generalized Stress-Strain Applications in Geotechnical Engineering session of the ASCE Annual Meeting, Hollywood, Florida, October, 1980.

6. D'Appolonia, D. J. and T. W. Lambe, "Method for Predicting Initial Settlement," *Journal of the Soil Mechanics and Foundations Division*, ASCE, Vol. 96, No. SM2, March, pp. 523-544, 1970.

7. Davis, E. H. and J. T. Christian, "Bearing Capacity of Anisotropic Cohesive Soil," *Journal of the Soil Mechanics and Foundations Division*, ASCE, Vol. 97, No. SM5, May, pp. 753-769, 1971.

8. Duncan, J. M. and C. Y. Chang, "Nonlinear Analysis of Stress and Strain in Soils," *Journal of the Soil Mechanics and Foundations Division*, ASCE, Vol. 96, No. SM5, September, pp. 1625-1953, 1970.

9. Duncan, J. M. and H. B. Seed, "Anisotropy and Stress Reorientation in Clay," *Journal of the Soil Mechanics and Foundations Division*, ASCE, Vol. 92, No. SM5, September, pp. 21-50, 1966a.

10. Duncan, J. M. and H. B. Seed, "Strength Variation Along Failure Surfaces in Clay," *Journal of the Soil Mechanics and Foundations Division*, ASCE, Vol. 92, No. SM6, November, pp. 81-104, 1966b.

11. Dunlop, P., J. M. Duncan, and H. B. Seed, "Finite Element Analyses of Slopes in Soil," Contract Report S-68-6, U.S. Army Engineer Waterways Experiment Station, Corps of Engineers, Vicksburg, MS, May, 1968.

12. Hansen, L. A., "Prediction of the Behavior of Braced Excavations in Anisotropic Clay," Dissertation submitted to the Department of Civil Engineering and the Committee on Graduate Studies of Stanford University in partial fulfillment of the requirements for the degree of Doctor of Philosophy, April, 1980.

13. Kinner, E. B. and C. C. Ladd, "Undrained Bearing Capcity of Footings on Clay," Proceedings of the 8th International Conference on Soil Mechanics and Foundation Engineering, Moscow, Vol. 1, Pt. 1, pp. 209-216, 1973.

14. Ladd, C. C., "Discussion - Main Session IV - Problems of Soil Mechanics and Construction on Soft Clays and Structurally Unstable Soils," Proceedings of the 8th International Conference on Soil Mechanics and Foundation Engineering, Moscow, Vol. 4, Pt. 2, pp. 108-115, 1973.

15. Ladd, C. C. and L. Edgers, "Consolidated-Undrained Direct Simple Shear Tests on Saturated Clays," Research Report R72-82, Department of Civil Engineering, Massachusetts Institute of Technology, Cambridge, Mass., July, 1972.

16. Ladd, C. C., R. Foott, K. Ishihara, F. Schlosser, and H. G. Poulos, "Stress-Deformation and Strength Characteristics," Proceedings of the 9th International Conference on Soil Mechanics and Foundation Engineering, Tokyo, Japan, Vol. 1, pp. 421-494, 1977.

17. Lo, K. Y. and J. P. Morin, "Strength Anisotropy and Time Effects of Two Sensitive Clays," Canadian Geotechnical Journal, Vol. 9, No. 3, August, pp. 261-277, 1972.

18. Norwegian Geotechnical Institute, personal communication to G. Wayne Clough, Prof. of Civil Engineering, Stanford University, 1978.

19. Simon, R. N., "Analysis of Embankment Construction by Finite Elements," Thesis submitted to Massachusetts Institute of Technology in partial fulfillment of the requirements for the Degree of Doctor of Philosophy, 1972.

20. Stille, H., A. Fredriksson, and B. B. Broms, "Analysis of a Test Embankment Considering the Anisotropy of the Soil," Numerical Methods in Geomechanics, Papers presented at the Second International Conference on Numerical Methods in Geomechanics, Blacksburg, VA, June, C. S. Desai, Editor, Vol. 2, pp. 611-622, ASCE,1976.

21. Vaid, Y. P. and R. G. Campanella, "Triaxial and Plane Strain Behavior of Natural Clay," Journal of the Geotechnical Engineering Division, ASCE, Vol. 100, No. GT3, March, pp. 207-224, 1974.

Uplift of Drilled Rigid Piers in Loose Sand

by

H.B. Poorooshasb*

INTRODUCTION

In the design of foundations for certain classes of structures, such as drilled piers supporting transmission towers or pile groups supporting sewage treatment tanks, the uplift behaviour of the foundation is a major design factor. This is especially so if the supporting soil is of a poor quality (loose sand or recently deposited clays) overlying a more competent bed.

In the present paper a theoretical approach to the solution of the uplift behaviour of rigid short piers or pile groups embedded in a layer of loose sand (density of the order of the critical density) is outlined. In obtaining the solution, which is expressed in the form of exponential power series, a nonlinear stress-strain law for the sand is employed and the soil below the level of the tip of the pier (or pile group) is assumed to act as a Winkler Foundation.

The analysis shows that for loads smaller than of about 50 percent of the ultimate, the butt displacements (or group heave) are almost linearly dependent on the magnitude of the uplift forces, that the mobilization of skin friction is linear with depth and that at the ultimate load the uplift coefficient is, to a very close approximation, equal to k_0, the at rest earth pressure coefficient.

ANALYSIS OF PILE GROUPS

In the analysis of pile groups the entrapped soil mass between the individual piles is assumed to move as a rigid body with the pile group. This is justified since the entrapped soil is work hardened during the driving operation.

If, in plan view, the length of the group is large compared to its breadth the deformation of the soil surrounding the system will be in plane strain. Here Cartesian coordinates (x,z) will be employed Fig. 1,a. In cases where the pile group has a length of

Key Words: Rigid Piers, Displacement Field, Critical Density, Butt Displacement, Ultimate Load, Pressure Distribution, Stress Relief, Uplift Coefficient.

---

*Professor, Concordia University, Dept. of Civil Engg., 1455 de Maisonneuve Blvd. West, Montreal, Quebec, Canada H3G 1M8.

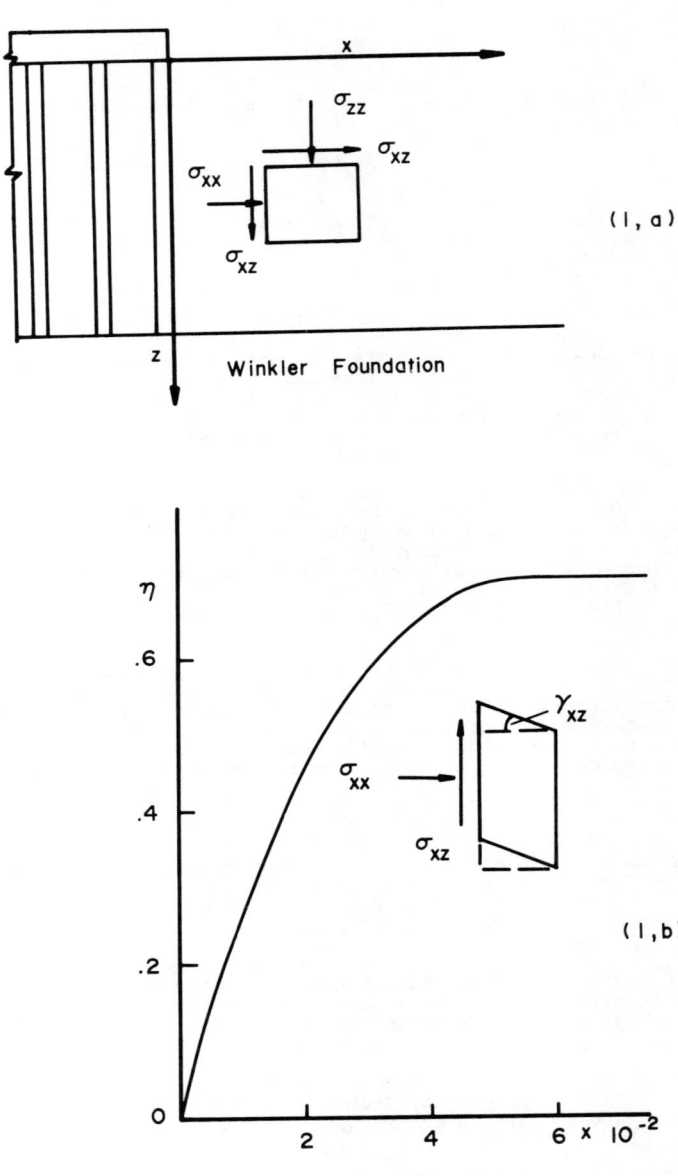

Fig. 1 - Configuration (a), "stress"- strain curve in simple shear (b) - no volume change.

the same order of its breadth or in cases where the piles are arranged to support a circular slab the system is treated as a short pier and cylindrical coordinates (r, θ, z) will be used.

Reverting to the plane strain problem the displacement field for the sand mass is prescribed in the form,

$$u_x = u_x(x) \quad , \quad u_z = u_z(x) \tag{1}$$

where u_x and u_z are the components of the displacement field vector $\vec{u}$. In this assumed mode of failure originally vertical planes remain vertical and preserve their original height. That is, the problem is treated as a "sand beam" on an elastic foundation.

With specification of (1) the volumetric strain of the soil is

$$v = u_{i,i} = \frac{\partial u_x}{\partial x} + \frac{\partial u_z}{\partial z} = \frac{\partial u_x}{\partial x}.$$

Now since the sand is at the critical density, by assumption, then

$$\frac{\partial u_x}{\partial x} = 0$$

which leads to u_x = Const. = 0 since at $x = 0$, $u_x = 0$. That is at the soil pile group interface no lateral movements take place. The engineering shear strain γ_{xz} is obtained from

$$\gamma_{xz} = \partial u_z/\partial x = \gamma_{xz}(x) \tag{2}$$

since u_z is a function of x only.

It is a well known experimental observation supported by most of deformation theories that the total strains encountered in a series of tests with simular stress paths are a function only of the "stress ratios" and not the magnitude of the stress components per se. In a simple shear test Fig. 1,b with shear stress parameter σ_{xz} and the normal stress component σ_{xx} the relationship may be expressed in the form

$$\gamma_{xz} = f(\frac{\sigma_{xz}}{\sigma_{xx}}) = f(\eta) \tag{3}$$

where η represents the value of σ_{xz}/σ_{xx}. For a derivation of this equation based on the incremental stress strain theory proposed by Poorooshasb et al (1966,1967) see Appendix I. Combining equations (2) and (3) a relationship is obtained which is of central importance in this analysis:

$$\eta = \eta(x) \tag{4}$$

It must be clearly understood that η is not a constant, rather a variable the magnitude of which is dependent on the magnitude of the deformation experienced by the sand mass. Combining equation

(4) with equations of equilibrium a set of equations obtain in the form

$$\frac{\partial \sigma_{xx}}{\partial x} + \frac{\partial \sigma_{xz}}{\partial z} = 0$$

$$\frac{\partial \sigma_{xz}}{\partial x} + \frac{\partial \sigma_{zz}}{\partial z} = \gamma \qquad (5)$$

$$\frac{\sigma_{xz}}{\sigma_{xx}} = \eta(x)$$

which must be solved with the following boundary conditions.

(i) Along ox the stress component σ_{xx} and σ_{zz} are equal to zero. This follows from the hypothesis that for a cohesionless media the failure surface in the stress space contains the origin where $\sigma_{ij} = 0$. Thus

$$z = 0 \qquad 0 \leq x \qquad \sigma_{xx} = \sigma_{zz} = 0 = \sigma_{xz} \qquad (6)$$

(ii) At large distances from the origin the state of stress remains unchanged. That is;

$$x \to \infty, \qquad \sigma_{xx} = k_0 \gamma z, \qquad \sigma_{zz} = \gamma z, \qquad \sigma_{xz} = 0 \qquad (7)$$

where k_0 is the coefficient of at rest earth pressure and γ is the soil density.

(iii) At the surface of the Winkler foundation the stress component σ_{zz} is related to the displacement component u_z through

$$\sigma_{zz} = -k_s u_z \qquad (8)$$

where k_s is the modulus of subgrade reaction of the foundation.

For convenience let $\xi = z + \int_x^\infty \eta(x) dx$. Then noting that $\frac{\partial \xi}{\partial z} = 1$, $\frac{\partial \xi}{\partial x} = -\eta(x)$ it may be directly verified that a solution in the form

$$\sigma_{xx} = \zeta(\xi)$$

$$\sigma_{xz} = \eta(x) \zeta(\xi) \qquad (9)$$

$$\sigma_{zz} = \gamma z + \eta^2(x) \int_0^z \zeta'(\xi) dz - \eta'(x) \int_0^z \zeta'(\xi) dz$$

satisfies the set of equations (5). The function $\zeta(\xi)$ must be so chosen as to satisfy the boundary conditions of the problem. In this paper the form of $\zeta(\xi)$ is assumed to be

It may be shown that if

$$\zeta(\xi) = \sum a_n e^{-\alpha_n \xi}$$

$$\sum a_n e^{-\alpha_n} \int_x^\infty \eta(x)\,dx \qquad (6,a)$$

and

$$\sum a_n e^{-\alpha_n z} = k_0 \gamma z \qquad (7,a)$$

then the boundary conditions (6) and (7) are satisfied. The two conditions (6,a) and (7,a) may be expressed by a simple equation

$$\zeta = \sum a_n e^{-\alpha_n \xi} = k_0 \gamma <\xi> \qquad (10)$$

To satisfy the boundary condition (8) it may be noted that at $z = L$ $\quad \xi = L + \int_x^\infty \eta(x)\,dx$. Furthermore

$$u_z = \int_x^\infty \gamma_{xz}(x)\,dx = \int_x^\infty f[\eta(x)]\,dx$$

in view of 3. Therefore

$$\eta^2 \sum a_n e^{-\alpha_n [L + \int_x^\infty \eta\,dx]} + \frac{d\eta}{dx}\left(\sum \frac{a_n}{\alpha_n} e^{-a_n [L + \int_x^\infty \eta(x)]\,dx}\right.$$

$$\left. - \sum \frac{a_n}{\alpha_n} e^{-\alpha_n \int_x^\infty \eta(x)\,dx}\right) = -k_s \int_x^\infty f[\eta(x)]\,dx. \qquad (11)$$

It is convenient to use normalized variables; $X = x/L$, $Z = z/L$, $\sigma_{ij}^N = \sigma_{ij}/\gamma L$, $\zeta^N = \zeta/\gamma L$ and $\xi^n = \xi/L$, where γ is the soil density and L is a characteristic length, the depth of the pile group say. With normalized notations equations (9), (10) and (11) reduce to

$$\sigma_{xx}^N = \zeta^N \qquad (a)$$

$$\sigma_{xz}^N = \eta \zeta^N \qquad (b)$$

$$\sigma_{zz}^N = z + \eta^2 k_0 \zeta^N + k_0 (\bar{\zeta}^N - \bar{\zeta}_{z=0}^N)\frac{d\eta}{dx} \qquad (c) \qquad (12)$$

$$\zeta^N = k_0 < \zeta^N > \qquad (d)$$

$$\eta^2 \zeta_{z=L}^N + \frac{d\eta}{dx}(\bar{\zeta}_{z=L}^N - \bar{\zeta}_{z=0}^N) = -K^N \int_x^\infty f[\eta(X)]dX \qquad (e)$$

where $\quad \bar{\zeta}^N = \frac{1}{\gamma L^2} \sum \frac{a_n}{\alpha n} e^{-\alpha_n \xi^N}$, $\bar{\zeta}_{z=0}^N = \frac{1}{\gamma L^2} \sum \frac{a_n}{\alpha n} e^{-\alpha_n} \int_x^\infty \eta(X)dX$

$$\bar{\zeta}_{z=L}^N = \frac{1}{\gamma L^2} \sum \frac{a_n}{\alpha n} e^{-\alpha_n}[1 + \int_x^\infty \eta(X)dX] \quad \text{and} \quad K^N = \frac{k_s}{k_0 \gamma}.$$

The set of equations (12) describes the stress and displacement field within the sand mass. From equation (12,c) the variation of η with x is obtained using a backward finite difference method by first replacing the upper limit of integral on the right hand side of equation (12,e) by x_i which assumes, tacitly, that $\eta(x)_{x_i < x} = 0$. That is, the presence of pile group has no effect beyond the distance x_i away from it. The coefficients to be used in equation (12,e), when written in its finite difference form, are obtained from Fig. 2 which shows the variation of ζ and $\bar{\zeta}$ with ξ. Once the form of $\eta(X)$ is obtained the stress components σ_{xx}, σ_{xz}, σ_{zz} are obtained from equations (12,a), (12,b) and (12,c) and the displacement of the pile group from equation

$$\delta = -L\delta^N = -\frac{k_s L}{k_0 \gamma} \int_0^{X_i} f[\eta(X)]dX \qquad (13)$$

ANALYSIS OF RIGID PIERS

Because of the axi-symmetrical nature of the problem the displacement field is defined by

$$u_r = u_r(r), \quad u_z = u_z(r) \quad \text{and} \quad u_\theta = 0 \qquad (14)$$

where u_r, u_z, and u_θ are the components of the displacement vector along the (r, z, θ) cylindrical coordinates. This assumed mode of displacement is equivalent to the shearing of concentric cylinders in the soil mass which has been used previously by many authors (e.g. Randolph & Wroth, 1978). The condition of no volume change requires that

$$\frac{\partial u_r}{\partial r} + \frac{u_r}{r} = 0 \qquad (15)$$

since $\frac{\partial u_z}{\partial z}$ is identically zero. Since at $r = r_0$ where r_0 is the pier radius, $u_r = 0$, then the only solution satisfying equation (15) is the trivial solution $u_r = u_\theta = 0$. Progressing in precisely

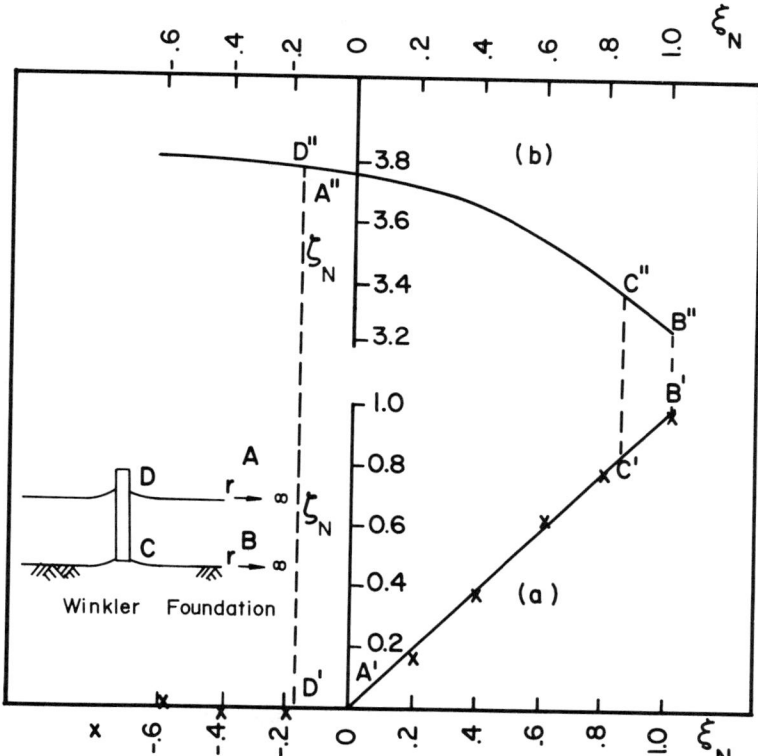

Fig. 2. (a) Transformation from (x,z) to (ζ^N, ξ^N) plane

(b) Variation of $\sum \dfrac{a_n}{\bar{a}_n} e^{a_n \xi}$ with ξ^N

the same manner as in the case of plane strain problem the set of equations describing the stress field σ_{rr} (assumed) equal to $\sigma_{\theta\theta}$), σ_{zz} and σ_{rz} obtain in the form

$$\sigma_{rr} = \sum a_n e^{-\alpha_n \xi} \tag{16}$$

$$\sigma_{rz} = \eta(r) \sum a_n e^{-\alpha_n \xi}$$

$$\sigma_{zz} = \gamma z + (\frac{d\eta}{dr} + \frac{\eta}{r}) \; [\sum \frac{a_n}{\alpha_n} e^{-\alpha_n \xi} - \sum \frac{a_n}{\alpha_n} e^{-\alpha_n} \int_r^\infty \eta(r) dr] + \eta^2 \sum a_n e^{-\alpha_n \xi}$$

provided that

$$a_n e^{-\alpha_n \xi} = k_0 \gamma <\xi>$$

and

$$\eta^2 \sum a_n e^{-\alpha_n [L + \int_r^\infty \eta(r) dr]}$$

$$+ (\frac{\eta}{r} + \frac{d\eta}{dr}) \{ \sum \frac{a_n}{\alpha_n} [e^{-\alpha_n (L + \int_r^\infty \eta(r) dr} - e^{-\alpha_n \int_r^\infty \eta(r) dr} \} \tag{17}$$

$$+ k_s \int_r^\infty f[\eta(r)] dr = 0$$

In terms of normalized coordinates $R = r/L$, $Z = z/L$ and variables ξ^N, ζ^N and $\bar{\zeta}^N$ as defined previously equations (16) and (17) reduce to

$$\sigma_{rr}^N = \zeta^N$$

$$\sigma_{rz}^N = \eta \zeta^N$$

$$\sigma_{zz}^N = Z + k_0 (\frac{d\eta}{dR} + \frac{\eta}{R}) \; [\bar{\zeta}^N - \bar{\zeta}_{Z=0}^N] + \eta^2 k_0 \zeta^N \tag{18}$$

$$\zeta^N = k_0 <\xi^N>$$

$$\eta^2 \zeta_{Z=1}^N + (\frac{d\eta}{dR} + \frac{\eta}{R}) [\bar{\zeta}_{Z=1}^N - \bar{\zeta}_{Z=0}^N] + K^N \int_R^\infty f[\eta(R)] dR = 0$$

and the displacement of the butt of the pier is given by

$$\delta = -LK^N \int_{R_0}^{\infty} f[\eta(R)]dR \tag{19}$$

APPLICATION

The foregoing analysis is applied to a pier with a diameter to length ratio equal to 0.1. Most of the conclusions apply to the pile group case also since they are fundamentally the same problem. The soil is assumed to have a density of 104 lb/ft^3 (1667 kg/m^3) with a stress strain curve as shown in Fig. 1,b. The modulus of subgrade reaction for a 1 x 1 ft plate is assumed to be 120 kips/ft^3 (Terzaghi (1955)) and $k_0 = 1 - \sin\phi$ [= $1 - \sin(\tan^{-1}.7)$, Fig. 1,b] is equal to .42. With these constants the value of K^N is evaluated to be 750 for piers of 1 meter in diameter or more.

In Fig. 3 is shown the variation of η with R for two cases, one, the upper curve, corresponding to state where the pier is experiencing its ultimate uplift capacity ($Q^N = Q_{ult}^N$) and the other, where it is supporting one half of this load. As expected, η, which is a measure of mobilized angle of friction decreases rapidly with distance from the face of the pier. Figs. 4 and 5 show the distribution of stress within the sand mass for the case of $Q^N = Q_{ult}^N$. In Fig. 4 is shown the variation of normal stress component σ_{rr} and the shearing stress component σ_{rz}. It is noted that the variation of both σ_{rr} and σ_{rz} are to a very close approximation, linear with depth. Of particular interest is the distribution of σ_{rr} and σ_{rz} at $R = R_0$ i.e. the interface of pier and soil. The ultimate uplift resistance of the pier due to skin friction is given by

$$Q_{ult} = 2\pi r_0 \int_{z_0}^{L} \sigma_{rr} \eta_0 dz \tag{20}$$

where z_0 is the theoretical no lateral pressure zone. (The theory presented here predicts a small zone at the pier interface of extent $\int_{r_0}^{r_i} \eta(r)dr$ below the surface for which $\sigma_{rr} = \sigma_{rz} = 0$. If the soil possesses a small amount of cohesion it would move away slightly from the pier up to a depth of z_0.) But

$$\sigma_{rr} = k_0 \gamma(z - z_0).$$

Therefore

$$Q_{ult} = 2\pi r_0 \eta_0 k_0 \gamma \int_{z_0}^{L} (z - z_0)dz$$

$$= 2\pi r_0 [\tfrac{1}{2} k_0 \gamma \eta_0 (L - z_0)^2] \tag{21}$$

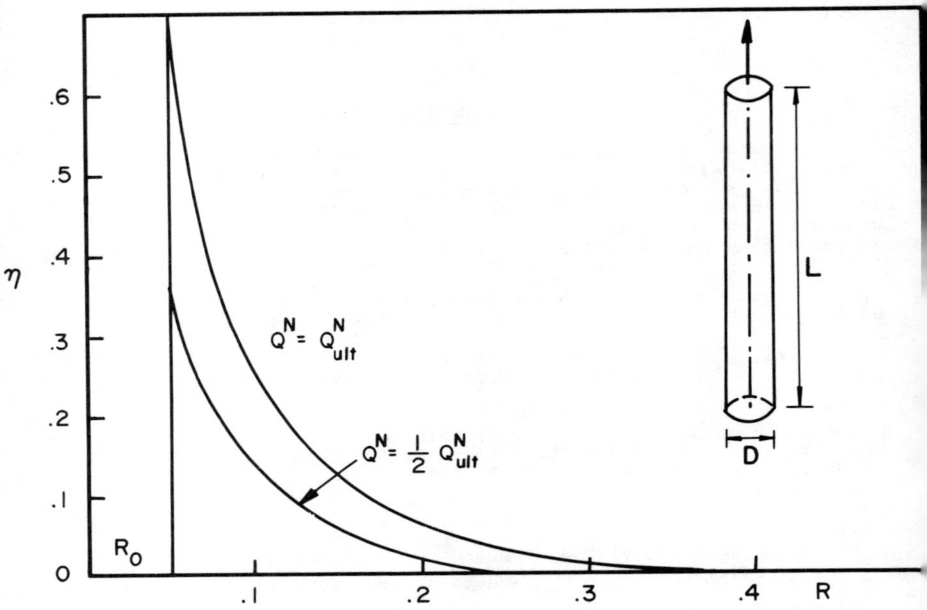

Fig. 3 - Variation of η with R for a pier D/L = 0.10.

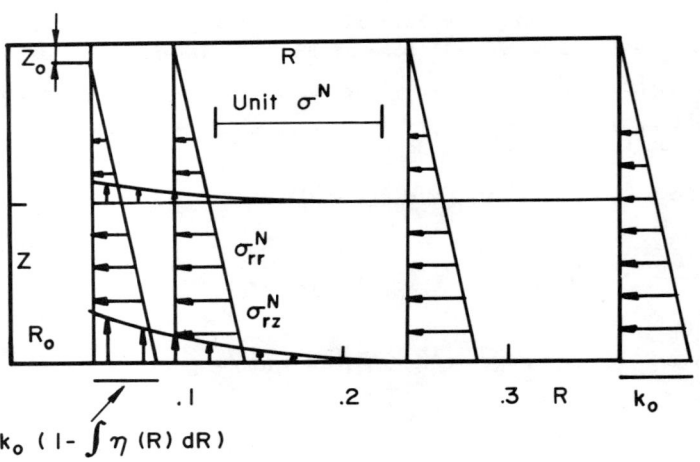

Fig. 4 - Distribution of σ_{rr}^N and σ_{rz}^N in (R,Z) domain.

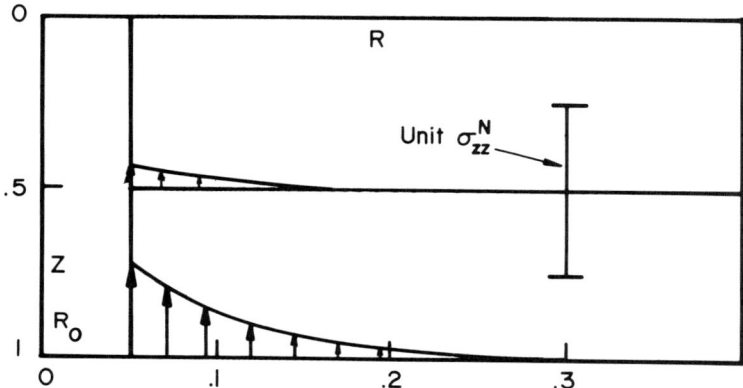

Fig. 5 - Normalized stress relief in σ_{zz} component.

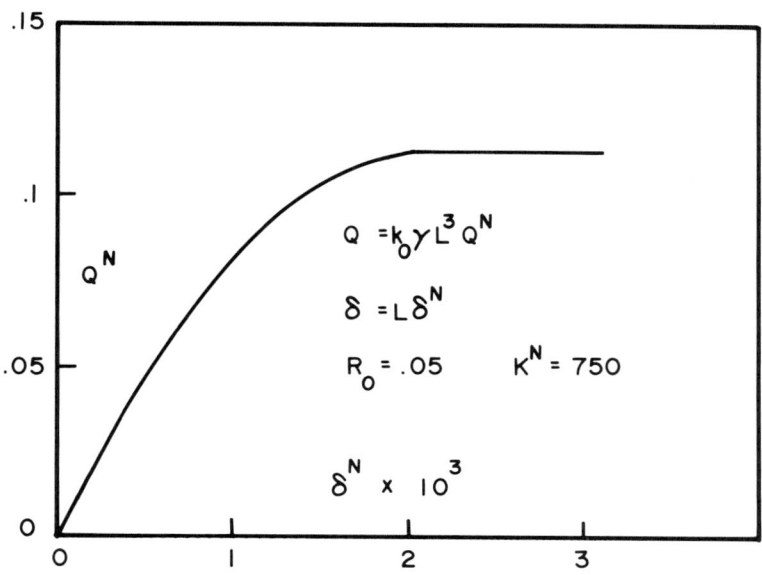

Fig. 6 - Normalized uplift load with Normalized Butt Displacement.

Since z_0 is usually quite small compared to L then

$$(L - z_0)^2 \approx L^2$$

and equation (21) reduces to the familiar form

$$Q_{ult} = 2\pi r_0 \left[\tfrac{1}{2} k_u \gamma L^2 \tan\phi'\right] \tag{22}$$

where

$$k_u \approx k_0 \tag{23}$$

Reese et al (1976), Meyerhof (1976) suggest a value for k_c, the coefficient similar in nature to k_u but used in analysis of compression piers, equal to k_0. Ismail and Klym (1979) suggest that, based on their field tests k_u is at most equal to k_c but indicate that the $k_u = k_0$ coefficient is too conservative. Kulhawy et al (1979), on the other hand, propose that for loose sand k_u is equal to k_a the coefficient of active earth pressure.

According to the analysis presented in this paper the use of $k_u = k_0$ is only justified if the soil density is of the order of its critical value. However, since sand deposits looser than critical density are intrinsically instable and not frequently encountered the use of k_0 as the uplift coefficient appears to be rational in the light of the analysis presented above.

In Fig. (5) the curves representing the variation of the stress relief in the normal component σ_{zz} are shown. The lower curve shows the stress relief at a depth $Z = 1$ (i.e. $z = L$, the elevation corresponding to the tip of the pier). The magnitude of this relief is a maximum at the immediate vicinity of the pier. There is a limit below which the soil can not tolerate a decrease in σ_{zz}, the limit depending on the mobilized $\eta(r_0)$. If σ_{zz} is reduced below this level the soil "flows" and the stress distribution along the soil-pier interface can no longer increase linearly with depth. The magnitude of the stress relief is sensitive to the compliance of the foundation soil (i.e. the soils below the tip of the pier) compared to the stiffness of the soil surrounding the pier. The larger the foundation compliance (lower k_s) the smaller would be the pressure relief and larger the pier movement under a given uplift load. This factor appears to have been ignored by many investigators who have concentrated on the effect of the properties of the soils surrounding the pier only.

Figure (6) shows the load-butt displacement curve for the pier. It is seen that a low stress levels, the magnitude of displacement is almost proportional to the uplift force. At higher loads there is a departure from this linearity until failure (pullout) condition is reached.

SUMMARY AND CONCLUSIONS

In this paper a method of analysis of the behaviour of vertical rigid bored piers (or pile groups) embedded in sand deposits with a density of the order their critical value is presented. The analysis shows that:

(a) Under most practical conditions the distribution of the normal stress as well as shearing tractions along the pier-soil interface is linear, increasing with depth of embedment (see item (c) below, however).

(b) For deposits no looser than the critical density the value of k_0 is a lower bound value to k_u where k_0 is the at rest earth pressure coefficient and k_u is the so called uplift coefficient.

(c) The compliance of the foundation soils (soils below the level of the tip of the pier) play an important role in the pier behaviour. For very low compliances (large k_s values) the pressure distribution will not be linear with the consequence that equations in the form of equation (22) above will not apply. The reason is that as soon as any vertical movement takes place the value of the vertical stress component σ_{zz} is significantly reduced resulting "flow" of the soils adjacent to lower parts of the pier.

REFERENCES

(1) Ismail, Nabil F. and Tony W. Klym. "Uplift and Bearing Capacity of Short Piers in Sand". Journal of Geotechnical Engineering Division, ASCE, Vol. 105, No. GT5, May 1979, pp. 579-594.

(2) Kulhawy, Fred H., David W. Kozera and James L. Witham. "Uplift Testing of Model Drilled Shafts in Sand". Journal of the Geotechnical Engineering Division, ASCE, Vol. 105, No. GT1, January 1979, pp. 31-47.

(3) Meyerhof, George Geoffrey. "Bearing Capacity and Settlement of Pile Foundations". Journal of the Geotechnical Engineering Division, ASCE, Vol. 102, No. GT3, March 1976, pp. 197-228.

(4) Poorooshasb, H.B., I. Holubec and A.N. Sherbourne. "Yielding and Flow of Sand in Triaxial Compression". Canadian Geotechnical Journal, Vol. IV, No. 4, 1967, pp. 376-397.

(5) Randolph, N.F. and Wroth, C.P. "Analysis of Deformation of Vertically Loaded Piles". Journal of the Geotechnical Engineering Division, ASCE, 1978, pp. 1465-1488.

(6) Tarzaghi, K. "Evaluation of Coefficient of Subgrade Reaction". Geotechnique, London, Vol. 5, No. 4, 1955, pp. 297-326.

(7) Reese, L.C., Touma, F.T., and O'Neill, M.W. "Behaviour of Drilled Piers Under Axial Loading". Journal of the Geotechnical Engineering Division, ASCE, Vol. 102, No. GT5, Proc. Paper 12135, May 1976, pp. 493-510.

ACKNOWLEDGEMENT

The financial assistance received from the National Science and Engineering Research Council of Canada is gratefully acknowledged.

APPENDIX I. DERIVATION OF EQUATION (3)

The constitutive law proposed by Poorooshasb et al (1966-67) may be modified to read, for simple shear,

$$d\varepsilon_{xz} = d\varepsilon_{xz}^{P} = \tfrac{1}{2} d\gamma_{xz} = h(\eta) \frac{\partial \Psi}{\partial \sigma_{xz}} <df> \qquad (A,1)$$

where

$$f = \frac{\sigma_{xz}}{\sigma_{xx}} = \eta \qquad (A,2)$$

and

$$\Psi = \sigma_{xx} \bar{\Psi}(\eta) \qquad (A,3)$$

and subscript p represents the plastic component [The elastic component of strain is assumed to be small in comparison with the plastic component]. Substituting from Equation A,2 and A,3 in A,1 results in

$$d\varepsilon_{xz} = h(\eta) \sigma_{xx} \bar{\Psi}'(\eta) \frac{\partial \eta}{\partial \sigma_{xz}} <d\eta> \qquad (A,4)$$

In a loading process where $d\eta$ is non-negative the singularity brackets of (A,4) may be dropped and the resulting equation integrated to yield

$$\varepsilon_{xz} = f(\eta) \qquad (3)$$

YIELDING LOAD OF ANCHOR IN SAND

By M. C. Wang[1] and A. H. Wu[2], Members ASCE

ABSTRACT

This study was undertaken with the aim of understanding the relative importance of the many variables associated with the pullout resistance of soil anchors. Variables investigated were anchor orientation, embedment depth, internal friction angle of sand, and anchor-soil friction. The study was conducted both analytically and experimentally.

The experimental phase was carried out using a test tank which was made of plywood with a pane of plexiglass on one side and strengthened with steel channels. The soil was Ottawa sand, and the test anchors were retangular plates of different materials. This phase of study provided anchor failure mechanisms for theoretical analysis and also test data for validation of the analytical results.

The theoretical analysis was conducted using the upper bound limit theorem together with the failure mechanism observed from the laboratory experiment. The results of the analysis agreed fairly well with the text results, and also compared satisfactorily with available information. The effect of each influencing factor on anchor capacity was examined in great detail.

Key Words: Anchor plates, laboratory tests, limit theorem, model tests, plasticity theory, sands, soil anchors, tests, yielding load.

---

[1] Associate Professor, Department of Civil Engineering, The Pennsylvania State University, University Park, PA 16802

[2] Soil Mechanics and Foundation Specialist, Naval Facilities Engineering Command, H. Q. Alexandria, VA 22332

INTRODUCTION

There is an increasing use of soil anchors both on land and off-shore to support structures subjected to tensile loading. On land, anchors are typically used with structures such as high-mast transmission towers to resist tensile forces due to wind and earthquake loading. Typical offshore applications of anchor are the anchorage of floating equipment in shallow and deep water.

As a result of the growing engineering applications of soil anchor, various methods for estimating anchor capacity have been proposed over the last two decades. Many of the methods have been developed for anchors at all depths (4, 7, 10, 11, 15), some for shallow anchors (2, 5, 8, 12), and others for deep anchors only (1). However, at present no generally accepted method exists for determining uplift resistance of soil anchors. The development of such a method requires thorough understanding of the relative importance of the many variables associated with pullout resistance of soil anchors.

In accordance with this need, a study was undertaken to investigate the effect of some important variables on the anchor capacity. The variables considered were anchor size, depth to the anchor, anchor inclination, and frictional characteristic of the anchor face. In this study, the plastic limit theorem was employed for theoretical analysis, and the analytical results were validated experimentally. This paper presents the method of analysis and the results obtained.

LABORATORY INVESTIGATION

A laboratory study was undertaken to assess the relative significance of some important factors influencing anchor behavior, to provide anchor failure mechanism for theoretical analysis, and to provide data for validation of the analytical results.

Test Setup. The experimental study was conducted in the laboratory using a plexiglass-sided test tank. The test tank had inside dimensions of 3.5 in. (88.9 mm) in width, 25.6 in. (650.9 mm) in length, and 33.3 in. (844.6 mm) in height. It was constructed of plywood with a pane of plexiglass on one side and strengthened with steel channels. The plexiglass on one side was used to aid in viewing slip line fields.

In the test tank, the test soil was compacted in layers using a vibrator. Thin vertical strips of colored sand were placed equally spaced along the plexiglass side. The test anchors were placed at different depths and various angles of inclination. They were pulled by a motor at a constant displacement rate of 0.006 in./s (0.152 mm/s). This displacement rate was chosen arbitrarily within the range of strain rate normally adopted for laboratory testing. The pullout load and displacement were monitored using a load cell and a 0.001 in. (0.025 mm) dial gage, respectively. Fig. 1 illustrates the schematic view of the test setup; Figs. 2(A) and (B) shows the color strips and failure mechanisms for shallow and deep anchors, respectively.

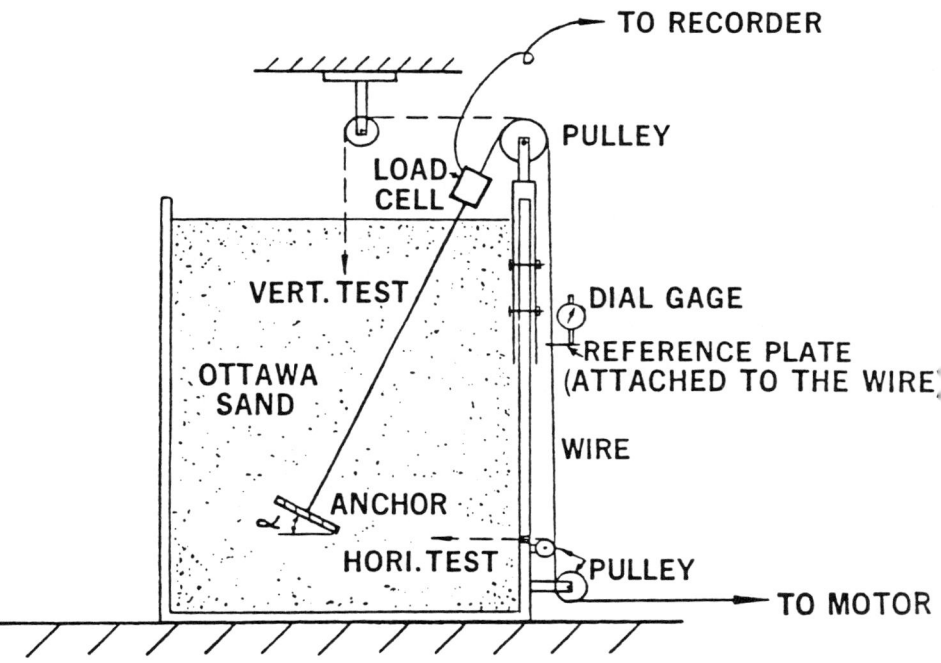

FIG 1 ANCHOR TEST SETUP

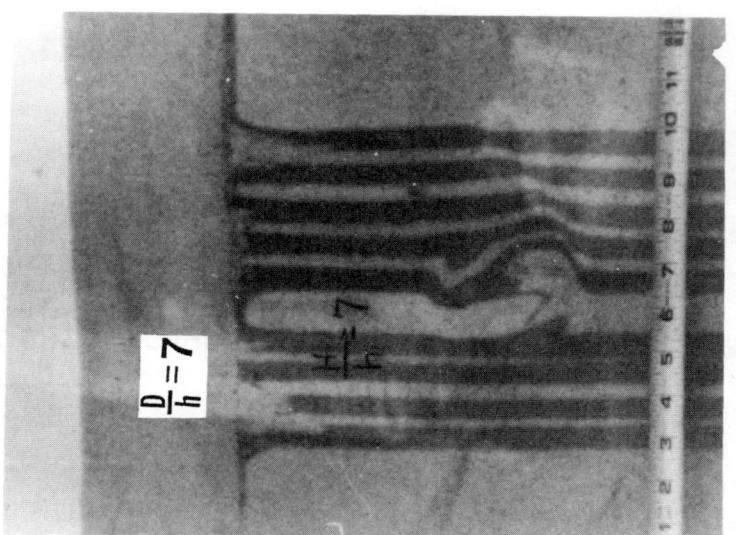

FIG 2(B) COLORED SAND STRIPS WITH TEST ANCHOR AND SLIP PLANE FOR DEEP ANCHOR CASE

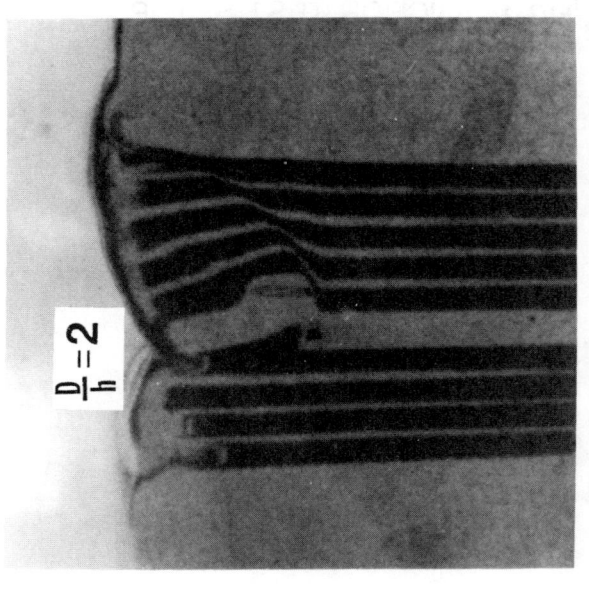

FIG 2(A) COLORED SAND STRIPS WITH TEST ANCHOR AND SLIP PLANE FOR SHALLOW ANCHOR CASE

Test Soil. The test soil was a commercially available Ottawa sand which had a D_{10} of 0.30 mm, a D_{50} of 0.54 mm, and a uniformity coefficient of 2.0. In the test tank, the sand was air dried and had two dry densities, 110 pcf (1780 kg/m^3) and 100 pcf (1618 kg/m^3). The strength property of the test sand was determined using the direct shear test. The angles of internal friction of the sand were approximately 30° and 35° for dry densities of 100 pcf (1618 kg/m^3) and 110 pcf (1718 kg/m^3), respectively.

Test Anchor. The test anchors were made of rectangular plates with a 0.25 in. (6.4 mm) diameter brass rod which was rigidly fastened at the center of the plate. The size of the plate was 3.38 in. (85.7 mm) wide, 1.5 in. (38.1 mm) high, and 0.25 in. (6.4 mm) thick. The width of the anchor plate fitted the test tank snugly so that the friction between the anchor plate and the tank walls was minimal. To study the effect of plate friction on anchor resistance, three different materials were used, namely, brass, plexiglass, and brass covered with sand paper. The friction angle between the plate and the test sand was measured using the shear test in which the plate material was overlaid by the test soil. The test soil was contained in a container (the upper half of a direct shear box was used) and was subjected to desired downward loads. Test results were the friction angles of 16°, 19°, and 25° for plexiglass, brass, and brass covered with sand paper, respectively.

Test Program. The test program was developed with the aim of studying the effect of soil density or internal friction angle of the test soil, anchor orientation, depth to the top of anchor, and anchor surface friction on the anchor resistance. For soil density study, the brass plate anchor was tested in two dry densities, 110 pcf (1780 kg/m^3) and 100 pcf (1618 kg/m^3). The anchor was placed horizontally ($\alpha=0°$) at various depths. For anchor orientation study, the brass plate was tested at various depths in a dry density of 110 pcf (1780 kg/m^3). Six levels of anchor orientation were used, i.e. $\alpha=0°$, 30°, 45°, 60°, and 90°. The effect of anchor surface friction was studied only for horizontal anchors which were embedded at various depths.

THEORETICAL ANALYSIS

The theoretical analysis was carried out using the upper bound limit analysis (3, 6). Because the anchor analyzed had a width such that the deformation of the surrounding sand in the direction normal to the anchor movement is negligible, it was treated as a plane strain problem. Fig. 3 shows the variables describing anchor size, anchor location, and slip line field. Based on the failure mechanism shown in Fig. 2(A), the slip line, ABCD, can be approximated by two straight lines and one segment of logarithmic spiral. Therefore, for a given anchor size (h), inclination (α), and depth (D), the slip line can be defined by two angles, ξ and θ. Where θ is the central angle of the radial shear zone, OBC, which is sandwiched between two rigid zones, OAB and OCDE, and ξ is the angle between the radial shear zone and the

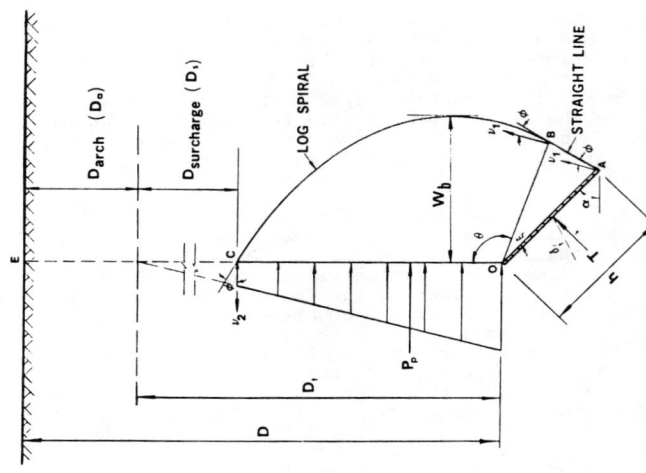

FIG 4 LOCAL SHEAR FAILURE FOR DEEP ANCHORS

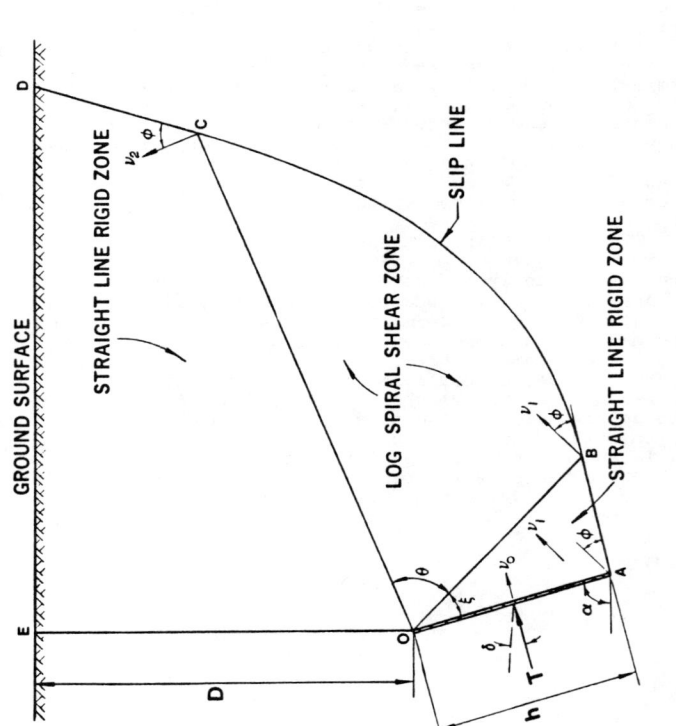

FIG 3 GENERAL SHEAR FAILURE FOR SHALLOW ANCHORS

anchor, OA. This figure also indicates the velocity field of each zone.

Using the slip line field and velocity field given in Fig. 3, the yielding load of the anchor (anchor resistance) was determined by equating the rate of work done by external forces to the rate of internal energy dissipation. The resulting equations for computing the yielding load are below:

$$\frac{T}{\gamma bh^2} = \frac{\cos \delta}{2 \cos^2 \phi [\cos(\xi+\delta) - \sin\xi \tan\delta]} (A+B+C) \quad \ldots \ldots (1)$$

where

$$A = \cos \phi \sin\xi \cos(\alpha-\xi) \cos(\alpha-\xi) \quad \ldots \ldots \ldots \ldots (2)$$

$$B = \frac{\cos^2(\xi-\phi)}{1+9 \tan^2 \phi} \{ \cos (\xi-\phi) [-3 \tan\phi + (3\tan \phi \cos \theta \quad (3)$$
$$+ \sin \theta) e^{3\theta\tan\phi} + \sin (\alpha-\xi)[1+(3 \tan\phi\sin\theta -\cos\theta)e^{3\theta\tan\phi}]\}$$

$$C = (\frac{D}{h})^2 e^{\theta\tan\phi} \cos (\xi-\theta-\alpha)\{ \cos^2\phi \cot (\xi+\theta-\alpha)$$
$$\div X^2[\tan(\phi+\alpha-\xi-\theta) -\cot (\xi+\theta-\alpha)]\} \quad \ldots \ldots \ldots (4)$$

and

$$X = \cos\phi - \frac{h}{D} e^{\theta\tan\phi} \sin(\xi+\theta-\alpha)\cos(\xi-\phi) \quad \ldots \ldots \ldots (5)$$

Also, ϕ and δ are the angle of internal friction of sand and the friction angle between the anchor face and the sand, respectively. γ is the unit weight of sand, b is anchor width which is equal to unity for the case under study, and T is yielding load. The other notations are defined in Fig. 3.

Fig. 3 illustrates a general shear failure mechanism in which the slip line field extends to the ground surface. This type of failure mechanism is typical for shallow anchors. For deep anchors, however, a local shear failure prevails. For a local shear failure, the slip line field develops in the vicinity of the anchor and does not extend to the ground surface as shown in Fig. 2(B). Accordingly, a different slip line field is required for determination of yielding load.

Fig. 4 shows the local shear failure mechanism adopted in this analysis. The slip line comprises of a straight line, AB, and a segment of logarithmic spiral, BC. The logarithmic spiral entends only to the vertical line through the top of the anchor, OC. As the anchor yields, passive earth pressure develops along OC. The passive earth pressure is estimated from the depth of arching (D_a) as shown. With the consideration of the work done by the passive earth pressure, the yielding load of the anchor may be computed using Eq. (1) with a replacement of C_1 for C, where C_1 is defined below:

$$C_1 = \tan^2 (45°+ \frac{\phi}{2}) \frac{2D_t}{h} \cos\phi \cos(\xi-\phi) e^{2\theta\tan\phi}$$
$$-\cos^2 (\xi-\phi) e^{3\theta\tan\phi}] \quad \ldots \ldots \ldots (6)$$

in which $D_t = D - D_a = D_s + \overline{OC}$. Both $\overline{OC}$ and depth of surcharge (D_s) are function of α, h, and ϕ. In this analysis, D_s was estimated using Terzaghi's theory of arching (14). The yielding load of the anchor was computed from a slip line field which gave the lowest resistance. This was accomplished by minimizing Eq. (1) with respect to ξ and θ. For deep anchors, however, only one variable, ξ or θ, was needed because of the following available relationship:

$$\xi + \theta - \alpha = 90° \quad\quad\quad\quad\quad\quad\quad\quad\quad\quad\quad\quad (7)$$

A computer program was developed to perform necessary computations. Using this computer program, the yielding load was determined for anchors with various values of α, h, D, and δ.

RESULTS AND DISCUSSIONS

Test results are given in terms of a dimensionless ratio $T/\gamma bh^2$; where T, γ, b, and h are yielding load, unit weight of the soil, anchor's width, and anchor's height, respectively. For the two soil densities studied, the anchor pressure vs. displacement curves of vertical anchors peak within 1 in. (25.4 mm) displacement when $\phi = 35°$. However, no peaks appear even up to 2 in. (50.8 mm) displacement when $\phi = 30°$; the curves approach constant slopes sooner for anchors with smaller ratio of anchor's embedment depth to anchor's height (D/h) so that the difference in $T/\gamma bh^2$ between 2 in. (50.8 mm), and 1 in. (25.4 mm) displacements increases with increasing D/h as shown in Fig. 5. Because of this difference in the anchor pressure-displacement behavior, the anchor capacity reported hereinafter is defined as the peak anchor pressure or the pressure at which the slope of the pressure-displacement curves becomes essentially constant as was adopted by Das (5).

Also demonstrated in Fig. 5 is that at $\phi = 35°$, the peak anchor resistance is significantly greater than the residual anchor resistance up to a D/h ratio of about 10. Above this ratio, all anchor pressure displacement curves exhibit no peaks and approach a maximum value. Furthermore, above this ratio, the anchor capacity becomes independent of the depth of embedment. This depth-height ratio referred to as the critical depth ratio separates the behavior of deep anchor from shallow anchor. Fig. 5 indicates that the critical depth ratio varies with soil density, being about 10 for $\phi = 35°$, and approximately 8 for $\phi = 30°$. This result is consistent with Meyerhof's finding (11).

Results of the theoretical analysis are compared with the experimental data in Figs. 6 through 10. The two sets of results agree fairly well and therefore, the theoretical approach was used to investigate anchor behavior within a broader range of conditions. The results of the study are presented in the following and are categorized according to the influencing variable. Along with the results of this study are some data available from the literature, which are included for the purpose of comparison.

ANCHOR IN SAND 299

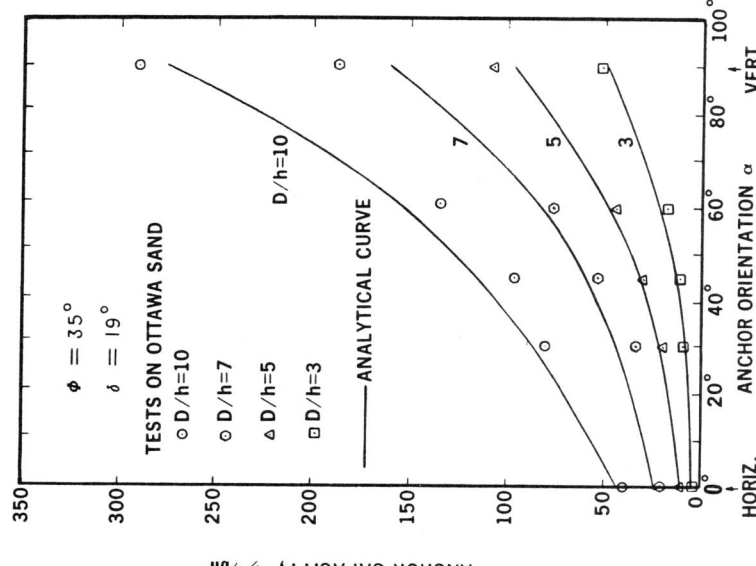

FIG 6 EFFECT OF ANCHOR ORIENTATION ON ANCHOR CAPACITY FOR A CONSTANT D/h RATIO

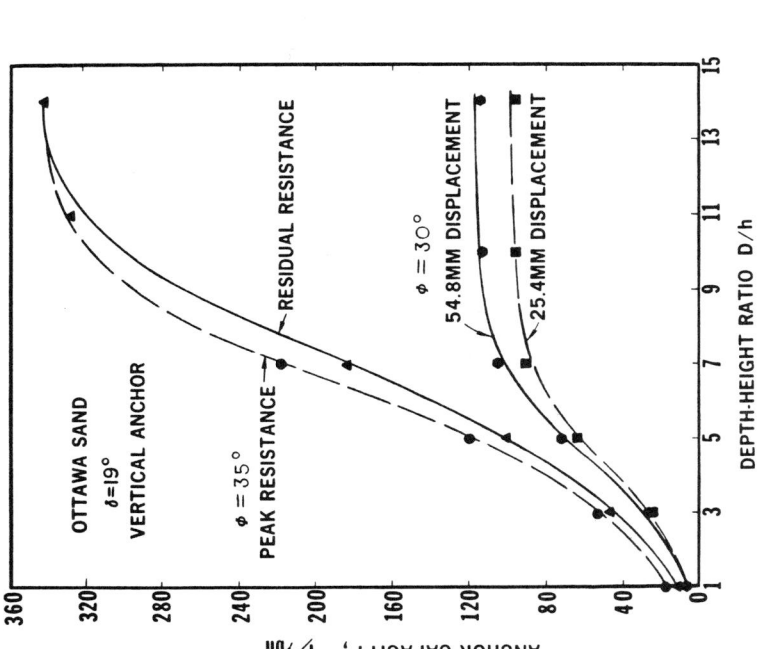

FIG 5 ANCHOR CAPACITY Vs EMBEDMENT-HEIGHT RATIO FOR TWO INTERNAL FRICTION ANGLES OF SAND

Effect of Anchor Orientation. Figs. 6 and 7 summarize the results of the analysis regarding anchor orientation effect for Ottawa Sand with $\phi = 35°$ and $\delta = 19°$. As would be expected, for a constant D/h ratio, anchor resistance increases with an increase in the angle of anchor inclination, being the minimum in the horizontal direction and the maximum in the vertical direction. The rate of increase is greater as the inclination angle becomes larger as illustrated by Fig. 6. This may be attributed to the difference in the extent of ruptured soil zone, becasue based on Fig. 3, for a given D/h, the ruptured soil zone enlarges as the anchor inclination angle increases; and the rate of enlargement depends on the value of D/h. According to Fig. 7, the critical depth ratio is not constant but decreases with a decrease in anchor inclination. This effect may also be accounted for based on the extent of local shear failure zone in front of the anchor plate. The local shear failure zone for a horizontal deep anchor is smaller than that of a vertical deep anchor; as a consequence, the critical depth ratio is greater for vertical anchors than for horizontal anchors.

Also included in Fig. 7 are some data interpreted from the findings of Meyerhof (11) and Das (5). Mayerhof's data agree with the results of this study surprisingly well for two angles of inclination (0° and 45°), but the data for vertical anchors are not close. Primary reasons for this broad discrepancy for one condition while excellent agreement for the others are not investigated. One point that should be stated, however, is that the comparison is made based on his data which are interpolated from a published figure which has been reduced in size considerably.

The results obtained by Das (5) are for shallow square and circular anchors with depth to height ratio up to about 4. In general, his data are higher than the results of this study. This is as expected because for both square and circular anchors, the slip line field emanates from four sides of the anchor rather than just from top and bottom as is the case for the anchor used in this study. The longer the total length of slip lines is, the greater the anchor resistance will be. Another possible cause for the discrepancy might be due to different anchor-soil friction and different sands. Das uses aluminum plates while brass plates are used to generate the results presented in Fig. 7. The effect of anchor-soil friction on anchor resistance is discussed in more detail in a later section. The sand used by Das is finer than the sand used in this study; his soil has 99% passing No. 20 sieve, 55% passing No. 40 sieve, and 0% passing No. 200 sieve, whereas the test sand has 95% passing No. 20 sieve, 29% passing No. 40 sieve, and 0% passing No. 200 sieve.

Effect of Internal Friction Angle of Soil. The effect of internal friction angle of soil on anchor capacity for horizontal anchors embedded at different depths is shown in Fig. 8. As expected, anchor capacity increases with increasing internal friction angle; the effect is more prominent at greater embedment-depth ratios. Also shown in Fig. 8 are some results obtained from Meyerhof and Adams (10), and Taylor and Lee (13). Although there is some dis-

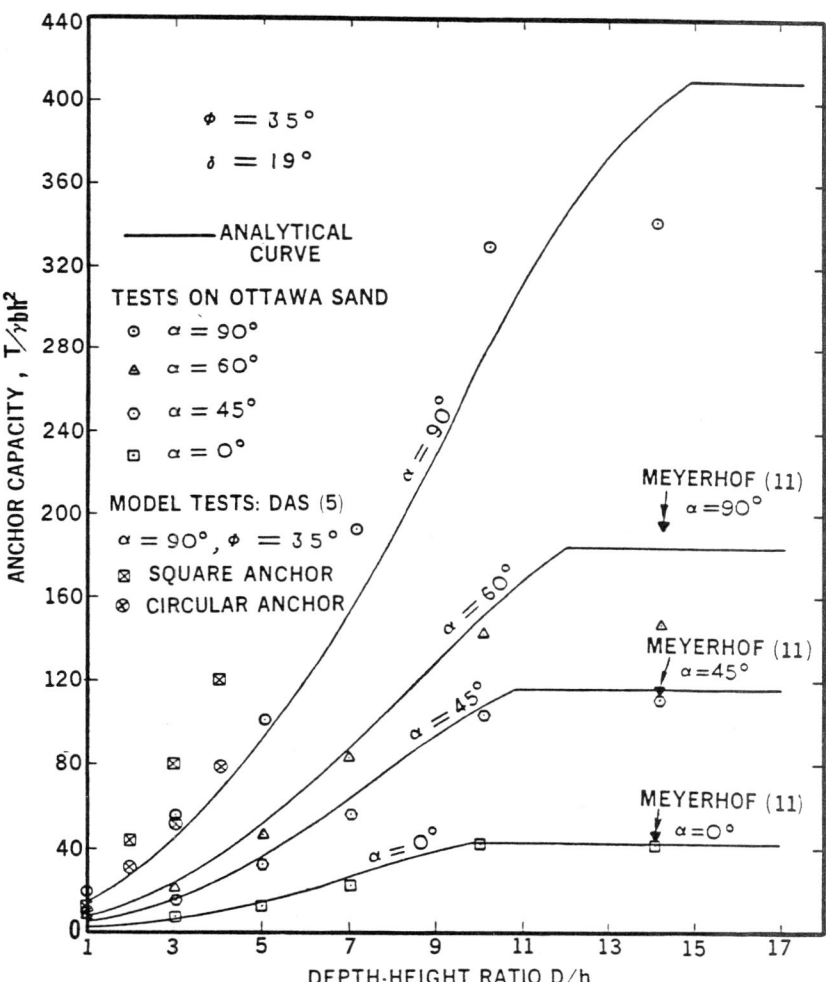

FIG 7 ANCHOR CAPACITY Vs EMBEDMENT-HEIGHT RATIO FOR A CONSTANT ANCHOR ORIENTATION

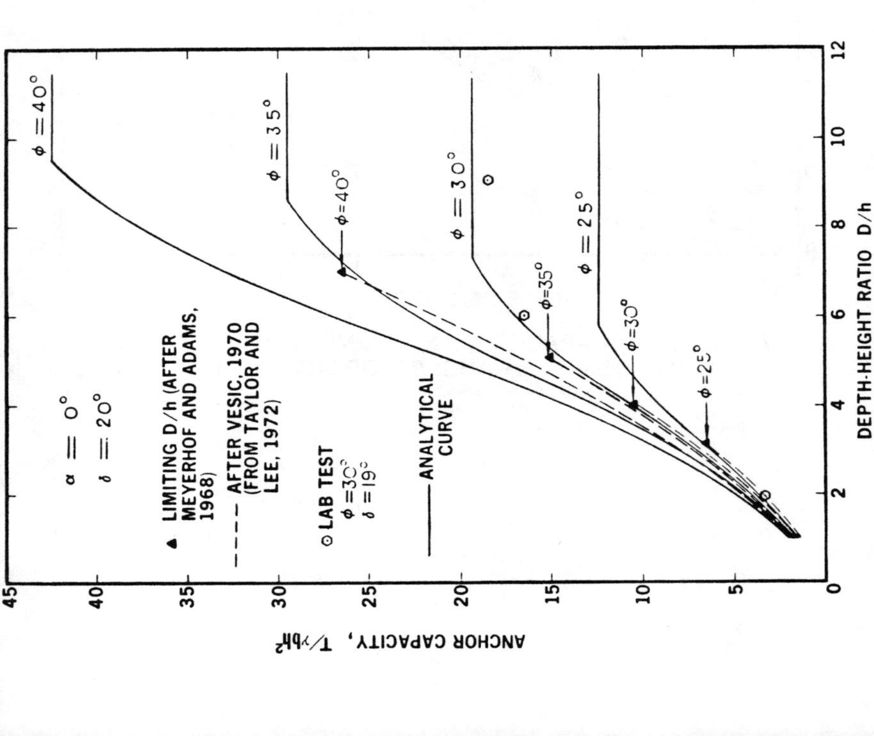

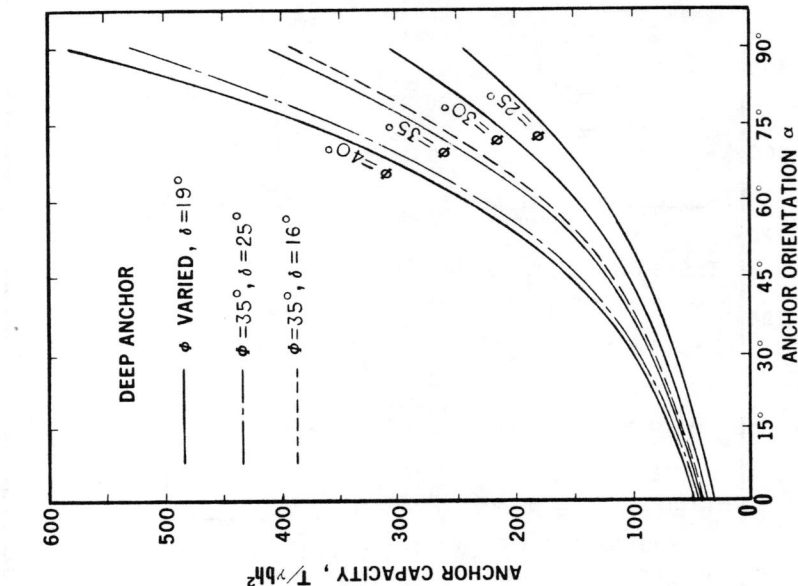

FIG 8 ANCHOR CAPACITY Vs EMBEDMENT-HEIGHT RATIO FOR A CONSTANT ϕ VALUE

FIG 9 ANCHOR CAPACITY Vs ANCHOR ORIENTATION

crepancy between the published data and the results of this study especially at higher angles of internal friction, the agreement in general is quite good. One possible cause of the discrepancy could be due to the friction between anchor surface and the surrounding soil. As will be discussed later, the effect of anchor-soil friction is very significant. The published data used in this comparison did not specify clearly the friction angle used.

Fig. 9 illustrates the variation of anchor capacity with internal friction angle for various anchor orientations. The figure indicates that the effect of internal friction of soil is greatest for vertical anchors and smallest for horizontal anchors. A possible reason for this result is that the extent of ruptured soil zone increases with increasing internal friction angle; it increases faster when the anchors are closer to the vertical direction.

Effect of Anchor-Soil Friction. Fig. 9 also demonstrates the variation of anchor capacity with three different angles of friction between anchor and the surrounding soil. It is seen that for a given soil condition, a greater anchor-soil friction results in a higher anchor resistance. This may be due to the fact that the amount of energy dissipated along the anchor-soil interface increases with increasing friction. As a result, a greater anchor capacity is needed to balance the dissipated energy.

More detailed information on anchor-soil friction effect is presented in Figs. 10 and 11. Fig. 10 gives the test results for one soil condition with two angles of anchor orientation and the test data obtained by Neely, et al. (12). Because the angles ϕ and δ used by Neely, et al. are slightly different from those used in this study, the agreement between these two sets of results are considered quite satisfactory. This figure indicates that the effect of δ on anchor capacity is more pronounced for deep anchors than for shallow anchors. Fig. 11 provides data for two anchor orientations and two embedment depth ratios. For a given D/h ratio, the anchor capacity increases with δ at an increasing rate.

To present the preceding results in a more general form, the anchor-soil friction angle is expressed in terms of the internal friction angle of the soil in Figs. 12(A) and (B). By taking the anchor capacity of a perfectly rough anchor plate as the reference value, the anchor capacity for other friction angles may be estimated from Figs. 12(A) and (B) through the use of reduction factor, R. Figs. 12 (A) and (B) are only for D/h = 2, and for α = 45° and 90°, respectively. However, more figures of this form can be developed for other boundary conditions to facilitate practical applications.

SUMMARY AND CONSLUSIONS

The behavior of anchor capacity in sand was investigated both analytically and experimentally. The analytical solution was developed using the upper bound limit analysis together with the failure mechanism observed from the laboratory tests. The

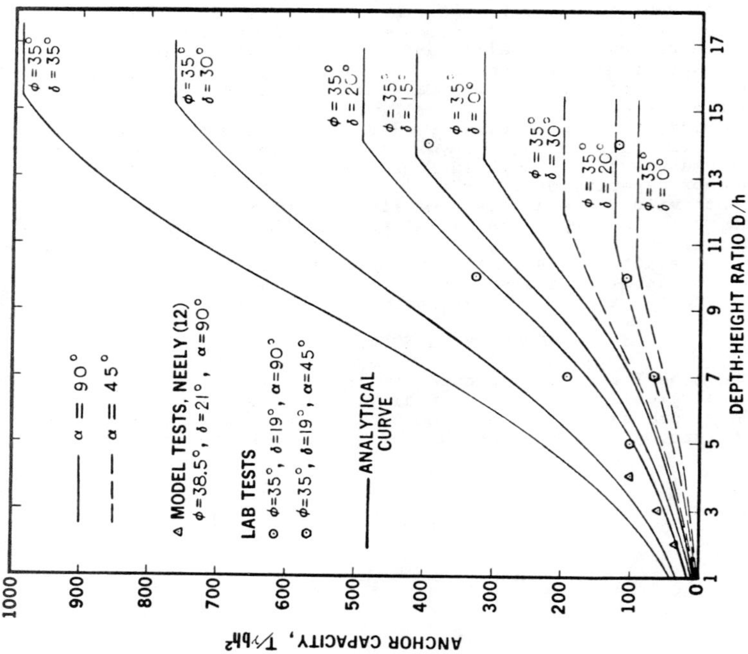

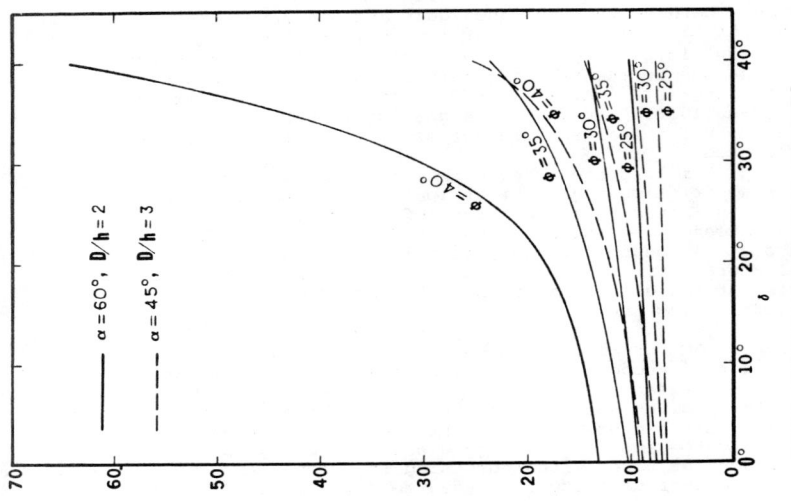

FIG 11 ANCHOR CAPACITY Vs δ VALUES FOR A CONSTANT φ ANGLE

FIG 10 ANCHOR CAPACITY Vs EMBEDMENT-HEIGHT RATIO FOR A CONSTANT φ AND δ VALUES

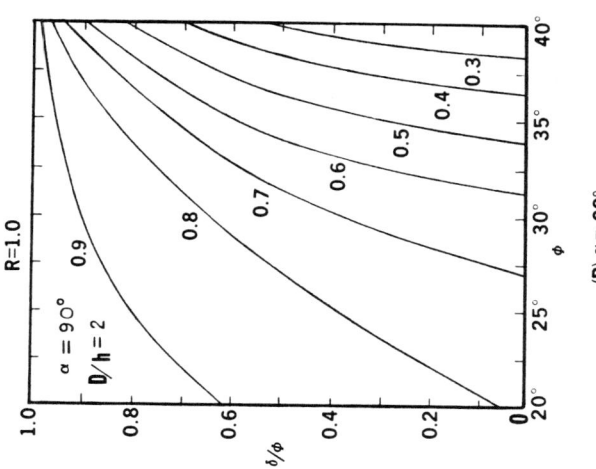

FIG 12(B) REDUCTION FACTOR (R) OF ANCHOR CAPACITY FOR VARIOUS RATIO OF δ/ϕ

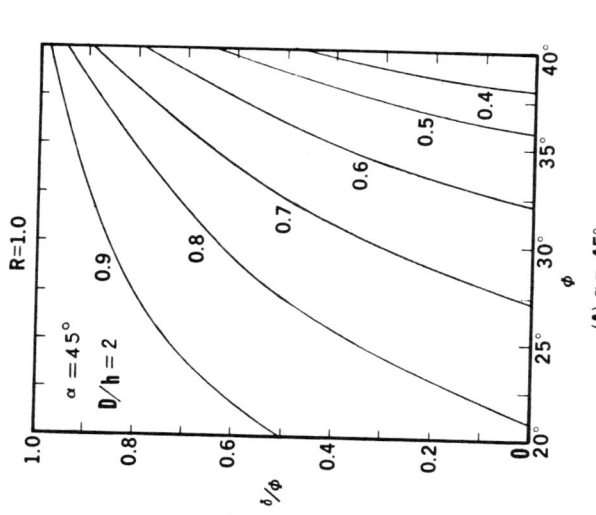

FIG 12(A) REDUCTION FACTOR (R) OF ANCHOR CAPACITY FOR VARIOUS RATIO OF δ/ϕ

results of the analysis agreed fairly well with the experimental results; also under most conditions the results compared satisfactorily with the information available in the literature.

Variables included in this investigation were anchor orientation, depth of anchor embedment, internal friction angle of soil, and anchor-soil friction. The effect of each of these variables on the anchor capacity was examined in great detail. It is hoped that the results of this study will be helpful for the development of a generally acceptable method for designing soil anchors.

Appendix I - REFERENCES

1. Ali, M.S., "Pullout Resistance of Anchor Plates and Anchor Piles in Soft Bentonite Clay," thesis presented to Duke University, at Durham, S.C. in 1968, in partial fulfillment of the requirements for the degree of Master of Science (also available in Duke Soil Mechanics Series No. 17).
2. Balla, A., "The Resistance to Breaking-out of Mushroom Foundations for Pylons," Proceedings, 5th International Conference on Soil Mechanics and Foundation Engineering, Paris, France, Vol. 1, 1961, pp. 569-576.
3. Chen W. F., Limit Analysis and Soil Plasticity, Elsevier Scientific Publishing Company, New York, 1975.
4. Colp, J.L., and Herbich, J.B., "Effects of Inclined and Eccentric Load Application on the Breakout Resistance of Objects Embedded in the Sea Floor," Sea Grant Publication No. TAMU-SG-72-204, Coastal and Ocean Engineering Division, Report No. 153, Texas A & M University, May, 1972.
5. Das, B.M., "Pullout Resistance of Vertical Anchors," Journal of the Geotechnical Engineering Division, ASCE, Vol. 101, GT1, January, 1975, pp. 87091.
6. Drucker, D.C., Prague, W., and Greenberg, J.J., "Extended Limit Design Theorems for Continuous Media," Quarterly of Applied Mathematics, Vol. 9, January 1952, pp. 381-389.
7. Mariupol'skii, L.G., "The Bearing Capacity of Anchor Foundations," Osnovaniya Fundamenty i Mekhanika Gruntov, Vol. 1, Jan. - Feb., 1965, pp. 14-18 (available in English translation from Consultants Bureau, New York, NY, pp. 26-32).
8. Matsuo, M., "Study on the Uplift Resistance of Footing (1)," Soil and Foundation, Tokyo, Japan, Vol. 7, No. 4, Dec., 1967, pp. 1-37.
9. Meyerhof, G.G., "The Ultimate Bearing Capacity of Foundations," Geotechnique, London, England, Vol. 2, No. 4, Dec., 1951, pp. 301-332.
10. Meyerhof, G.G., and Adams, J.I., "The Ultimate Uplift Capacity of Foundations," Canadian Geotechnical Journal, Ottawa, Canada, Vol. 5, No. 4, Nov., 1968, pp. 225-244.

11. Meyerhof, G.G., "Uplift Resistance of Inclined Anchors and Piles," Proceedings of the 8th International Conference on Soil Mechanics and Foundation Engineering, Vol. 2, Part 1, Moscow, 1973, pp. 167-172.
12. Neely, W.J., Stuart, J.G., and Graham, J., "Failure Loads of Vertical Anchor Plates in Sands," Journal of the Soil Mechanics and Foundations Division, ASCE, Vol. 99, No. SM9, Sept., 1973, pp. 669-686.
13. Taylor, R.J., and Lee, H.J., "Direct Embedment Anchor Holding Capacity," Naval Civil Engineering Laboratory Technical Note, TN-1245, Port Hueneme, CA, December, 1972.
14. Terzaghi, K., Theoretical Soil Mechanics, John Wiley & Sons, Inc., NY.
15. Vesic, A.S., "Expansion of Cavities in Infinite Soil Mass," Journal of the Soil Mechanics and Foundations Division, ASCE, Vol. 98, No. SM3, Proc. Paper 8790, Mar., 1972, pp. 265-290.

Appendix II - NOTATIONS

The following symbols are used in this paper:

D = Depth of anchor plate from ground surface to top of plate;
D_a = depth of sand which provides arching effect
D_s = depth of sand which provides surcharge load to passive lateral earth pressure
b = width of anchor plate
h = height of anchor plate
T = pullout force
V_i = velocity field
α = anchor plate orientation measured from horizontal direction
δ = anchor plate-soil friction angle
γ = initial unit weight of sand
ξ = angle between anchor plate and radial shear zone
θ = central angle of radial shear zone
ϕ = internal friction angle of sand

ANALYSIS OF SOIL COMPACTABILITY BY ROLLERS

by

R. N. Yong[1], M.ASCE and E. A. Fattah[2]

ABSTRACT

Historically, the problem of specification of input work for soil compaction under moving rollers, has generally been addressed in terms of empirical relationships and techniques. In this study, the finite element method (FEM) of analysis is used as an analytical/computer tool for evaluation of the amount of compaction energy required to compact a particular soil to a certain depth (thickness) under rigid or pneumatic tyre rollers. The compaction effort is essentially viewed in terms of the amount of compaction energy required to produce a resultant unit deformation of the soil layer. The influence of varying soil stress-strain relationships can be demonstrated by the analysis and has been confirmed from corresponding laboratory tow-bin tests.

KEY WORDS: soil compaction, F.E.M. analysis, roller compaction, stress-strain, rebound modulus, compaction efficiency.

---

[1] Director, and William Scott Professor of Civil Engineering and Applied Mechanics, Geotechnical Research Centre, McGill University, Montreal, Canada.

[2] Division Chief, Mechanics Division, Geotechnical Research Centre, McGill University, Montreal, Canada.

SOIL COMPACTABILITY ANALYSIS

INTRODUCTION

The objective of soil compaction in the field by moving rollers is to produce a competent soil-bearing stratum. The increased densification of the soil, together with the reduction in permeability of the compacted soil, combine to insure that the compacted soil develops greater soil strength which can be maintained for a long time duration.

In field soil compaction, a layer of loose soil is compacted until its density and strength reach specific values established a priori from laboratory tests, and until the density and strength are approximately similar to the previously compacted underlying layers of soil. The objective of field soil compaction is to obtain specified mechanical properties and soil characteristics with a minimum number of passes of the roller or with a minimum amount of roller input energy (work done).

In general, the compaction of layers of soil is accelerated by increasing the number of the magnitude of roller-imposed normal stresses established at the roller-soil interface, i.e. roller weight. However, if the roller-imposed normal stresses far exceed the local bearing capacity of the soil, i.e. if excessive local shear failure occurs, the soil will tend to extrude or flow under the rollers instead of compacting. Note that local shear failure of the compacting soil under initial compaction of loosely placed soil constitutes a special case of study. In this present study, attention is directed towards soil compaction of soil layers beyond the first "pass" loose soil compaction.

The magnitude of imposed normal stress can be increased by increasing roller load in the case of rigid rollers, and increasing tyre load or increasing inflation pressure in the case of rubber-tyre rollers. Excessive sinkage and roller immobilization can occur if the general bearing strength of the soil is far exceeded. The required work done (input energy) by a roller to produce a specific degree of compaction is a function of:

(a) <u>Roller Type</u>: solid (rigid) roller, pneumatic tyre roller, number of tyres in the roller, arrangement, roller load and speed.

(b) <u>Tyre Characteristics</u>: diameter, width, cross-sectional shape, inflation pressure, carcass stiffness and shape, tyre structure and tyre material mechanical properties.

(c) <u>Soil</u>: all the factors which control stress-strain behaviour of the soil during loading and unloading, such as - soil type, density, moisture content, soil structure, confining pressure, etc.

In this study, an analytical model is established (a) to calculate the amount of work done per unit distance for the moving roller for every pass, and (b) to determine the total work as required to produce a certain permanent deformation (i.e. compacted depth) of the soil layer.

METHOD OF ANALYSIS

The method of analysis for treatment of the problem is based on the principle of energy conservation. It is assumed that the input energy required to keep the roller in constant motion during a certain pass is equal to the sum of the useful output energy and the following parasitic energy components:

(a) energy spent in compacting the soil,
(b) energy dissipated at roller-soil interface due to slip,
(c) energy dissipated in distorting the tyre. Note that in the case of a rigid roller, this parasitic energy component becomes vanishingly small and can be ignored.

Figure 1 shows a schematic representation of the roller-soil interaction phenomenon. The various energy components participating in the interaction are identified in the Figure.

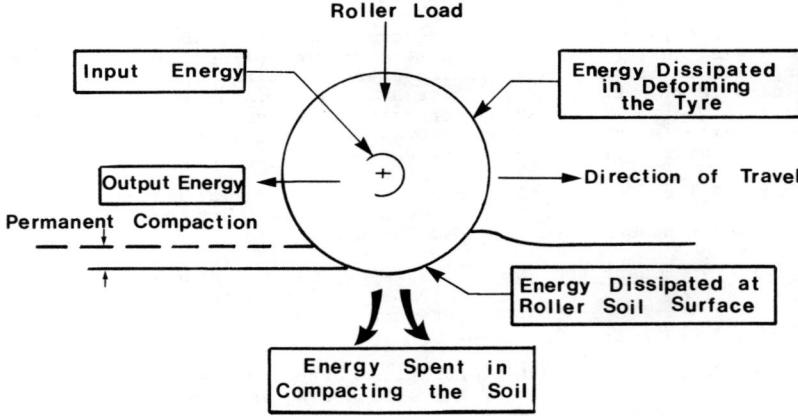

Fig. 1   Roller-soil energy system

Since the rollers or tyres used in compaction will be relatively wide, a two-dimensional analysis can be adopted and all calculations can be performed in terms of a unit width of the roller or tyre. It is assumed that at any stage of roller compaction, the soil continuum being compacted consists of two layers, as shown in Fig.2. The bottom layer consists of the previously compacted layer (or layers) where the density and strength have reached, or nearly reached, the specified values. The top layer is the current soil layer to be compacted to a specified or desirable density.

The model chosen and shown in Figs. 1 and 2 utilizes the Finite Element Method (FEM) for solution.

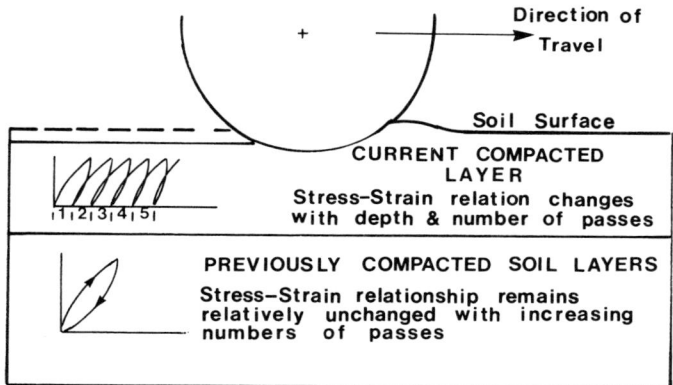

Fig. 2   Idealized soil continuum showing a developed two-layer system

ELEMENTS OF THE PROBLEM

The finite element analysis adopted herein is an extension of the analysis previously reported by Yong et al. (5, 6). The requirements for solving the problem can be summarized as:

(a) Specification of loading boundary at roller-soil interface in terms of contact area dimensions and normal and tangential stress distributions.

(b) Soil stress-strain relationships for state of loading and unloading at different confining pressures (i.e. depth and prior compaction effects).

Analytical Relationships

To briefly develop the model, the governing relationships consider a body occupying a space S subjected to traction forces q. Using the principle of virtual work and equating the external and internal work, the (approximate) equations of equilibrium in updated coordinate form [7] may be obtained as follows.

Writing the external work done as:

$$\int_{\overline{V}} \overline{P}\{q\}\{dU\}d\overline{V} + \int_{\overline{A}} \{\overline{P}\}\{dU\}d\overline{A} \tag{1}$$

and internal work done as:

$$\int_{\overline{V}} \{\overline{\sigma}\}^T d\{\overline{e}\}d\overline{V} \tag{2}$$

The relationship between strain and displacement can be written as:

$$d\{\overline{e}\} = [B] d\{\delta\} \tag{3}$$

Equating relationships (1) and (2), one obtains the equilibrium equations as follows:

$$\{\psi\} = \{R\} - \int_{\overline{V}} [B]^T \{\overline{\sigma}\} d\overline{V} = 0 \tag{4}$$

where

$$\{R\} = \int_{\overline{V}} \overline{P}[N]^T \{q\} d\overline{V} + \int_{\overline{A}} [N]^T \{\overline{P}\} d\overline{A} \tag{5}$$

and where

[ ]	=	matrix form;
{ }	=	column vector;
{R}	=	equivalent external nodal forces;
{ψ}	=	nodal forces - required to bring the assumed displacement pattern into nodal equilibrium;
{$\bar{P}$}	=	surface forces per unit area of the deformed body;
{q}	=	body forces per unit mass;
[N]	=	shape function;
$\bar{V}$ and $\bar{A}$	=	volume and area of the deformed body;
$\bar{\rho}$	=	density in the deformed body;
{U}	=	displacement at any point within a finite element;
{δ}	=	displacement at nodal points;
[B]	=	displacement function; and
{$\bar{\sigma}$} and {$\bar{e}$}	=	vector forms of stress and strains.

Load Boundary Conditions

For a generalized roller situation, the pneumatic tyre is considered since this permits one to incorporate the tyre energy loss term in the analysis. This term can be ignored in the case of the rigid roller. The tyre-soil interfacial stress distribution is a function of : (a) tyre geometrical characteristics, (b) tyre stiffness and (c) soil mechanical properties. The tyre-soil interfacial stresses can be determined from continuum mechanics; however, the solution technique is quite complex because of (a) transient type of soil loading, (b) non-linearity of the soil stress-strain relations, (c) tyre distortion under load and in motion, and (d) relative movements at the tyre-soil interface due to slip. In order to simplify the problem, the interfacial stress distribution is specified in terms of known distributions based on previously available or reported measurements [3].

Constitutive Relations

Since the applied roller load is a transient type of loading, any point in the soil continuum is subjected to a state of loading or unloading according to its position with respect to the roller, i.e. for each roller pass the soil is subjected to a complete stress cycle (load and unload). Note that complete soil rebound (i.e. recovery) is not a necessary requirement since the intent of the roller load application is to compact the soil. The selected constitutive relationship should represent the soil behaviour for a complete stress-reversal cycle. A nonlinear elastic approach can be adopted to represent the soil during loading and an elastic approach during unloading. For loading and unloading of a soil element the following code is used in the analysis:

$$\Delta W = \sigma_{ij} dE_{ij} \qquad (6)$$

If $\Delta W < 0$ the element is unloading.

If $\Delta W > 0$ the element is loading.

where

σ_{ij} = state of stress;

dE_{ij} = incremental state of strain.

Method of Solution

To apply the method of solution, the soil continuum is idealized as shown in Fig. 3 with the finite element mesh. For simplicity in generalization of the problem, it is easier to express the dimensions of the interacting components of the problem (i.e. roller and soil) in terms of the contact patch length of the roller-soil interface. This is given as the footprint length - FPL. This procedure also allows one to minimize the amount of input data and computing time.

Implementation of the analysis requires one to determine the stresses and displacements in the soil layer being compacted under the roller load with the roller centre in position 1. (Fig. 3). This computational procedure is repeated by moving the roller centre to position 2. Note, however, that in moving the roller from position 1 to 2, unloading at position 1 occurs as loading at position 2 is implemented. The procedure is essentially applied in the continuous roller travel on the ground surface by taking successive roller positions and is required in order to account for the transient motion of the roller - i.e. the solution format is a pseudo kinematic procedure, since the state of stress in the soil at the end of the increment depends on the initial state of stress established at the beginning of the travel increment. The incremental roller travel computational procedure is terminated automatically when the state of stress at any point in the soil with respect to one roller position does not change significantly with the imposition of the next increment.

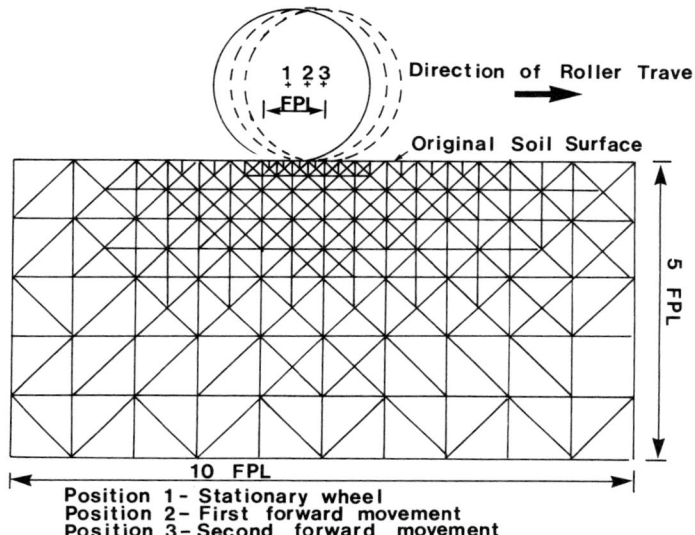

Fig. 3 Finite element idealization of the roller roller compaction problem

The output is obtained in terms of stresses and strain in the subsoil at any point. Strain-rates at any point in the subsoil are automatically computed from a knowledge of the time utilized in moving the roller one travel increment. Defining e_d as the compaction energy utilized in deforming a unit element of soil per unit time, it is possible to define the distribution of consumed energy in the subsoil in the compaction process.

LABORATORY TEST PROGRAM

In the laboratory test program developed to provide information to serve as input and to verify the predicted results of the analytical model, three types of test information and results were obtained: (a) tyre-deformation tests results and contact area measurements (INPUT), (b) tyre-soil interfacial characteristics, (VERIFICATION), and (c) soil properties (INPUT).

Tyre Types and Tests

A model rigid wheel and two types of model pneumatic tyres were used in the testing program. These were :

(a) aluminum wheel   - 34 cm diameter,   9.5 cm width.
(b) 3.00-8.00 2PR   - 34.3 cm diameter, 7.7 cm width.
(c) 4.10/3.50-4.00 2PR - 26 cm diameter,   7.0 cm width.

The dimensions and type of tyres were essentially dictated by the size of the soil tow bin. The width of the 11 m long soil bin was adjusted to allow the tests to be conducted as a two-dimensional towed or driven wheel test. The 120 cm depth of the soil tow bin served to minimize and prevent any bottom boundary effects.

The wheels and tyres used in the laboratory test program were chosen to reflect the effects of wheel diameter, carcass shape and stiffness. In the towed or driven roller tests performed on the soils, the following measurements were made:

(1) Measurement of contact areas on rigid and soil surfaces under different loadings and inflation pressures.

(2) Measurement of tangential stresses using a spinning wheel on the soil surface. The torque required to spin the wheel is a measure of the tangential stresses developed.

The drawbar pull tests performed in the 11 m long soil tow bin allowed the selection of various wheel slips by controlling both the dynamometer carriage velocity and the driven wheel (roller) rotational velocity. This was possible since the dynamometer carriage was driven independently from the wheel. By unlocking the driven wheel or roller, a towed wheel condition with some negative slip was obtained. For a no-slip condition, both the carriage and the roller (wheel) were driven such that no relative displacement between the soil and wheel surfaces occurred. The ability to drive the roller independent from the dynamometer carriage permitted one to examine roller and compaction behaviour as a function of slip. The various above-ground (surficial) items measured in the tow bin tests other than the tyre properties included: input torque, drawbar pull, dynamic (instantaneous) and permanent soil deformation, roller or wheel rotational and translational speeds [4].

By using a flash X-ray system, positioned to record embedded lead markers in the test soil below the moving roller, it was possible to obtain information on the time-displacement response performance of the loaded soil (Fig. 4). Since the X-ray pulse sequence was in the order of 10^{-8} seconds, a multitude of recordings could be obtained during the passage of the roller. In a general test sequence, only 5 X-ray pulses were required [3]. Recording of the pulses on fresh individual exposure plates permitted one to trace the time-movement sequence of individual lead markers previously embedded in an ordered matrix format in the test soil. Reduction of the lead marker information permitted one to view soil deformation in the entire soil mass in terms of (a) "stream line" flow akin to soil moving past a stationary roller, (b) strain-rate and velocity fields, and (c) stress fields, - provided a proper constitutive relationship could be formulated.

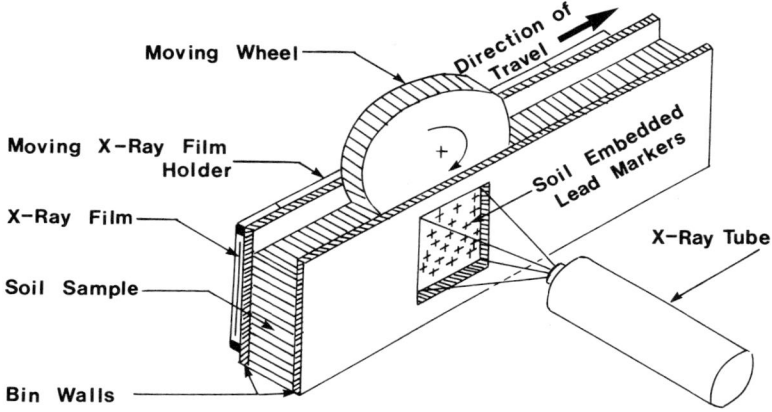

Fig. 4   Schematic diagram of the experimental system showing the flash X-ray tube and the roller moving in the soil tow bin

Soil Properties

Two types of soil were used in the tow bin tests:

(a) A soft kaolinite clay with liquid limit = 54%, plastic limit = 37%, moisture content 52 to 56%, and dry density of 1.25 ton/m^3. This type of soil was used for the rigid aluminum wheel (roller) tests.

(b) A mixed soil consisting of fine silica sand passing sieve number 30 and kaolinite proportioned in terms 30:70 respectively. The soil mixture was mixed with water and compacted in the soil bin to a total dry density of 1.88 ton/m^3 and moisture content of 13%.

To obtain the relevant soil mechanical properties, soil samples were taken from the tow bin and tested in plane strain compression. The corresponding stress-strain curves obtained are showed in Figs. 5 and 6.

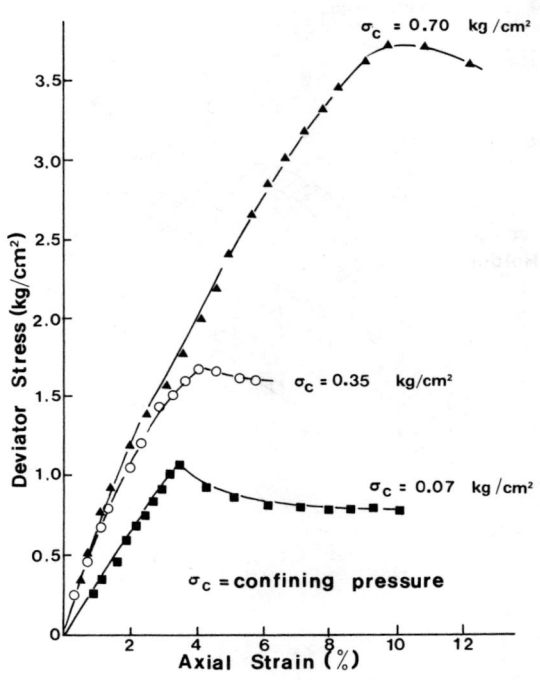

Fig. 5   Stress-strain curves of kaolinite clay from plane-strain triaxial tests

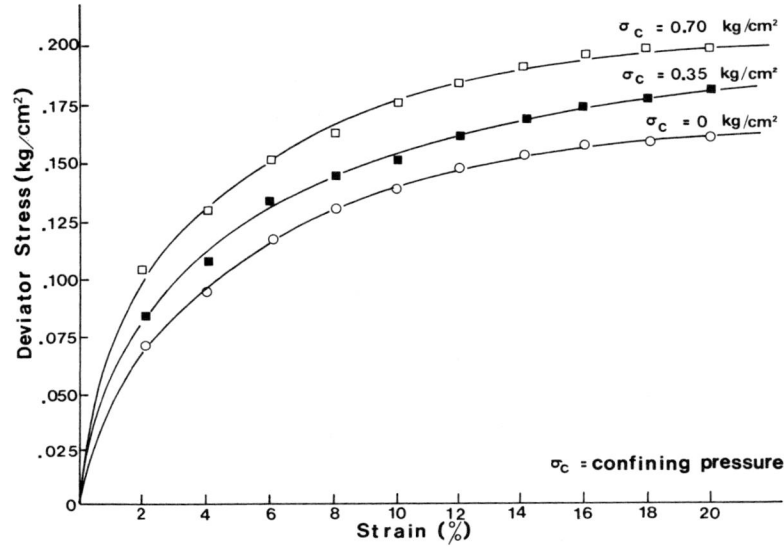

Fig. 6   Stress-strain curves of clayey sand from plane-strain triaxial tests

RESULTS AND DISCUSSION

To simplify comparison of predicted and measured values, the presentation of results can best be shown for the cases of a single soil layer subject to one roller pass. Note that the underlying previously compacted layer (or layers) is already factored into the consideration. In addition, the single pass under consideration implicitly accounts for the previous passes through the use of the current stress-strain relationship in the continuing solution of the problem - i.e. each successive compaction pass of the roller will use a different set of stress-strain curves to account for prior passes of the roller.

Figure 7 shows measured soil "stream lines" beneath a moving rigid roller on soft soil. Reduction of test data from the lead markers in the soil permits one to, in essence, "freeze" the roller and visualize the soil moving past the "frozen" roller.

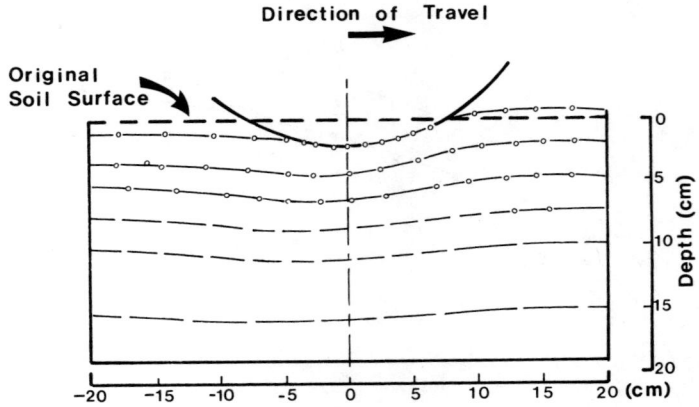

Fig. 7 Soil "stream line" flow under a moving rigid wheel. Wheel load = 24.6 kg; 0 degree slip; dynamic sinkage = 1.6 cm.

The soil surface instantaneous and permanent deformation can be seen directly below and at a distance apart from the roller centre respectively.

Since the multiple X-ray exposures of the embedded lead markers recorded displacements under the transient roller load, the recorded displacements at various points and times in the subsoil can be used to compute the "experimental" compaction energy e' expended at the various points. This is calculated by the following relationship [4]:

$$e'_d = 2\tau\sqrt{I} \qquad (7)$$

where

τ = shear strength of the soil, and

I = second strain rate invariant obtained from evaluation of the lead marker movements in the subsoil.

Computations performed at various points in the subsoil will thus permit one to produce a set of contours which define the expended compaction energy distribution in the subsoil.

From the finite element analysis, computations for the FEM predicted expended compaction energy e_d can be obtained for a unit width of soil element as follows:

$$e_d = \sigma \dot{\varepsilon} \tag{8}$$

where

σ = FEM computed stress acting on the soil element;
$\dot{\varepsilon}$ = incremental strain rate.

The FEM predicted e_d values at the centers of the subsoil idealized elements can also be plotted in sets of contours and compared with the e'_d values obtained from experimental observations as shown in Fig. 8. The case considered in Fig. 8 is a rigid roller operating at a zero slip condition. The correspondence between the e_d and e'_d is good near the rigid roller, but is relatively weak at distances away from the roller. The shape of the contours, however, appear to compare well.

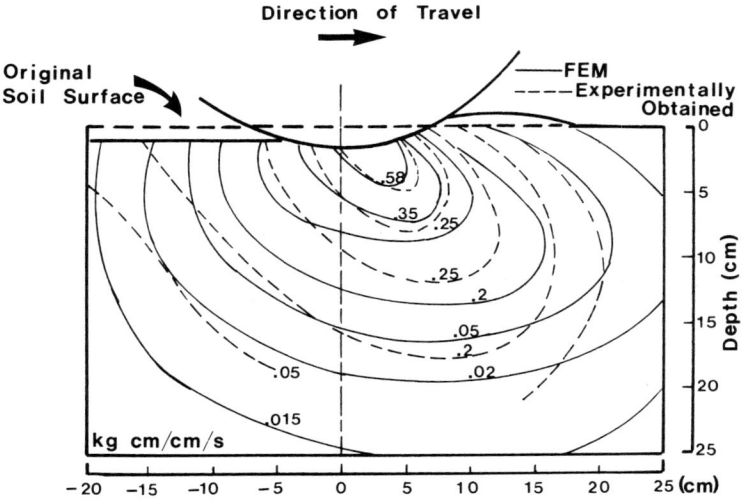

Fig. 8 Deformation energy contours (0 percent slip) developed due to passage of a rigid roller on kaolinite clay

In the case of the pneumatic tyre roller, Fig. 9 shows the effect of slip on the production of e_d contours. Although the experimentally computed e'_d values are not shown, the same regions of close correlation as with the rigid roller have been identified.

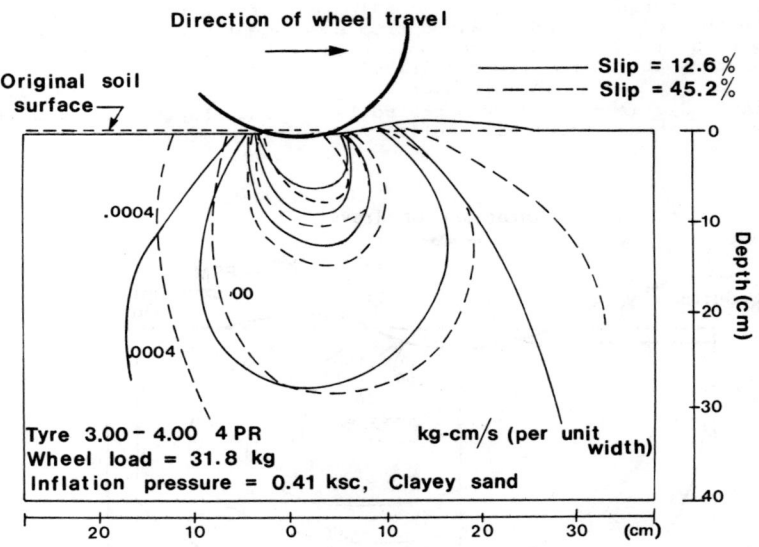

Fig. 9   Deformation energy contour
(Tyre 4.10/3.50-4.00 2PR)
Developed as a result of passage
of a pneumatic tyre on clayey sand.

The total energy consumed in compacting the soil (E_d) can be obtained as [4]:

$$E_d \text{ (experimental)} = \sum_I^N \int e'_d \, dxdy \tag{9}$$

where

N = number of subsoil idealized elements;
dx, dy = soil elemental dimensions.

In the case of the FEM prediction, the theoretically computed E_d values can be obtained as:

$$E_d \text{ (theoretical)} = \sum_I^N \int e_d \, dA \tag{10}$$

where

N = number of finite elements;
A = area of each element.

The slip test results for the different tyres surface conditions using the spinning wheel test are shown in Fig. 10. As noted previously, these results are used to estimate the roller-soil or wheel-soil tangential stress distribution. Note that the intercepts on the ordinate are the torques required to initiate slip. Expressing the wheel or roller-soil interface performance in energy relationships, one obtains:

$$E_i = P + E_f + E_d + E_t \tag{11}$$

where

E_i = input energy = torque T multiplied by ω, where ω = rotational velocity of wheel;
P = useful output energy;
E_f = interfacial energy loss = $\frac{T}{r}(\omega r - V)$;
r = rolling radius;
V = translational velocity;
E_d = energy utilized in deforming or compacting the soil;
E_t = energy dissipated in deforming the pneumatic tyre. Note that in a rigid roller, this term is essentially zero.

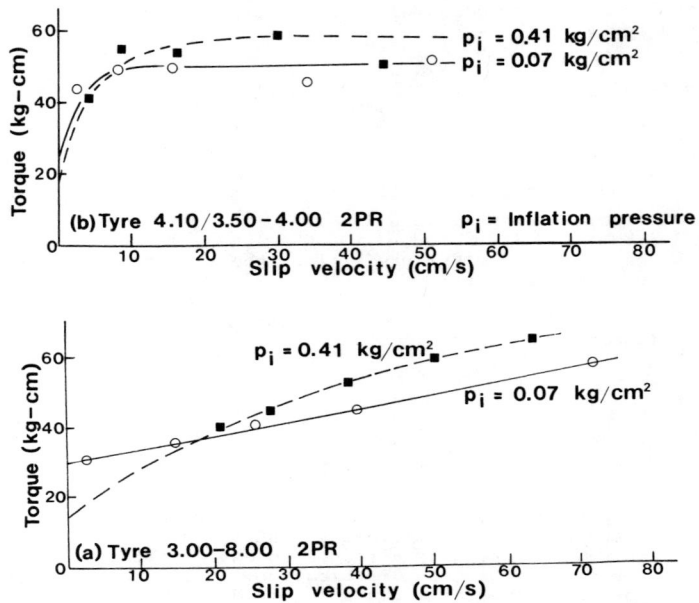

Fig. 10 Relationship between torque and slip velocity, (kaolinite clay). Note that intercepts on ordinate are torques required to initiate slip.

Since the useful output energy P and torque T are measured in the laboratory experiment, together with ω, the input energy E_i and P are known quantities. From the spinning wheel tests, one can determine E_f the interfacial slip energy loss term. The term E_t can be taken as zero for a rigid roller, or can be computed as the product of tyre motion resistance and the travelling velocity on rigid surface. Thus, the experimentally determined energy used for soil compaction, E_d (experimental) can be verified by comparing the FEM predicted E_d (theoretical) with that obtained by applying Eq. (11) or by Eqs. (9) and (10).

Figure 11 shows the energy balance relationship demonstrated in Eq. (11). The graph shows the energy relations established in roller-soil compactive interaction as a function of slip. This type of plot shows the performance of the roller at different degrees of slip from towed roller or self-driven roller.

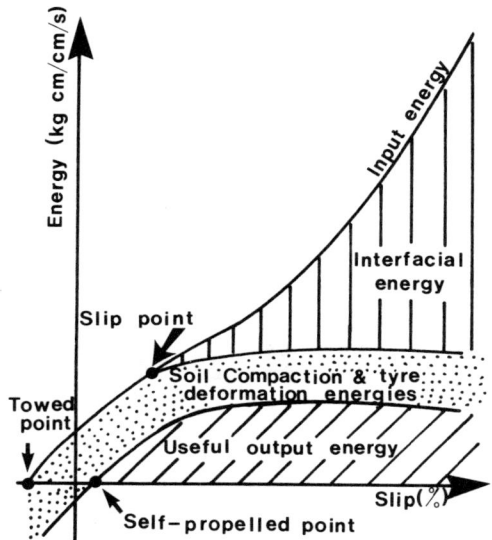

Fig. 11 Typical energy balance developed for a tyre moving on soft soil

The relationships between the energy spent in compacting the soil and the slip for the rigid roller and the pneumatic tyre are shown in Figs. 12 and 13 for the two kinds of soil tested. Note that a certain amount of slip generates a greater compactive effort. It should also be observed that the correspondence between FEM predicted and experimentally computed values is good.

The instantaneous and permanent soil surface deformations due to the application of the moving roller is shown in Fig. 14. It should be noted that after a certain degree of slip, the tangential stresses at the roller tyre-soil interface are fully mobilized. At that time the instantaneous and permanent deformations remain constant. The maximum stress difference ($\sigma_1 - \sigma_3$) in the stress-strain curve chosen for the FEM analysis will control the magnitude of the instantaneous surface deformation obtained. When the roller moves on, the surface rebounds. The amount of rebound is controlled by the unloading modulus in the stress-strain load-unload curve. From previous experience and correlations, [6], it has been noted that in using the FEM analysis, the rebound modulus can be taken as 1.5 times that of the initial tangent modulus defined by the stress-strain curve chosen - for the kinds of soils used in this test program. Note that in working with all kinds of soils, it

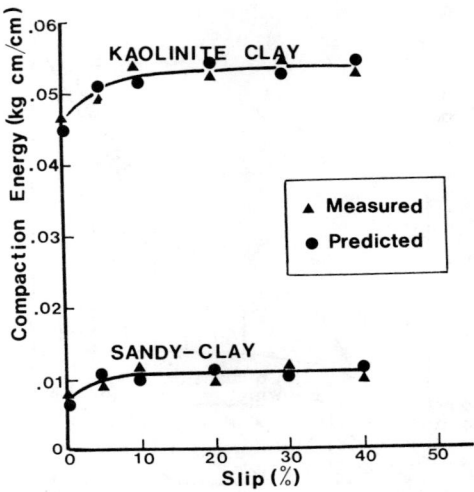

Fig. 12 Effect of type of soil on compaction energy - rigid roller

is necessary to physically determine the actual rebound modulus. The influence of a proper representative stress-strain relationship which reflects the compactive properties of the soil is significant. It should again be noted that the correspondence between predicted and experimentally computed values for both the instantaneous and permanent soil surface deformations are good. Note that the permanent deformation is obtained by subtracting the rebound from the instantaneous deformation. The results also show that the sinkage and energy spent in compacting the soil are higher for a driven tyre (identified because of slip-generation), than that of the corresponding towed tyre.

Figure 15, which is obtained from the previous figures, shows the energy spent per unit depth of permanent deformation for a unit travel distance for the two different soils. This is defined as the specific compaction energy. The specific compaction energy required for the clay soil is seen to be considerably less than that expended for the sand-clay mixture. The factors and parameters controlling specific compaction energy include (a) the inherent stiffness of the soil, (b) its rebound capability, (c) the load applied to the roller, and (d) the stiffness of the roller. A greater use of the data or predictions can be made to assist in evaluation of compaction effort and depth of layer compacted for various other design parameter requirements if desired. To demonstrate the utility of the analysis developed, an actual test example is presented to show the effect of

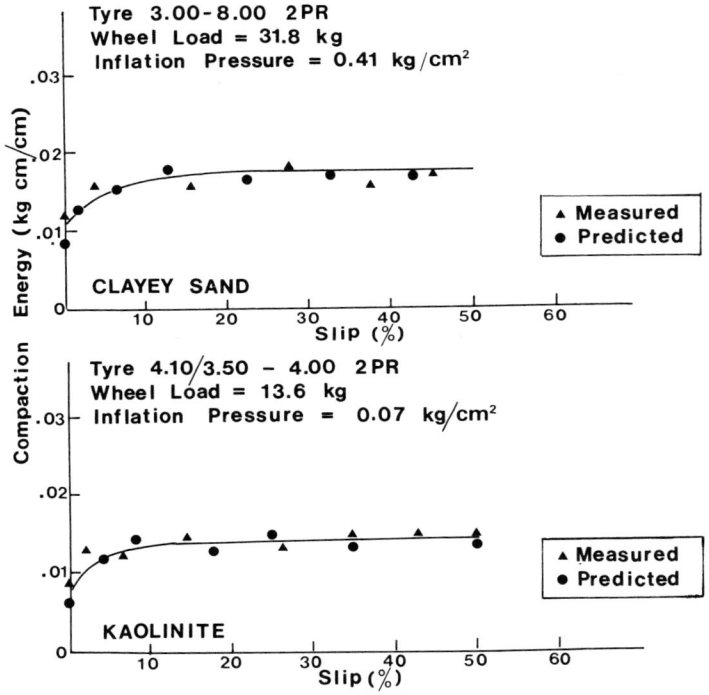

Fig. 13 Effect of type of soil on compaction energy - pneumatic tyre

soil type and soil layering on pneumatic roller performance. In this example the roller load is 515 kg. Contact area of roller on rigid surface can be characterized as -

width = 15.8 cm; length = 20.8 cm.

The soil stress-strain curves are presented in Fig. 16. As noted previously the tangential stresses at the tyre-soil interface can be determined from laboratory experiments. In the three test examples treated, it is assumed that the two separate cases for the single-layer system have the same soil depth of 104 cm, and that the two-layer system has layers of 20.8 cm and 83.2 cm thick, with the thinner layer on top. In the two separate considerations of the single-layer system, a soft soil and a stiff soil are considered separately.

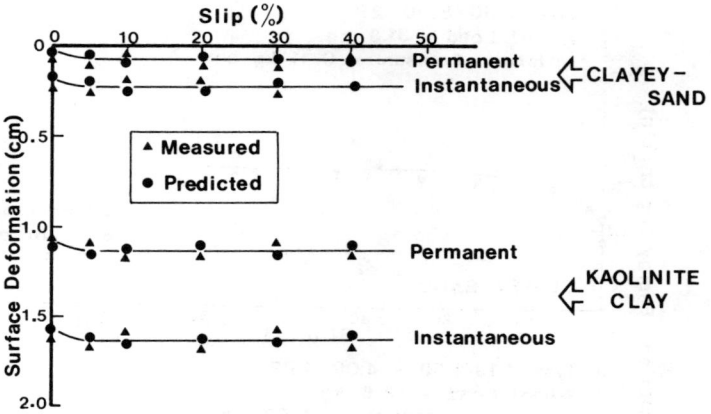

Fig. 14 Effect of type of soil on developed surface deformation due to passage of a rigid roller

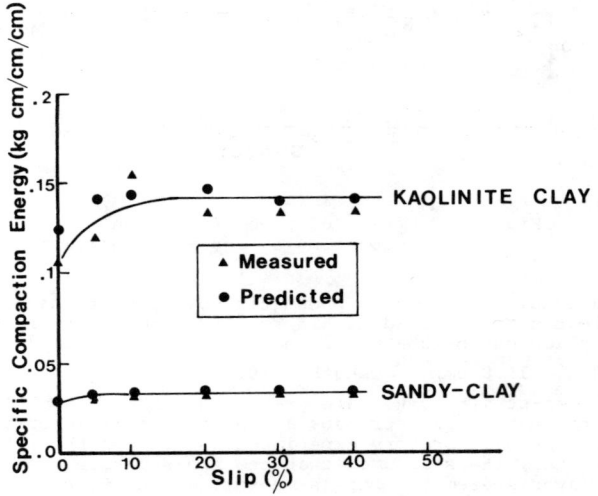

Fig. 15 Effect of type of soil on specific compaction energy requirements for a rigid roller

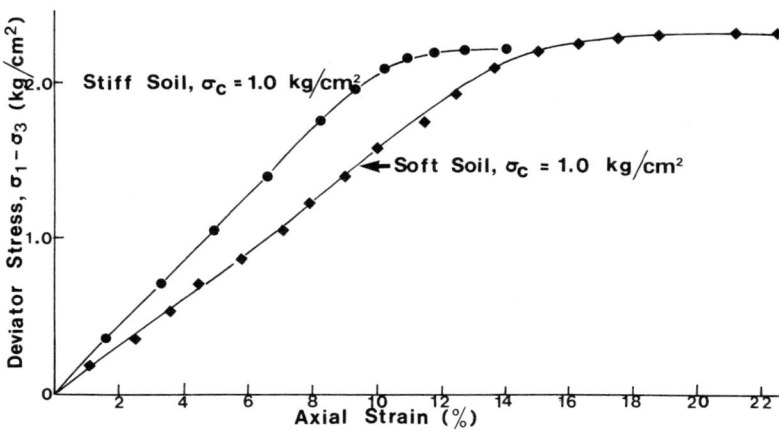

Fig. 16 Stress-strain curves of silty clay from triaxial tests

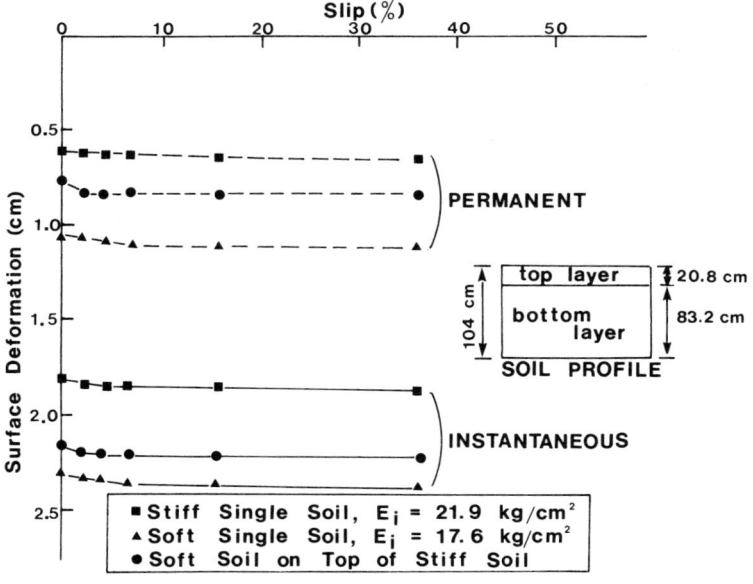

Fig. 17 Effect of type of soil on developed surface deformation due to passage of a pneumatic tyre - Load = 525 kg

Figure 17 shows the computed surface deformations for the three cases presented, whilst Fig. 18 shows the energy spent in compacting the soil as a function of slip. Dividing the results given in Fig. 18 by the deformations shown in Fig. 17, one obtains the specific compaction energy curves shown in Fig. 19. As expected, Fig. 19 shows that the stiff single soil layer requires more specific energy than that utilized for the soft single soil layer, - because one requires more energy to compact a stiffer soil to the same permanent unit deformation as compacted to a softer soil. Thus, even though Fig. 19 shows that the soft single soil layer absorbs less specific compaction energy than the stiff layer, the total permanent deformation sustained is larger.

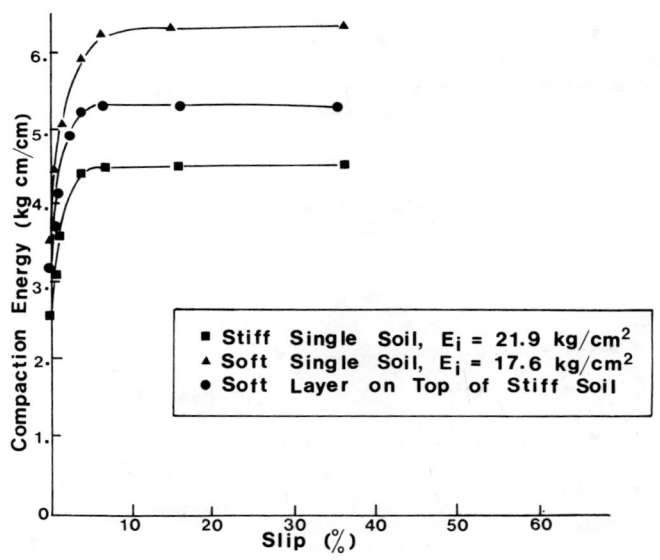

Fig. 18 Effect of type of soil on developed compaction energy due to passage of a pneumatic tyre - load = 525 kg

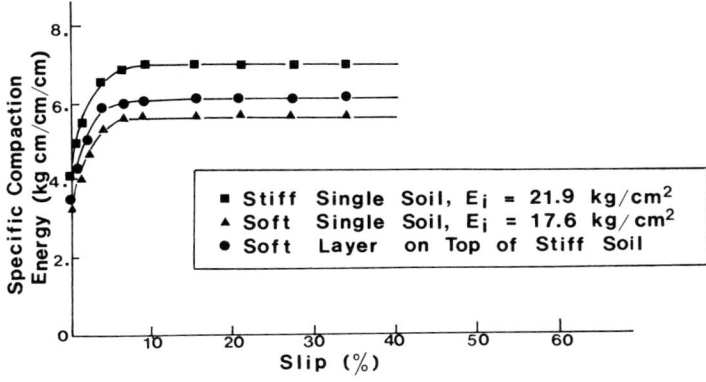

Fig. 19 Effect of type of soil on specific compaction energy requirements for a pneumatic tyre - load = 525 kg

CONCLUDING REMARKS

The comparisons between predicted and experimentally computed values of compaction energy show that the finite element method can be used to evaluate the compactibility of the soil in terms of the amount of energy required for producing a permanent unit surface deformation. Comparison between the predicted and the measured values confirm the validity of the proposed analytical model. This model takes into consideration the nonlinear behaviour of soil, the effect of large deformations and the effect of stress reversal due to the nature of roller movement.

This model is a first step towards establishing a general model which considers the case of multi-layered system and successive number of passes. The ability to evaluate the roller performance with respect to a specific type of soil in terms of the energy required to produce a unit depth of permanent deformation of the soil surface serves to provide an insight into the effect of roller parameters on roller efficiency.

ACKNOWLEDGEMENTS

The authors wish to acknowledge the experimental assistance and input provided by Dr. P. Boonsinsuk. This study was jointly supported by the Natural Sciences and Engineering Research Council of Canada under Grant No. A-882, and the Department of National Defence (Canada).

REFERENCES

1. Freitag, D. R., Green, A. J. and Murphy, N. R. Jr., *Normal Stresses at the Tyre-Soil Interface in Yielding Soils*. Highway Research Record 74, 1-18, 1964.
2. Krick, G., *Radial and Shear Stress Distribution under Rigid Wheels and Pneumatic Tyres operating on Yielding Soils with consideration of Tyre Deformation*, J. Terramechanics 6, 73-98, 1969.
3. Windisch, E. and Yong, R. N., *The Determination of Strain-Rate Behaviour beneath a Moving Wheel*, J. Terramechanics 7, 55-67, 1970.
4. Yong, R. N. and Webb, G. L., *Energy Dissipation and Drawbar Pull Prediction in Soil-Wheel Interaction*, Proc., Third Int. Conf. ISTVS, Essen, Vol.1, pp. 93-142, 1969.
5. Yong, R. N. and Fattah, E. A., *Prediction of Wheel-Soil Interaction and Performance using the Finite Element Method*. J. Terramechanics 13, 227-240, 1976.
6. Yong, R. N., Fattah, E. A. and Boonsinsuk, P., *Analysis and Prediction of Tyre-Soil Interaction and Performance using Finite Elements*. J. Terramechanics 1, 43-63, 1978.
7. Zienkiewicz, O. C. and Nayak, G. C., *A General Approach to Problems of Plasticity and Large Deformation using Isoparametric Elements*, Proc., Third Conf. on Matrix Methods in Structural Mechanics, WPAFB, Ohio, pp. 881-928, 1971.

Comments by J. M. Duncan *
Session 2 - "Generalized Stress-Strain Applications
   in Geotechnical Engineering"

My own experience with the use of generalized stress-strain applications in Geotechnical Engineering has been concerned mostly with the use of the hyperbolic model in static finite elemnt analyses. In the past 10 years some lessons have emerged from this experience which appear to have a degree of generality, and which are perhaps worthy of comment.

Whenever finite element analyses are applied in specific practical cases, there are inevitable uncertainties about what the field conditions will be when the structure is built. These uncertainties frequently include such questions as these: What will be the average gradation of a fill? What will be its average field density? What will be its average water content? These questions need to be answered before laboratory tests can be performed to measure stress-strain properties. In some cases (such as with coarse rockfills) no laboratory tests can be performed, and properties must be estimated. Other questions also arise, such as: Are samples of natural soils really "undisturbed"? If not, to what degree has disturbance affected the test results?

Each of these questions deserves careful consideration, because each can have a large influence on the behavior in the field and the results of the analysis. In the writer's experience more than half of the time and effort involved in typical stress-strain applications in Geotechnical Engineering is devoted to consideration of such questions. For example, in a recent analysis of stresses and movements in an earth dam, about

*Prof., Univ. of California, Berkeley, CA

250 hours of effort, out of a total of 400 hours, was devoted to answering such simple questions.

Even after the best possible answers have been developed for these questions, it must be realized that there are still uncertainties. As a result, soil stress-strain properties can only be estimated within a range of possible values. In this circumstance it seems appropriate to employ fairly simple stress-strain relationships, because a high degree of precision in matching field behavior is unlikely even with the most sophisticated relationship. (The writer admits that his point of view may be somewhat prejudiced by the fact that his experience has been concerned primarily with application of the simple hyperbolic stress-strain relationship).

The inevitable uncertainties involved in defining conditions for analysis makes it more desirable to perform a number of comparatively simple analyses rather than a single very detailed analysis. Parameter studies can then be used to evaluate the importance of the uncertainties about properties and other conditions.

At this session a question was raised concerning the use of generalized stress-strain relationships for _design_ in Geotechnical Engineering. It is worth noting in this regard that these procedures have been applied to a number of different problems: analyses of embankment dams, excavation bracing, U-frame lock structures, buried culvert structures. In most cases the analyses have been performed to predict or study movements and stresses, and they have not had extensive impact on design. In circumstances where adequate and soundly based design procedures are available (embankment dams, excavation bracing, small culverts under deep fills) the use of new, untried procedures has no advantage. Only in cases where design procedures have not been developed (U-frame lock structures, long-span culverts with shallow cover and heavy live loads) have finite element analyses and generalized stress-strain relationships had significant impact on design.

STRESS-STRAIN MODELS FOR DEEP EXCAVATION
(Panel Discussion)
By Zdenek Eisenstein[1], M.ASCE

The problem of magnitude and distribution of lateral pressure acting on tangent pile walls supporting deep excavation in glacial till has been studied by integrating field measurements with a finite element analysis.

An agreement between the field and analytical results in terms of displacements is accepted as a criterion of validity for analytical results of stresses, where direct in-situ stress measurements are difficult to obtain and interpret. Of special importance is the calculated lateral pressure against the wall and its relation to the stiffness of the wall and to the magnitude of associated ground movements.

The finite element analysis employed several stress-strain models and comparisons were made as to their efficiency in predicting the observed field behaviour. Special attention has been devoted to the effects of stress paths.

The following laboratory tests were performed on block samples of the till:
- Triaxial Passive Compression (TPC) Test - Conventional triaxial test in which the vertical stress is increased while the lateral stress is kept constant (this test does not model any stress path around excavation, but has been included in the study for the purposes of comparison);
- Triaxial Active Compression (TAC) Test - Triaxial test in which the lateral stress is reduced while the vertical stress remains constant (this test represents stress path beside the excavation);
- Triaxial Active Extension (TAE) Test - Triaxial test in which the lateral stress remains constant while the vertical stress is reduced (this test represents stress path below the excavation);
- As a major part of the excavation deformed under a condition of plane strain, in addition to triaxial testing, tests in a plane strain apparatus were performed to simulate the actual field conditions even more realistically. The plane strain experiments of the following type were performed:

---

[1] Professor of Civil Engineering, University of Alberta, Edmonton, Alberta, Canada, T6G 2G7

- Plane Strain Passive Compression (PSPC) Test - A test in a plane strain apparatus, with vertical stress increasing and lateral stress constant;
- Plane Strain Active Compression (PSAC) Test - The lateral stress is reduced while the vertical stress remains constant.

Nonlinear stress path dependent stress-strain curves obtained from these tests were used as input in the following analyses:
1) Nonlinear elasticity with conventional triaxial data - TPC tests;
2) Nonlinear elasticity with stress path dependent triaxial data - TAC and TAE tests;
3) Nonlinear elasticity with plane strain testing data - PSAC and PSAE tests.

In addition to the nonlinear models a fourth analysis was performed for the sake of comparison:
4) Linear elasticity - constant values of parameters used throughout (elastic moduli derived from pressuremeter in-situ testing).

The results of tangent pile wall movement with different stress-strain assumptions are in Figure 1. The use of linear elasticity results in reduced displacements. This is caused by a constant value of the modulus of deformation being used even for elements with high stresses. Results of the nonlinear conventional triaxial data (TPC) exhibit displacement significantly higher than field measurements. This is due to the more rapid weakening of the soil along the TPC stress path than in reality. There is no significant difference between results from the triaxial TAC/TAE analysis and the plane strain analysis. They both indicate an improved agreement with the field behaviour as obtained by a slope indicator embedded in the pile.

Comparing the analytical results with the field measurements it appears that the closest agreement is obtained for the analyses using the plane strain or the triaxial stress path dependent data. This is not surprising since these two analyses employ soil moduli derived from tests with stress paths most closely resembling the actual behaviour.

The final goal of this study was to analytically determine a realistic pattern of lateral pressures along the tangent pile wall, which is practically impossible to measure directly.

The lateral stress distribution along the wall obtained analytically using the four different stress-strain models is shown on Figure 2. Also indicated in the same Figure is the distribution recommended by Peck (2) for stiff soils. Apart from the assumption of linear elasticity, which shows an unreasonable distribution, there is a reduction of lateral stress in zones

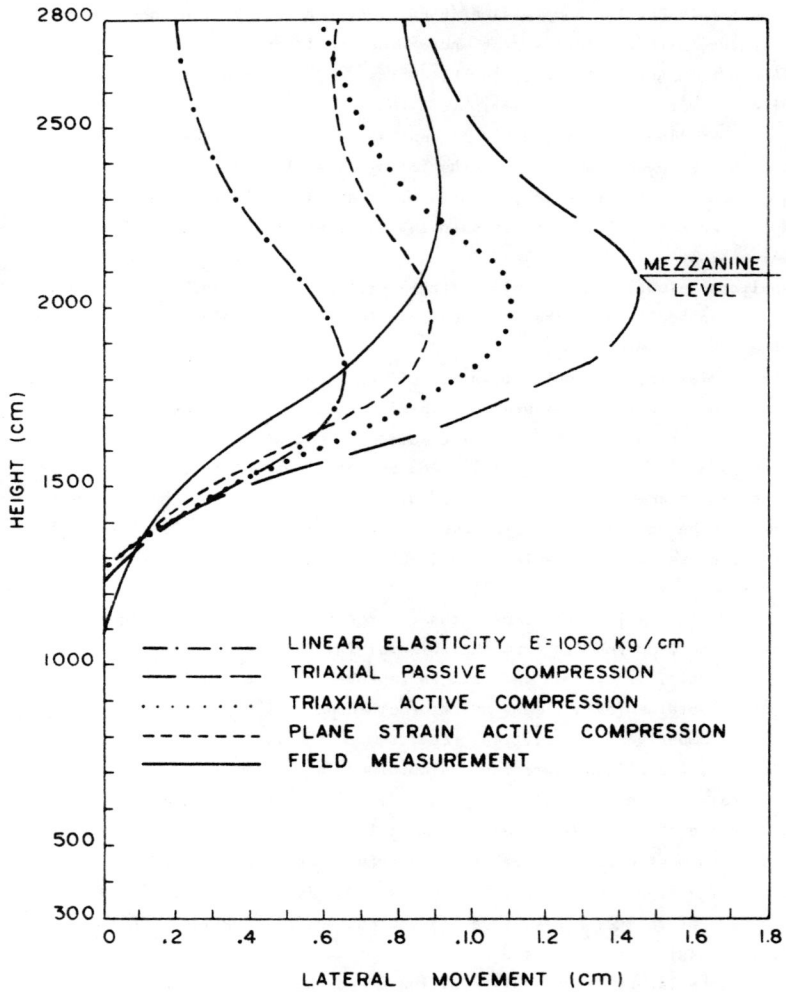

COMPARISON OF FIELD MEASUREMENT OF
SLOPE INDICATOR S12 AND FINITE ELEMENT PREDICTIONS

FIGURE 1 Comparison of Field Measurements with Analytical Results for Tangent Pile Wall

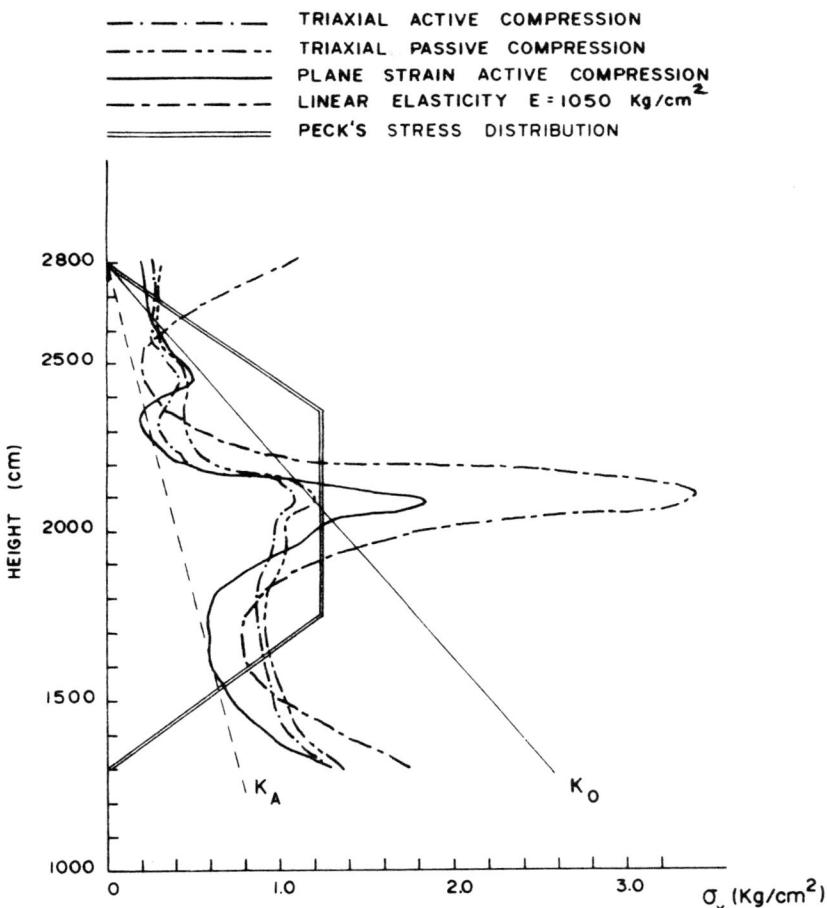

LATERAL STRESS DISTRIBUTION ALONG THE WALL

FIGURE 2 Lateral Stress Distribution along the Wall ($1 \text{ kg/cm}^2 = 98 \text{ kPa}$)

where the retaining structure yields and an increase on the supporting points. Below the bottom of the excavation the stresses increase rapidly towards the at rest state of stress. The back-calculated lateral pressure distribution can be approximated by a linear increase with depth, exhibiting peaks at the levels of the struts.

Another interesting point is that the differences between the various stress-strain models in Figure 2 are much less pronounced than for similar comparisons of displacements. This is because analyses of stresses are known to be less dependent on stress-strain characterization than analyses of displacements. From this follows that if an analysis predicts movements reasonably close the stresses calculated by the same analysis can be assumed to be at least equally reasonable.

More detailed description of the study is given elsewhere (1).

Conclusions and answers to panel questions:
1) A stress-strain analysis was used to interpret field displacement data and to calculate lateral pressures;
2) Several stress-strain models were used. Best agreement between field data and analysis was obtained for models simulating the actual stress paths around the excavation. Conventional triaxial tests yielded results departing from the field behaviour.
3) The commercial cost of this analysis would be in the order of Cdn. $20,000. This figure does not include the cost of specialized testing, which would add another Cdn. $30,000. Direct savings resulted due to reduction of lateral support and exceeded these costs many times.
4) A stress-strain analysis to be credible and reasonably accurate has to be linked with field observation program.
5) Complex structures, such as deep excavations, with complicated stress change patterns, require nonlinear, stress-path dependent stress-strain models.
6) The presented approach is based within the framework of the conventional elasticity, trying to take into account some important non-elastic soil features (nonlinearity, stress path dependency). The next step would be to apply a truly elasto-plastic model.

REFERENCES

1. Eisenstein, Z., and Medeiros, L.V., "Supported Deep Excavations in Stiff Soil", Proceedings 33rd Canadian Geotechnical Conference, Calgary, Alberta, Canada, September 1980, pp. 931-941.
2. Peck, R.B., 1969. "Deep Excavations and Tunnelling in Soft Ground", 7th ICSMFE State-of-the-Art Volume, Mexico, pp. 215-290.

A DISCUSSION ON THE
STRESS-STRAIN APPLICATIONS IN GEOTECHNICAL ENGINEERING

By George Y. Baladi,* Member, ASCE

For those of you who were not present at the Montreal workshop, I would like to point out that I participated in the predictions using an elastic-plastic model. Therefore, in addressing the following assigned questions as a panel member, my comments will be centered primarily on elastic-plastic models. The questions are:

1. What models are suitable for which types of problems?
2. What are the limitations of the model or models that you use?
3. Have you used these models in practice and at what cost?
4. What testing improvements are needed to expand our models or to verify them?
5. What have we learned from the three meetings?

The first question can be best answered by examining the characteristics of soil behavior and pointing out those characteristics that can be simulated by the model. Questions 2, 3, and 4 can be answered directly. The last question can be best answered by stating the basic requirements and conditions of constitutive models so that a practicing engineer can use them in solving engineering problems.

Characteristics of Soil Behavior to be Displayed by the Model

Each constitutive model has its own range of application and displays a limited number of the major responses of the stress-strain behavior of soil. Therefore, the adequacy of a constitutive model for the solution of a given engineering problem can be assessed by first determining which characteristics are relevant to the particular engineering problem of interest, and second selecting the model that best reflects these characteristics. The following are the major features of the stress-strain behavior of earth materials:

---

* Research Civil Engineer, U. S. Army Engineer Waterways Experiment Station, Vicksburg, Mississippi, 39180.

1. Soils exhibit a nonlinear compacting behavior under hydrostatic states of stress.
2. The compressive volumetric strain is bounded.
3. Under cyclic isotropic compression, soils display permanent volumetric strains which are cumulative but bounded.
4. The shearing strength of soils is bounded.
5. The shearing strength of soils depends generally on the mean normal stress.
6. Soils exhibit shear-induced volume change.
7. The slope of the shear stress-shear strain curve always decreases as the shear strain increases.
8. Under cyclic shear loading, soils display permanent deformations and exhibit cumulative but bounded irrecoverable volume changes. In addition, the unloading-reloading shear stress-shear strain curves generate hysteresis loops which change with the number of cycles.
9. Soil behavior depends on past loading conditions.
10. For dynamic problems, soil stress-strain properties are generally affected by rate of loading.
11. Soils dissipate energy even for very small amplitudes of stress changes.
12. Stress increments are almost parallel to strain increments for small deviations from unstressed states; however, for large stresses, the strain increments are eventually parallel to the total stresses.
13. Under undrained conditions, soil is a multiphase material.
14. Soils display some degree of anisotropy.

Some of these characteristics are demonstrated qualitatively in Figures 1 through 6. Figure 1 shows a typical behavior of soil subjected to an isotropic state of stress (isotropic consolidation test). It is clear from this figure that soils, in general, exhibit a nonlinear compacting hydrostat. It is also clear that soils display permanent volumetric strain under cyclic isotropic compression-type loading. Figure 2 shows typical shear stress-shear strain behavior of soil. The dependency of the shearing stress of soils on the confining pressure P_c or mean normal stress is obvious in this figure.

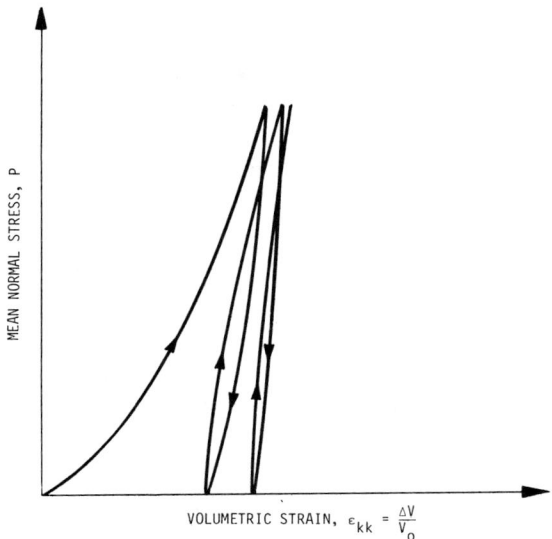

Figure 1. Typical behavior of soil under isotropic consolidation test conditions

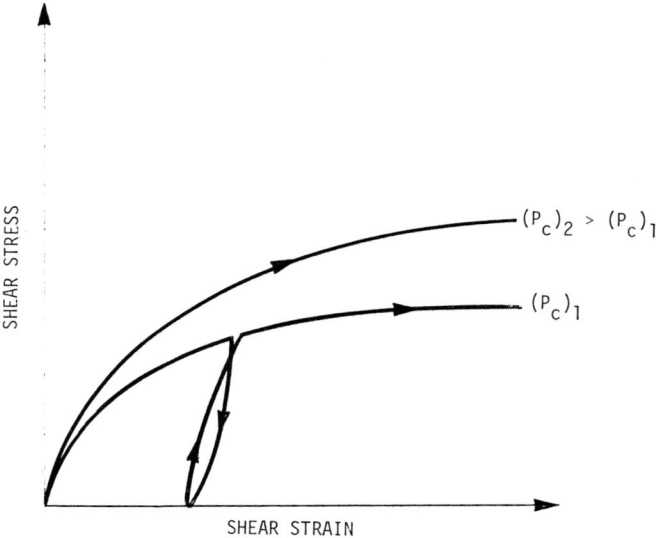

Figure 2. Typical shear stress-shear strain curves for soil

Figure 3 shows typical (qualitative) stress-strain response curves for soils sheared under drained triaxial compression conditions; i.e., the curves marked "1" represent dense sand or overconsolidated clay, while the curves marked "2" depict response typical of loose sand or normally consolidated clay.

Figure 4 shows a typical variety of stress-strain-pore pressure response curves manifested by saturated soils tested in undrained shear in a triaxial compression device. The three specimens were first isotropically consolidated to the same effective mean normal stress level (point 2), then sheared undrained. The shear curves marked "2 → 3" show the typical response of a normally consolidated clay or a very loose sand. The curves marked "2 → 5" show behavior typical of an overconsolidated clay or a very dense sand. Within the extreme limits of these loose and dense soil responses, there is a graduated response, typified herein by the curves marked "2 → 4." The latter response depends on the state of compaction (consolidation) of the material. It is clear from this figure that the effective stress is the only part of the total stress that affects soil shear strength.

Figures 5 and 6, respectively, show the typical behavior of an isotropic and a transverse-isotropic soil subjected to hydrostatic stress. For the isotropic soil (Figure 5), all strains are equal under hydrostatic states of stress; however, as indicated in Figure 6, in the case of transverse-isotropic soil, the strain in the plane of isotropy, ε_r, is different from that in the axial (symmetry axis) direction, ε_z.

The elastic-plastic model that I presented at the workshop in Montreal can duplicate quantitatively almost all these characteristics in the compression region. In the extension region, however, the model can only qualitatively describe the behavior of the material. In addition, the model cannot produce hysteretic behavior under small-amplitude cyclic loading conditions.

Advantages and Disadvantages of Elastic-Plastic Models

Advantages

The models used by the author at the Montreal workshop (1) satisfy all rigorous theoretical requirements, (2) can fit reasonably well most characteristics of soil behavior in the compression region, and (3) are suitable for both finite-element and finite-difference numerical techniques.

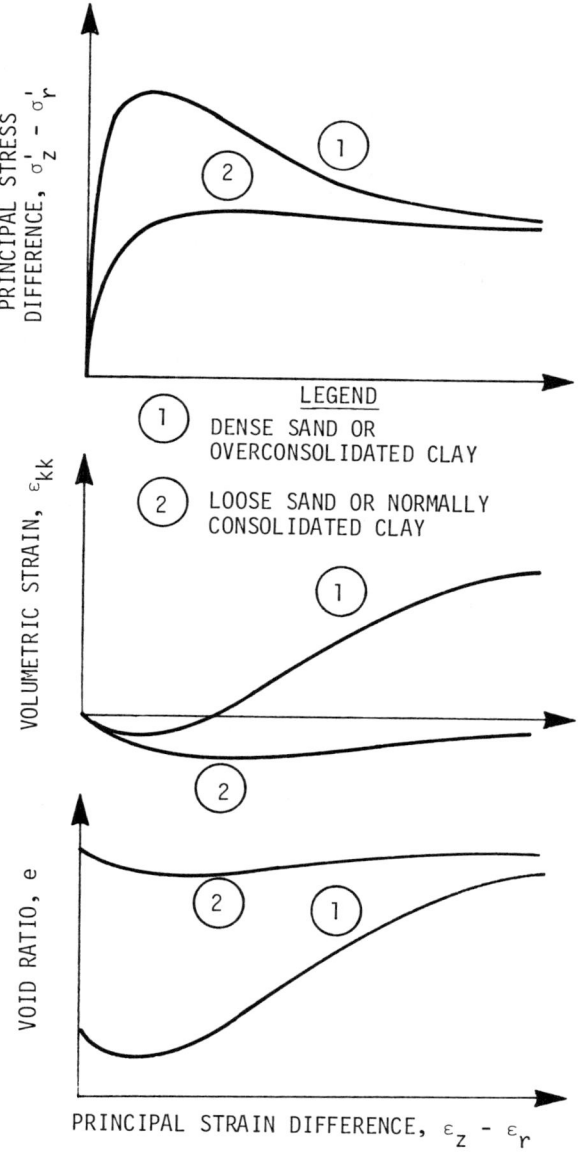

Figure 3. Typical behavior of saturated soils tested under drained triaxial test conditions

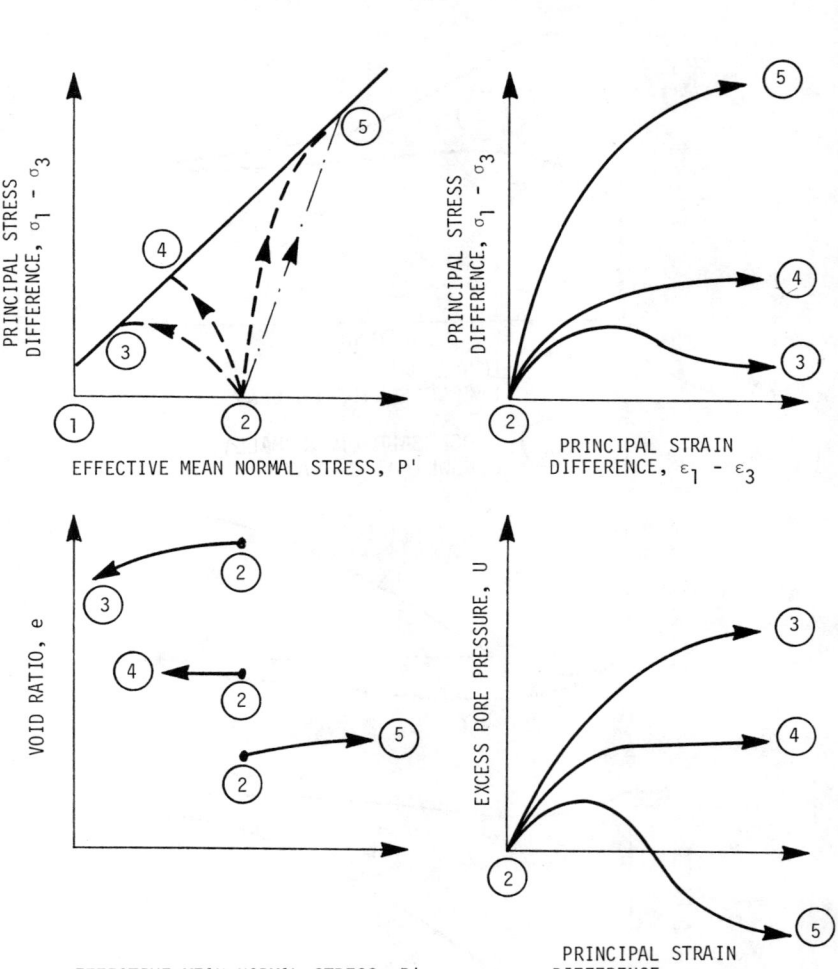

Figure 4. Typical behavior of saturated soil tested under undrained triaxial test conditions

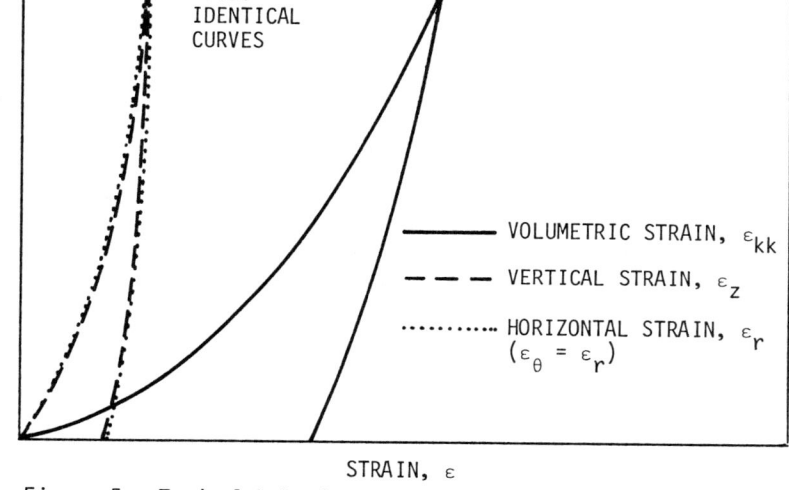

Figure 5. Typical behavior of a dry or undrained isotropic soil under hydrostatic loading and unloading

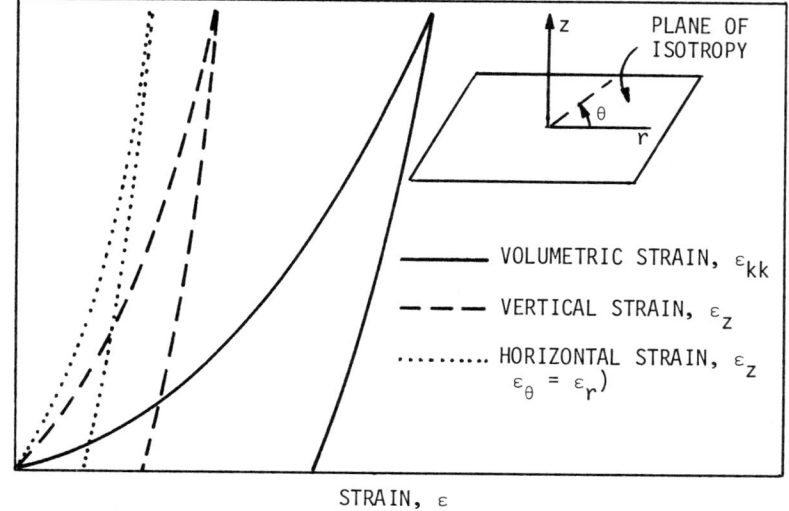

Figure 6. Typical behavior of a dry or drained transverse-isotropic soil under hydrostatic loading and unloading

Disadvantages

The limitations of the model are (1) it duplicates very poorly the behavior of soil in the extension region or under a simple shear circular loading, and (2) it requires an indirect approach to fit soil data.

Application of the Model and Cost

The model presented by the author at the Montreal workshop has been used successfully for the solution of ground shock problems and for analysis and interpretation of laboratory data. The computer cost for solutions of ground shock problems depends on the physical dimension of the problem of interest. The cost for the analysis and interpretation of the laboratory data, however, is less than a dollar per calculation.

Types of Tests Required to Fit an Elastic-Plastic Model

The following tests are required to fit an elastic-plastic model:

1. Isotropic consolidation test (IC).
2. Drained uniaxial strain test (DUX).
3. Consolidated-drained triaxial shear test (CDTX).

Under undrained conditions, the following additional tests are needed:

1. Isotropic compression test (UIC).
2. Undrained uniaxial strain test (UUX).
3. Consolidated-undrained triaxial shear test (CUTX).

The first step in the fitting procedure using the results of the above tests is to employ the unloading portion of these essential tests to determine the appropriate elastic prescription of the model, since the model behaves elastically during initial unloading in these tests; for example, initial unloading behavior in isotropic consolidation tests is indicative of the bulk modulus K and in triaxial stress tests is indicative of the shear modulus G. Other material property tests or stress paths, if available, may be used to check or adjust the overall fit of the elastic portion of the model.

The next step in the fitting procedure is to establish the failure envelope, i.e., that portion of the yield surface which absolutely limits the shearing stresses that the material can withstand. While the failure envelope could be chosen as a work- or strain-hardening yield surface, it is generally adequate and much simpler to assume it to be ideally plastic. The failure envelope is generally obtained using failure data from triaxial stress and/or proportional loading shear tests. This envelope is fit as a function of the stresses, which is usually assumed to involve only the first stress invariant and the second invariant of the stress deviator tensor.

The remaining step in the fitting procedure is the most difficult, involving the hardening portion of the model which is obtained by a trial-and-error procedure in which a hardening surface and hardening rule are assumed and the behavior of this assumed model computed and compared to the representative data. If the fit requires improvement, a new set of parameters is tried and the procedure is repeated. The computation of the model behavior can be based on the equations describing the relations between the stress and strain increments during the common laboratory loading paths.

Obviously, the success of such trial-and-error fitting procedures and the rapidity with which they converge are strongly dependent on the experience of the modeller. Knowledge of the effect on the model behavior of changes in the model parameters is important for rapidly obtaining a satisfactory fit. For example, the fitting procedure is greatly simplified by the knowledge (obtained through experience in fitting several models by trial and error) that the hardening rule strongly affects the stress-strain curves for uniaxial strain and hydrostatic stress paths while the shape of the hardening surface plays an important role in determining the stress-strain behavior for triaxial stress situations and the stress path for uniaxial strain. In fact, the hardening rule has been obtained for the most recent elastic-plastic models by using a separate program to compute the plastic volumetric strain during isotropic consolidation tests.

In summary, to construct elastic-plastic type models the following (minimum) tests are required: (1) loading and unloading in isotropic consolidation, and (2) loading and unloading in triaxial shear. Additional tests, such as uniaxial strain tests, proportional loading tests, direct shear tests, etc., are useful for verification of the model fits.

Basic Requirements and Conditions of Constitutive Models

Theoretical restrictions, such as those imposed by uniqueness and continuity considerations, should be satisfied to the maximum practicable extent. It should be kept in mind, however, that earth materials are almost invariably nonhomogeneous, anisotropic, stress-history dependent, chemically entangled mixtures of gases, liquids, and solid particles. Therefore, it is unreasonable to expect that a single or universal constitutive model can be developed which satisfies all the mathematical restrictions imposed by the assumptions of formal continuous media theories while accurately mirroring all of the detailed responses observed during various laboratory tests. Hence, for a constitutive model to be useful in the solution of practical problems, it can, at best, apply only to limited ranges of strain, strain rate, etc. Therefore, we must resist the temptation of trying to apply a specific constitutive model beyond its range of applicability.

The number of coefficients in the response functions of the model should be kept to a minimum. In addition, the numerical values of these coefficients should be readily derivable from laboratory test data. It is most important that the coefficient values not be merely a set of numbers generated through a trial-and-error "black box" routine to fit a given set of data, but that they have physical significance in terms of compressibility, shear strength, etc., so that when extrapolating to different materials, rational engineering judgments can be made as to their relative magnitudes based on geologic descriptions, mechanical properties, and other conventional indices.

The model should be capable of qualitatively and quantitatively matching the salient nonlinear and hysteretic response characteristics of earth media as determined from a variety of laboratory test boundary conditions. If confidence is to be placed in the solutions of practical problems, the material model used should at least have the capability to reasonably duplicate the response of relatively homogeneous specimens of simplified geometries loaded in the laboratory along known stress paths under carefully controlled conditions.

Finally, the model should be easily incorporated into existing finite-element and/or finite-difference computer codes.

The above requirements and conditions are admittedly more pragmatic than theoretical. However, the author strongly believes that any critical analysis of material models should include examination in the light of all these criteria in order to maintain a balance in perspective. The elastic-plastic models, presented by the author in Montreal, are based on these requirements and conditions.

SUBJECT INDEX

Page numbers refer to first page of paper

Analytical models, 25
Anchor plates, 291
Anisotropic plasticity theory, 96
Anisotropy in clays, 253

Bearing capacity of foundations, 7
Bounding surface, 78
Braced excavation, 205
Butt displacement, 277

Cavity expansion, 153
Clays, 78
Cohesionless soils, 226
Compaction efficiency, 308
Constitutive models, 96
Constitutive relations, 78, 96
Cyclic loading, 96, 240

Deep excavation, 336
Deformation analysis, 205

Elastic-plastic failure states, 166
Elastic-plastic models, 241

Failure conditions, 7
Finite element analysis, 226, 333
Finite element models, 166, 205

Lateral pressure, 336
Limit equilibrium, 53
Liquefaction phenomenon, 25
Load-unload behaviour, 25

Marine foundations, 139

Non-linear constitutive relations, 182
Nonlinear elastic models, 253

Offshore engineering, 139
Offshore installations, 96

Passive earth pressures, 7
Plasticity models, 115
Plasticity theory, 7
Pore-water pressure model, 240
Pore-water pressures, 240
Pullout resistance, 291

Rigid piers, 277
Roller compaction, 308

Sands, 153
Shallow penetration, 139
Soil anchors, 291
Soil behavior, 241
Soil compaction, 308
Soil modulus, 205
Soil response, 115
Soil strength, 205
Stability problems, 53
State-of-the Art, 7
Stress-strain applications, 241, 333
Stress-strain curves, 226
Stress-strain models, 336
Stress-strain relations, 182

Undrained strength, 253
Uplift forces, 277

Yielding load, 291

AUTHOR INDEX

Page numbers refer to first page of paper

Abbas, M.H.B., 205

Baker, R., 25
Baladi, George Y., 341
Bhatia, S.K., 240

Chang, C.S., 205
Chen, W.F., 115
Chowdhury, R.N., 53
Christian, J.T., 182
Clough, G.W., 253

Defailias, Y.F., 78
Duncan, J.M., 333

Eisenstein, Zdenek, 226, 336
Evgin, E., 226

Fattah, E.A., 308
Finn, W.D.L., 96, 240
Frydman, S., 25

Galil, J., 25

Hansen, L.A., 253
Herrmann, L.R., 78
Hjorth, B.E., 166

Lievre, B., 153

Martin, G.R., 96
Meyerhof, G.G., 7
Miller, T.W., 139
Mizuno, E., 115
Murff, J.D., 139

Poorooshasb, H.B., 153, 277
Prevost, J.H., 166

Selig, E.T., 1

Wang, M.C., 291
Wu, A.H., 291

Yong, R.N., 1, 308

ISBN 0-87262-294-0